THE LIMESTONES AND CAVES OF NORTH-WEST ENGLAND

THE LIMESTONES AND CAVES OF BRITAIN

Series Editor: T. D. Ford

in preparation

Limestone and caves of Mendip
Limestone and caves of the Peak District
Limestone and caves of South Wales

THE LIMESTONES AND CAVES OF NORTH-WEST ENGLAND

compiled and edited by

A. C. Waltham

assisted by

M. M. Sweeting

published by

DAVID & CHARLES : NEWTON ABBOT

for

The British Cave Research Association

0 7153 6181 3

Set in 11 on 13 pt Lectura
and printed in Great Britain by
Bristol Typesetting Company Limited for
David and Charles (Holdings) Limited
South Devon House Newton Abbot Devon

Contents

List of Illustrations

PLATES

TEXT-FIGURES

LIST OF TABLES

Foreword

Much of the limestone scenery of Britain is of high scenic, recreational and educational value. Two of our main National Parks, in the Peak District and Yorkshire Dales, and our main nature reserves are in limestone country, and, by its very nature, most of our caves are in these areas. Yet no comprehensive account of the limestone scenery and caves of each of these areas is to be found. The general review *British Caving—an Introduction to Speleology* was written a generation ago by members of the Cave Research Group of Great Britain, and it stimulated so much interest that a whole series of volumes is necessary to summarise the present state of knowledge. Once again, members of the Cave Research Group have given freely of their time in writing the chapters within.

Though of the same geological age, the limestones of our different areas, of the Yorkshire Dales, Peak District, Mendip Hills and South Wales, each have their own subtleties, and each has a distinctive pattern to the landscape partly resulting from the characters of the limestone itself. The histories of that limestone through subsequent geological periods, not least the Ice Ages of the Pleistocene, have stamped further variations and contrasts on the regions. These factors have controlled the nature and distribution of the cave systems and underground drainage. In turn these differences are reflected in the degree of prehistoric habitation by early man and contemporary animals, and by the present-day occupation of the cave environment by insects and other life specially adapted to permanently dark and damp conditions.

The present state of knowledge of all these aspects of our limestone areas is reviewed within. One purpose of this book is to bring together the essentials of our knowledge, but, in so doing, the authors hope they have high-lighted the gaps in our knowledge, gaps which must be filled by future investigation, whether it be more accurate measurements of underground water flow, or limestone solution rates, further study of the life of prehistoric man, or the metabolism of modern insects underground, or, more simply, the discovery of new caves.

It is hoped that this series of books will provide a survey of facts for the landowner, quarry industry and water authorities alike, where a conflict of interests is sometimes manifest. The books should provide guides for teachers and students who use these areas so much to gain an understanding of the weathering and other processes affecting our environment. It is hoped that the cave descriptions will encourage the sporting caver to take a greater understanding of the features visible in his favourite caves, and the cave scientist will have a clearer background into which to fit his own piece of research.

As Series Editor for these books, I wish to express my sincere gratitude to all those who have contributed, in particular to the co-ordinating authors of each volume. The sense of their feelings towards their respective areas comes through vividly in the pages that follow. I hope that you, the reader, will enjoy reading this book as much as we have enjoyed preparing it for you.

TREVOR D. FORD
Series Editor

I

Introduction

A. C. Waltham

The great white cliff of Malham Cove and the awe-inspiring pot-hole of Gaping Gill are two of the finest natural spectacles in the whole of Britain. Yet they are integral features of the landscape that surrounds them. For this small part of north-west England, centred on Ingleborough Hill, is a region of bare white cliffs, disappearing rivers, limestone pavements and cave systems. It is a magnificently developed area of karst—a topography cut into lime-stone and therefore showing all the features typical of erosion by natural solution of the rocks.

The scenery is invigorating, the karst features are classic and the caves are fascinating. Consequently the area attracts country-lover, scientist and caver alike, and tends to generate a type of person who fulfils at least two of these roles.

It is nearly 70km from the northern shore of Morecambe Bay to Pateley Bridge in Nidderdale, and the landscape in a 15km wide belt between these points is nearly all dominated by lime-stone. The area straddles the axis of the Pennines; this line of highlands—the backbone of England—is formed of sandstones, limestones and shales of Carboniferous age. The thick limestones of the northern half of the Pennines are exposed on the surface to form three distinctive karst regions (see Fig 1). The most specta-cular comprises the dissected mountains of the Ingleborough area; here, the limestone plateaux stretch from Casterton Fell in the west to Grassington Moor in the east and are broken by deep

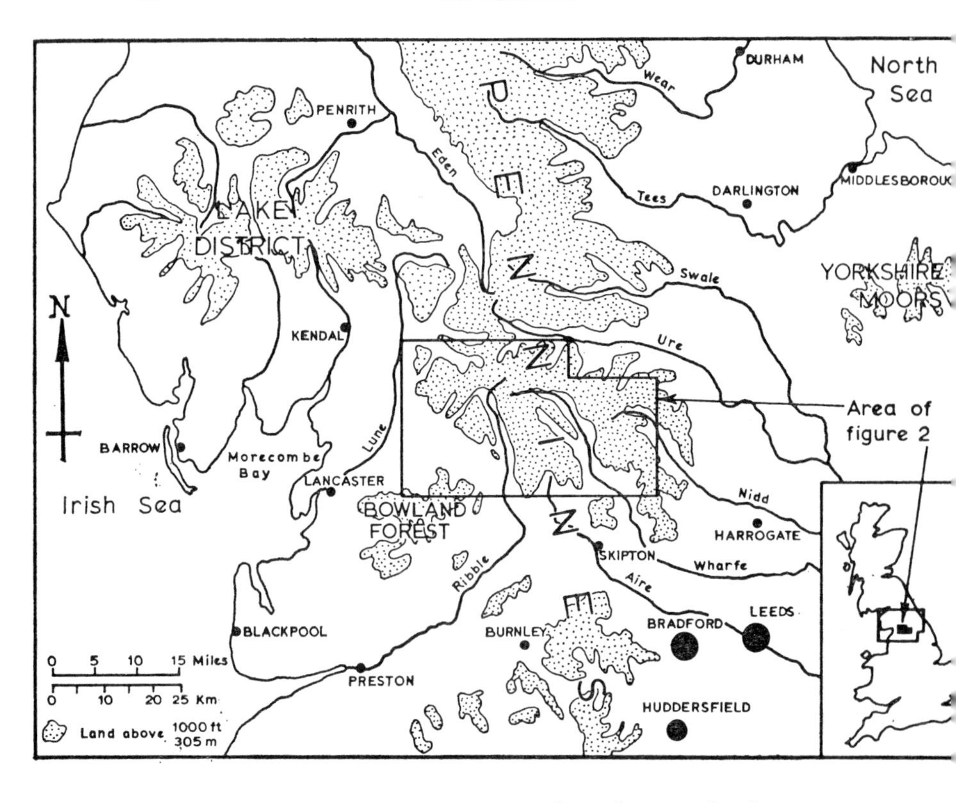

Fig 1 Location map of northern England

valleys such as those of the Rivers Ribble and Wharfe. Immediately to the west the same limestones provide a completely contrasting topography where they form isolated blocks projecting through the alluvial plains that border Morecambe Bay. These two regions are the subjects of this book.

The northern boundary of the area covered by this volume is the southern watershed of Wensleydale. Though this is not the limit of karst development, it provides a convenient boundary between the Ingleborough district and the third of the Pennine karst regions. The Dales of the North Riding of Yorkshire lack the thick limestone of Ingleborough, but are cut into a number of thinner limestones giving a distinctly different type of karst geomorphology.

GEOGRAPHY OF THE AREA

The topography shows a general variation from east to west across the limestone outcrops (see Fig 2). Perhaps the bleakest area is the high moorland of Great Whernside—Grassington Moor lying between Nidderdale and Wharfedale. This forms one single roadless block never lower than an altitude of 400m and culminating on Great Whernside at 704m (2,310ft).

Further west the highlands are much more dissected around the Three Peaks—the adjacent summits of Ingleborough, Penyghent and Whernside, each about 700m high. Here the relief is greatest, the limestone best exposed, the scenery the most spectacular, and the caves and karst the best developed.

Between the River Lune and the shores of Morecambe Bay, the relief is again much more moderate. The valleys are nearly at sea level, but few of the limestone ridges have summits at altitudes above 200m (650ft).

The climate of the karst region is essentially mild and wet. It is generally described as a 'Marine West Coast' climate and falls into group *Cbf* on the Köppen classification. Table 1 summarises the climatic data at Malham Tarn, and this must typify the limestone outcrops except those close to Morecambe Bay. Precipitation is generously and fairly evenly distributed throughout the year, being measurable on an average of 220 days per annum; there are only slightly marked extremes of a late spring minimum and an early winter maximum. Monthly mean temperatures are all above freezing point, and are probably just above freezing even on the highest summits. In the valleys, snow lies on the ground for an average of only about twenty mornings per year, though this figure rises to over forty on many of the higher slopes; the daily minimum temperature is below freezing on an average of ninety-one days per year at Malham Tarn.

The relief and climate of the region have strictly dictated its

TABLE 1 *Climatic statistics (30-year averages) (after Manley, 1957)*

(Station: Malham Tarn House Altitude 382m (1,297ft))

	J	F	M	A	M	J	J	A	S	O	N	D		
Average temperature:	1·5	1·7	3·5	5·7	8·7	11·6	13·6	13·1	10·9	7·5	4·1	2·2	7·0 °C	44·6 °F
Average daily maxima:	3·8	3·9	6·5	9·4	13·0	15·8	17·3	16·6	14·1	10·1	6·3	4·2	10·0 °C	50·1 °F
Average daily minima:	—0·8	—0·6	0·3	2·1	4·5	7·4	9·8	9·6	7·6	5·1	1·9	0·2	4·0 °C	39·1 °F
Average precipitation:	144	112	125	89	85	89	119	147	108	146	142	155	1.461mm	573·8in

natural vegetation and pattern of farming. Most of the limestone outcrops are high and windswept, supporting only grassland. Fescue grasses dominate in the typically short springy turf. Thin birch woods have developed on some of the more sheltered slopes, and ferns are important in the grykes of the bare pavements. Much of the limestone is covered by boulder clay and here the natural vegetation is similar to that on the adjacent outcrops of the non-calcareous rocks. There, acid peat bogs are covered by sedge masses, though the steeper, better drained, slopes support a thick growth of heather.

Farming on the high limestones is therefore almost totally restricted to sheep grazing, though more mixed farming activity takes place on the floors of the lower Dales and in the Morecambe Bay area. A more conspicuous form of land use is quarrying, both of the limestone and of the harder lithologies within the basement rocks. Present activity is most extensive in the Horton–Settle area and around Grassington and the majority of the quarries are unsightly scars on the natural beauty of the Yorkshire Dales National Park. Vein mining has now ceased in the region, and the removal of limestone clints, for use as rockery stone, is now fortunately minimal.

Farming, quarrying and service industries provide most of the employment in the Dales—that is, excluding the larger towns in the Morecambe Bay district. Consequently most of the population is concentrated along the southern edge of the limestone plateaux; only a few smaller villages lie in the higher parts of the Dales. However, access to the karst is very easy. Nowhere is more than 6km from a tarmac road and much of the limestone is still closer.

DISTRIBUTION OF CAVES AND KARST

The geological map (Fig 3) shows the extent of the limestone outcrop, but neither caves nor surface karst features are spread uniformly over the area. Known caves are most abundant and spectacular in the Ingleborough–Gragareth region, while the surface karst features are best developed to each side—in the

Malham area, and west of the River Lune—and the largest sinking river is far to the east in Nidderdale. The distribution of the surface features is mainly related to the differing long-term erosional histories of the separate parts of the karst. In contrast, the cave distribution is mostly a result of variation in the limestone geology; admittedly, future discoveries of further cave systems will certainly add to the present pattern, but should not unbalance it significantly.

Discovery and exploration of the caves and potholes of north-west England is of course a continuing process, and is one with a long and interesting history. The nineteenth century saw the first really serious explorations and the conspicuous holes such as Gaping Hill, Alum Pot and Ingleborough Cavern received the attentions of the first explorers. From the turn of the century onwards, speleology became a slightly more popular pastime and new caves were discovered regularly year after year. The open shafts, such as Rowten Pot and Marble Steps, were soon explored, and around 1930 some of the more difficult caves, like Swinsto Hole and Washfold Pot, gained their first visits. Then relatively small-scale digging, mainly removing the glacial debris, revealed caves such as Bar Pot and Lancaster Hole, until the year 1950 saw the start of a lull in the pace of exploration. Finally the evolution of new equipment, and indeed new attitudes, has led to many further discoveries since about 1965; extensive excavation and the passing of 'impossibly' tight passages has revealed the more difficult systems such as Pippikin Hole and Marble Sink.

It is therefore true to say that the underground karst of the North West is not yet completely known—and indeed it never will be. Both this fact and the uneven distribution of caves and karst, mentioned above, have partly determined the pattern of the regional descriptions which comprise chapters 11–23 of this volume (see Fig 2). These chapters are intended to be comprehensive, but not exhaustive, for the latter would lead to tedious reading. The degree of detail in each chapter is also partly related to the local intensity of research, but the chapters have been

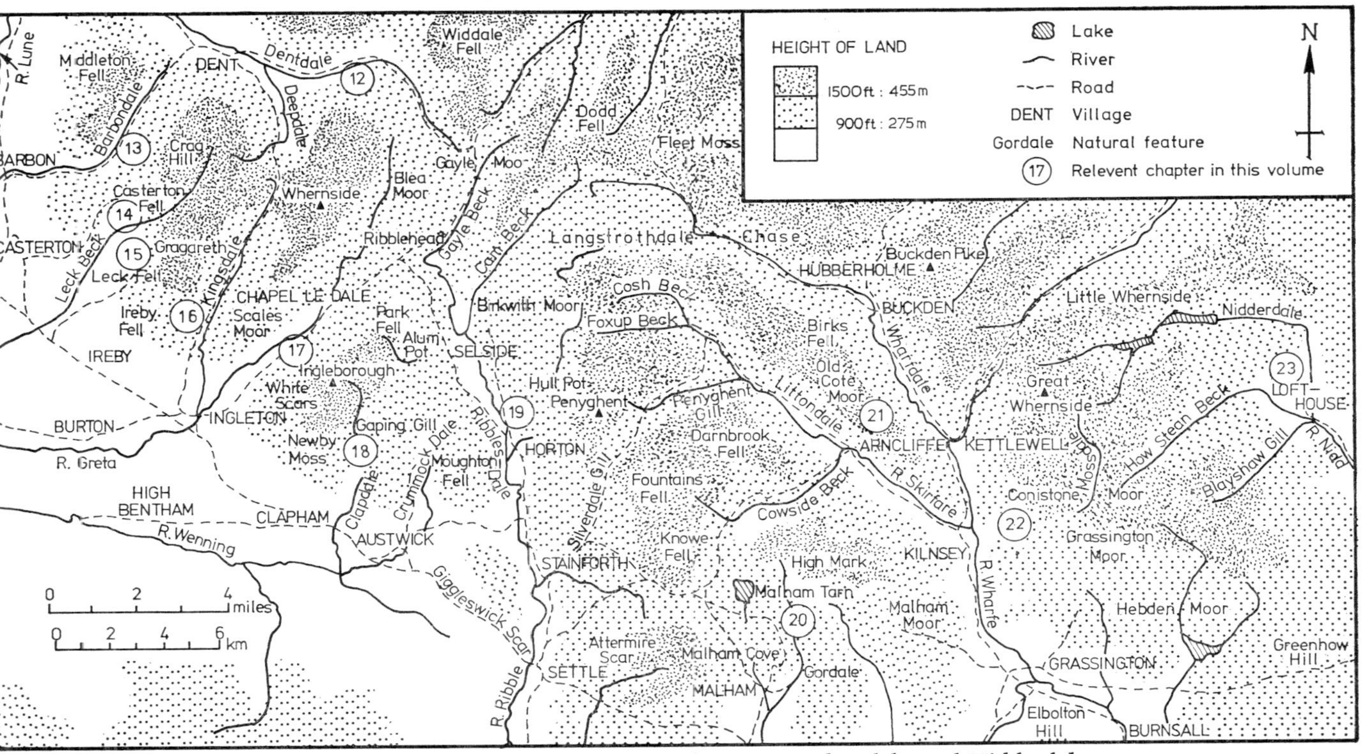

Fig 2 Location map of limestone areas between Barbondale and Nidderdale

designed to describe adequately the contrasting types of karst and cave development present across north-west England.

KARST RESEARCH IN NORTH-WEST ENGLAND

The karst features on the surface of the Pennines must clearly be related to the long and complex geomorphic development of the region. Most important of the phases in this erosive history have been the Pleistocene glaciations, but it is not known even how many glaciations crossed the area, nor what were their different effects. The environments of Tertiary erosion are similarly almost unknown.

Underground, the sporting caver is continually revealing more caves and potholes, and often his contribution to research is limited to a topographic survey. Few of the sporting cavers turn to detailed research in the caves; there is enough exploration to occupy them fully. Furthermore the majority of caves are far too difficult for the non-caving scientist to visit casually.

Largely for reasons such as these, the history and extent of karst research in north-west England is limited. The first ten chapters of this volume review the present state of knowledge related to this region within each of the karst sciences. There is clearly immense scope for future work both above and below ground; the problems are complex but fascinating.

A study of karst landscapes and features is one sphere where scientist and sportsman must work together to achieve success; the results to date for north-west England are reviewed in the following pages.

2

The Geology of the Southern Askrigg Block

A. C. Waltham

Over much of north-west England the Lower Carboniferous Limestones are topographically prominent. They reach their culmination where the Great Scar Limestone facies forms the high plateaux of the Ingleborough–Malham region. The limestone is here situated on the elevated southernmost part of the Askrigg Block, a major structural feature of the basement rocks, whose uplifted form has influenced both the deposition and present topography of the limestone. Along its southern and western margins the Askrigg Block is bounded by the Craven and Dent Faults, which are therefore taken as limits of the area described, together with the watershed to Wensleydale, and Nidderdale, as the northern and eastern limits (see Fig 3). The Askrigg Block, and consequently the limestone, dips gently towards the north and east, and inliers of the older basement rocks are therefore exposed where the deeper valleys cross the Craven Faults. Most of the hills above about 420m (1,400ft) are formed of the Yoredale Series and Millstone Grit, both overlying the limestone and therefore obscuring its continuation further downdip to the north.

The whole of the Great Scar Limestone and a number of the limestones within the Yoredale Series are well known for their karst features, and contain the caves which make the area famous.

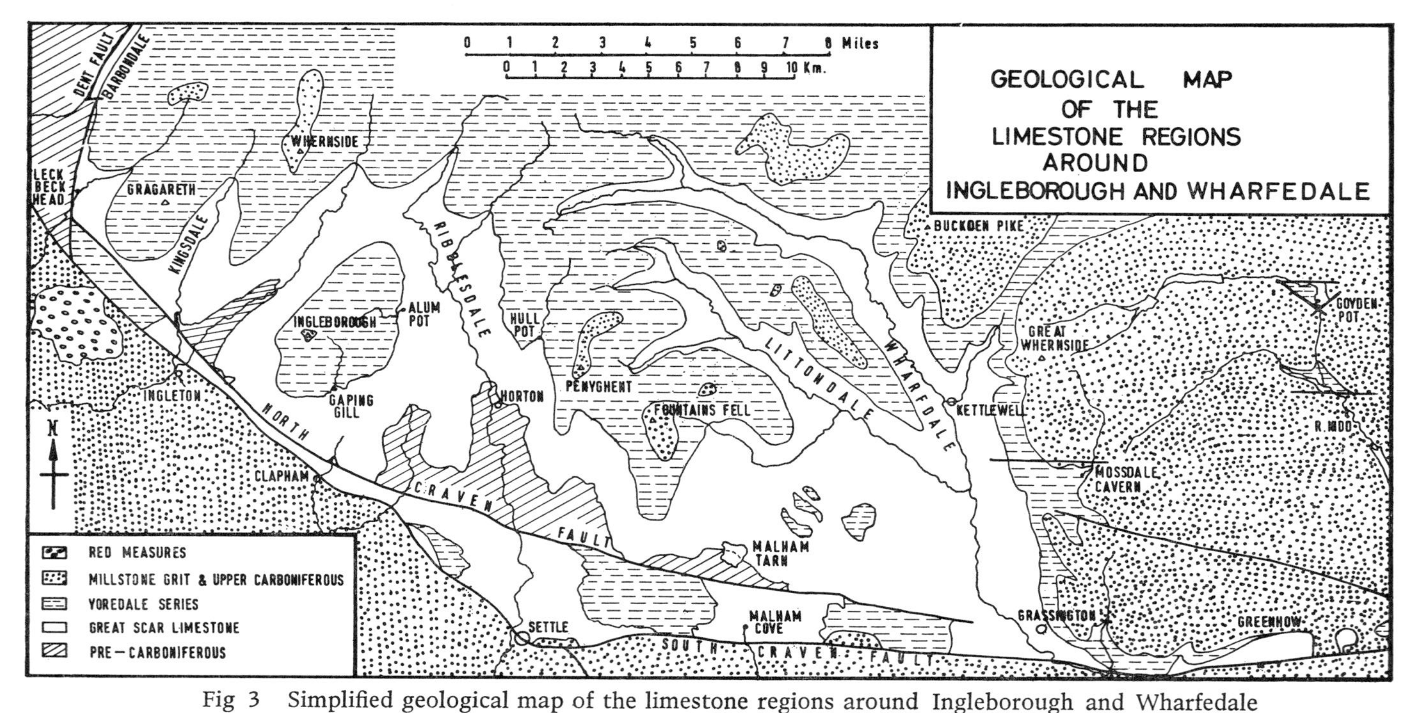

Fig 3 Simplified geological map of the limestone regions around Ingleborough and Wharfedale

PREVIOUS GEOLOGICAL RESEARCH

Partly due to the inspiring natural exposures, the karst regions of north-west England have been the subject of an extensive geological literature centred on the Ingleborough and Wharfedale districts.

Late in the last century the Geological Survey opened the modern phase of research by detailed mapping of the whole area, and the publication of a long account (Dakyns *et al,* 1890). Still almost the standard reference on the Great Scar Limestone is the detailed stratigraphic account by Garwood & Goodyear (1924). However, Dunham *et al* (1953) provided a useful more modern review, and Rayner (1953) described the Carboniferous rocks in a wider context.

Hughes (1909) produced the first detailed lithological description of the Great Scar Limestone and the same subject was investigated by Schwarzacher (1958). Recent more specific studies of the limestone lithology include those by Sweeting & Sweeting (1969) and Waltham (1971 *a*). The Yoredale Series have their own descriptive literature, but of particular significance are the papers by Moore (1958), Hicks (1959) and Wilson (1960). Joints are an important feature of the Great Scar Limestone and have been described at length by Wager (1931) and Doughty (1968).

The above papers are only the more important of many, and do themselves refer to a wider literature. There are also a number of published accounts of small parts of the karst region, and of particular value is the description of the Grassington area (Black, 1950).

PRE-CARBONIFEROUS ROCKS

Large areas of pre-Carboniferous rocks crop out to the west of the Dent Fault, but to the student of karst the relatively small inliers of these rocks, within the confines of the Askrigg Block, are more important. These inliers are exposed where the main valleys have been cut right through the elevated edge of the Great Scar Limestone immediately north of the Craven Faults, and the main ones are situated at Ingleton, Crummock, Horton and Malham (see Fig 3).

All these inliers reveal relatively similar successions of Ingletonian (probably pre-Cambrian), Ordovician and Silurian slates and coarse sandstones, distorted into a series of tight east–west folds and with the finer sediments cleaved to form slates. Also exposed are some lamprophyre dykes, a bed of very coarse arkose locally known as the 'Ingleton granite', and the impure Ordovician Crag Hill Limestone (exposed near Horton) which is not significantly permeable. It is not known what rocks underlie the limestone north of these inliers, but recent geophysical surveys indicate the presence of a granitic mass (Myers & Wardell, 1967), structurally similar to that known to underlie the Alston Block, further north.

The general effect of the pre-Carboniferous rocks is to provide a completely impermeable basement to the limestone. Though

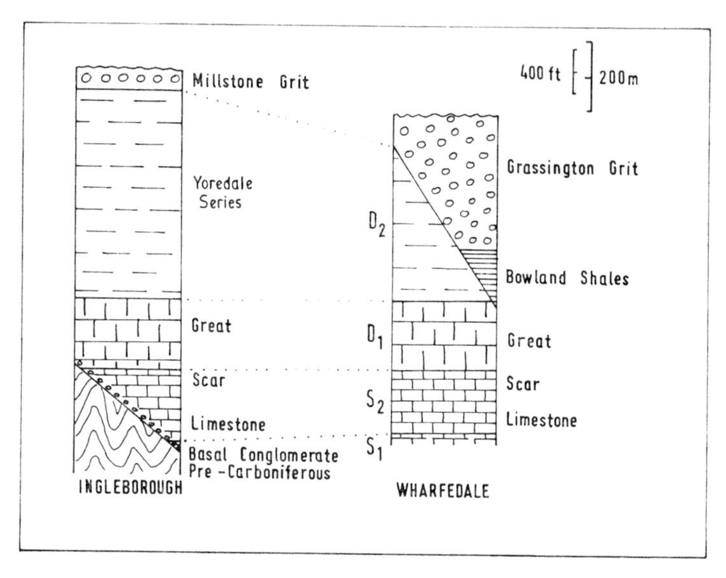

Fig 4 The stratigraphy of the area, demonstrated by simplified sections in the Ingleborough and Wharfedale areas

these rocks were much eroded down before deposition of the limestone, the resultant early Carboniferous topography was still very irregular. Detailed mapping of the basal Carboniferous unconformity reveals considerable relief in its morphology and the gently transgressive nature of this unconformity results in the limestone being very much thicker over the pre-Carboniferous valleys where limestone deposition was initiated at an earlier date. The scale of this relief is generally in the order of 30–50m though locally up to 120m, and the consequent subdivision of the lowest limestone beds into horizontally disconnected compartments must have an effect on the development of the present karst hydrology. For example, a marked ridge in the basement, visible in Crummock Dale, if continued to the north-west, probably marks the watershed between the catchment areas of Clapham and Austwick Beck Heads.

GREAT SCAR LIMESTONE

Nearly all the major cave systems of the karst of north-west Yorkshire and adjacent counties are formed in the thick limestones, known as the Great Scar Limestone, which form the lowest part of the Carboniferous succession in the area. This limestone belongs mainly to the S_2 and D_1 zones, but locally the lowest beds are of S_1 and C_2 ages. The irregularity in the age of the lowest Great Scar beds is due to the uneven topography, which was transgressed by the Carboniferous seas, thereby restricting the early phases of deposition to bays in depressions of this surface; furthermore the earliest stages of the Carboniferous are unrepresented on the Askrigg Block.

Consequently there is a considerable variation in the thickness of the limestone and around Ingleborough alone it varies between 100 and 205m thick. The depth of most of the cave systems is restricted therefore to these figures, hence there are many caves around 150m deep. In some cases, though, the caves are slightly deeper than the thickness of the limestone due to the passages running for some distance down dip. For example, the hydro-

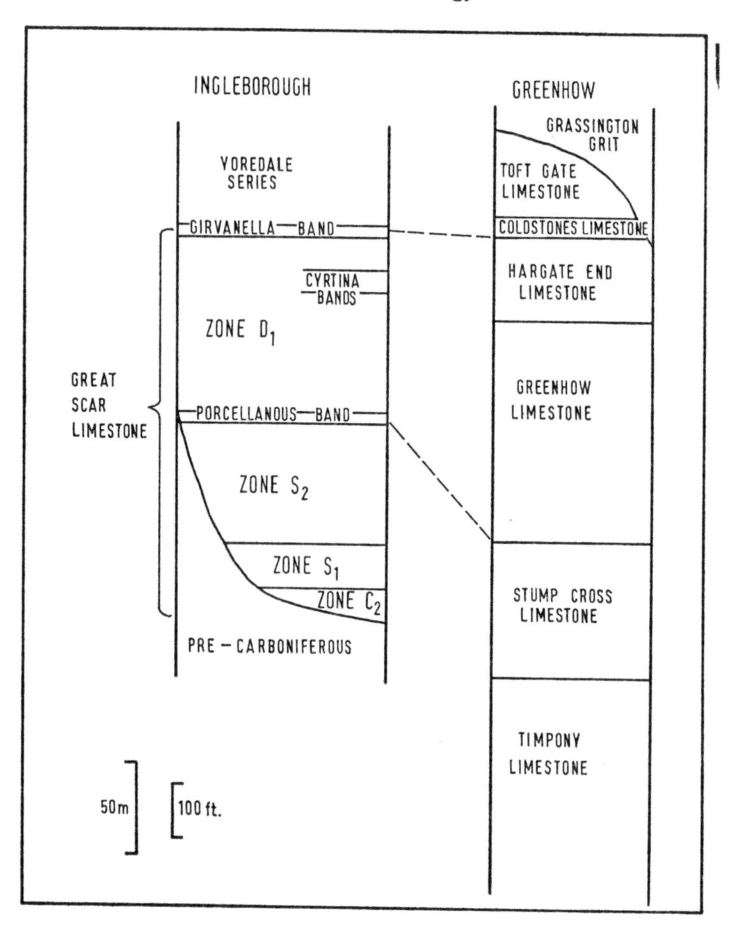

Fig 5 Stratigraphy of the Great Scar Limestone, a typical section at Ingleborough, and the stratigraphic equivalents at Greenhow

logical systems of Ireby Cavern–Leck Beck and Gingling Sink–Brants Gill are both over 200m deep though formed in limestone of less than maximum thickness. There are also two important cases where even deeper hydrological systems are known. In Wharfedale the Black Keld rising drains a continuous limestone block over 230m thick and this exceptional thickness of limestone is due to three different features: first there is a local development of the C_2 limestone (with the base not seen) and consequent maximum development of the S limestones; secondly, each zone of the limestone is here unusually thick; and, thirdly, the Great Scar Limestone passes up without a break into the Hardraw Limestone, itself about 30m thick (see Fig 7). The Long Kin West–Moses Well connection has the exceptional vertical range of 250m, but this is due to the water passing through a number of limestone fault blocks stepped down towards the Craven Faults (see Fig 8).

Lithologically the Great Scar consists mainly of fine-grained bioclastic limestones; they are generally pale grey to cream in colour though the uppermost D_1 limestones and those of the lower S_2 beds are significantly darker. Table 2 shows some representative analyses of the limestone. A matrix of fine calcite mud

TABLE 2 *Chemical analyses of Great Scar Limestone*

	(1)	(2)	(3)	(4)
$CaCO_3$	99·05	93·67	98·00	85·52
$MgCO_3$	0·46	4·38	12·18	1·42
Fe_2O_3	0·06	—	—	—
Al_2O_3	0·06	—	—	—
$CaSO_4$	0·05	—	—	—
SiO_2	0·15	—	—	—
Insoluble	—	1·95	5·52	0·58

(1)=Pure limestone from Horton in Ribblesdale (Anderson & Vernon, 1970).
(2)=Means, (3)=Maxima, (4)=Minima } of ten limestones from various localities (analyst, R. J. Bowser).

or coarser sparry calcite makes up about half of the volume, and contains foraminifera and fragments of shells and crinoid ossicles making up the other half. The fragments range in size from microscopic to 5cm or more, while the matrix grain size varies from less than 0·01mm to about 1mm. Limestones with the extremely fine grained groundmasses are known as micrites, or as biomicrites if they have a significant fossil content, and the coarser matrix varieties are termed sparites. The variations in grain size of both the matrix and the organic fragments are partly responsible for the fine laminations in the limestone so commonly visible on water-polished walls of the potholes; the lighter bands are generally of finer grain. Also patchy recrystallisation of the calcite is responsible for the mottled texture of the 'pseudobreccias'.

Schwarzacher (1958) suggested that beds of micrite and sparite alternated through the Great Scar Limestone in rhythmic successions. It was claimed that these alternations were primary sedimentary features, and that each cycle, including a micrite and a sparite horizon, was about 10m thick. Undoubtedly the micrites and sparites do alternate stratigraphically, but examination of very clean underground sections does reveal lithological variations far more complex and irregular than 10m cycles. Both this and the lack of correlation with the shale bed sequences (Waltham, 1971 *a*) must question the cyclic concept of the limestone sedimentation.

The horizontal continuity of the Great Scar Limestone is generally very uniform, except adjacent to the Craven Faults, but vertically there are numerous interruptions in the calcareous succession. Through the thickness of the limestone exposed in the caves there are about twenty significant shale beds, which being impervious have considerably influenced cave development. As they are also easily weathered they are rarely exposed at the surface, but correlation of the underground outcrops show that many of the shales are of very great horizontal extent (Waltham, 1971 *a*). Fig 6 shows the shale bands revealed in five major potholes spaced at varying intervals. Within the confines of the three westernmost sections, the shale beds are easily correlated and also reveal variation in the thickness of the limestone beds, while the greater distances to Juniper Gulf and Pasture Gill Pothole make any correlation very dubious.

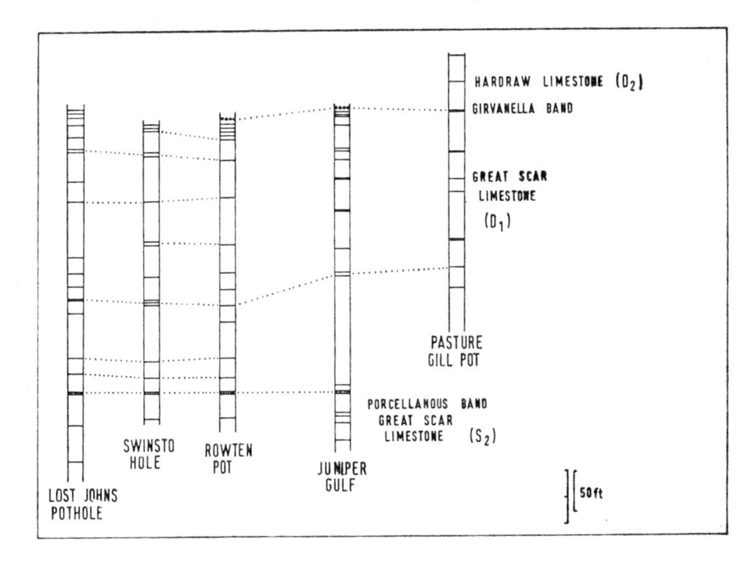

Fig 6 Distribution of shale bands in five representative cave sections through the Great Scar Limestone

Most of the shale beds range in thickness between 1 and 50cm though there are many which consist only of paper-thin partings in the limestone. Also some beds are much thicker and the extreme is represented by the beds 2m thick in Shale Cavern, Lost Johns Pothole, and Shale Pitch, Birks Fell Cave. Petrologically the shale beds are more accurately described as a fairly pure, light grey mudstone, completely unfossiliferous but containing some nodules of pyrite which rapidly alter to brown limonite. G. M. Walkden (1972, personal communication) has found a significant proportion of bentonite in some of the shales indicating some volcanic parentage; at present it is only conjecture as to whether the bentonite derives from Derbyshire or Scottish volcanoes, or erosion of the Lake District or Cheviot volcanic rocks. In two places, Meal Bank Quarry, Ingleton, and Four Ways Chamber in Notts Pot, coal seams have developed to a thickness of up to 50cm at the tops of shale bands. Their existence confirms the hypothesis that the Great Scar Limestone was formed in very shallow water subject to periodic uplift and consequent subaerial erosion.

Fossils are abundant and conspicuous in the Great Scar Limestone, and particularly notable are various types of brachiopod (eg *Productus, Gigantoproductus*), corals (eg *Lithostrotion, Caninia, Dibunophyllum*) and fragments of the crinoid *Actinocrinus* (for further details on the palaeontology see Garwood & Goodyear, 1924). There are also two widespread fossil horizons. At the very top of the Great Scar is the *Girvanella* band, which is often very difficult to locate as it is only about a metre thick but, when weathered, the elliptical algal nodules about a centimetre across sometimes stand out very conspicuously, for example, at the entrances to Birks Fell Cave and Loose Pot, Newby Moss. Lower in the succession the *Cyrtina septosa* band was first described by Garwood & Goodyear (1924), but it appears to be unreliable as an accurate marker horizon as locally there are two or more bands, occurring over a range of nearly 15m.

The most useful marker horizon, however, is probably the Porcellanous Band, a conspicuous bed of fine, white, slightly less pure limestone occurring at the base of the D_1 zone. It is usually about 70cm thick but may split into two leaves, as in Gaping Gill Main Chamber. Also in Gaping Gill it is seen to be replaced by a shell-bed before it feathers out to the south-east.

At the very base of the limestone is a layer of conglomerate which, due to the irregularity of the surface on which it was deposited, varies between the S and C_2 zones in age. It is generally no more than 15m thick, and often is much less. The texture of this basement bed varies considerably and includes layers of flagstone; most is coarser, and boulders of older rocks within it reach a diameter of 50cm, though, passing upwards, the boulders get fewer and smaller so that the calcareous matrix slowly grades into the pure limestone.

Approaching the Craven Faults the Great Scar Limestone undergoes a series of changes, not surprisingly, for it is absent in the Craven Lowlands to the south. Reef knolls have developed in the limestone along the fault zone, and two fine examples just south of Grassington now form Elbolton and Thorpe Kail Hills. The knolls consist mostly of a porcellanous limestone containing an abundance of thick shelled brachiopods, grastropods and lamellibranchs, and were formed on the very edge of the Askrigg Shelf overlooking the deeper water of the Bowland Trough to the south. Their structure is generally massive and the limestone is only bedded in small patches within the knolls. There is considerable unresolved debate regarding the contribution of organic structures to the morphology of the knolls, and also regarding the degree to which the present rounded hills—the reef knolls—represent exhumed primary structures.

In the Greenhow area, the limestone changes in character, again due to the influence of the Bowland Trough, and the uniform nature of the Great Scar Limestone gives way to a more variable succession (shown in Fig 5) which is well exposed in the small anticline containing the caves of Stump Cross and Mongo Gill. The massive white Greenhow Limestone contains all the large chambers of Mongo Gill Caverns, in contrast to the more densely jointed, grey Stump Cross Limestone in which has formed the small low-level passages of the cave of the same name, while the grey, thin bedded Timpony Limestone merely contains the risings of Dry Gill. At Greenhow the Toft Gate Limestones are completely cut out by the unconformity of the overstepping Grassington Grit (Dunham & Stubblefield, 1945).

YOREDALE SERIES

Overlying the Great Scar Limestone is a repetitive cyclic series of limestones, shales and sandstones collectively known as the Yoredale Series. Although best developed in their type area to the north, Wensleydale, they are still 300m thick on Ingleborough.

A single Yoredale cyclothem consists of a limestone bed overlain by first shale and then sandstone, this threefold unit, or cycle, being controlled by the depth of water in which the sediments were orginally deposited. There are nine major cycles recognised in the Yoredales (Moore, 1958), though these commonly include other minor cycles. The Yoredale limestones are generally fairly similar to the Great Scar Limestone except that they are mostly darker in colour and commonly more thinly bedded and slightly less pure. Also the shales are generally dark grey or black, containing in places thin seams of coal, while the sandy beds vary in lithology between fine flagstones and coarse cross-bedded sandstones. Fossils in the Yoredales are very varied but the lowest 30m of beds are particularly noteworthy for their abundant fauna of corals, brachiopods and even some goniatites.

South of Wensleydale the two lowest cyclothems, named after the Hawes and Gayle Limestones, are not represented, as the deltaic facies at this time was restricted to the more northerly regions, so the successions at Ingleborough and Conistone Moor–Kettlewell are as shown in Fig 7. Only the lowest third of the Yoredale stratigraphy is shown in these sections as the limestones above the Middle Limestone contain no known significant caves within the confines of the main karst region. However, it should be noted that many of the higher limestones are thick and therefore should contain at least minor cave development, especially where their outcrops are crossed by significant streams.

West of Wharfedale, the Yoredale limestones contain only extremely restricted underground drainage systems, and the most important influence on the karst is exercised by the 10–15m of shales, immediately above the Great Scar Limestone, as these efficiently collect all the water on the surface to feed the major sinks and caves of the area. In contrast, however, there are numerous interesting and important caves in the Yoredale limestones east of the River Wharfe.

Over much of this eastern area, the Hardraw Limestone is hydrologically attached to the Great Scar Limestone owing to the absence of any intervening non-calcareous sediments. Consequently the upper 20m of some major potholes, for example Pasture Gill Pot, are in this limestone, and the cave streams pass freely into the Great Scar Limestone. This is a very local feature

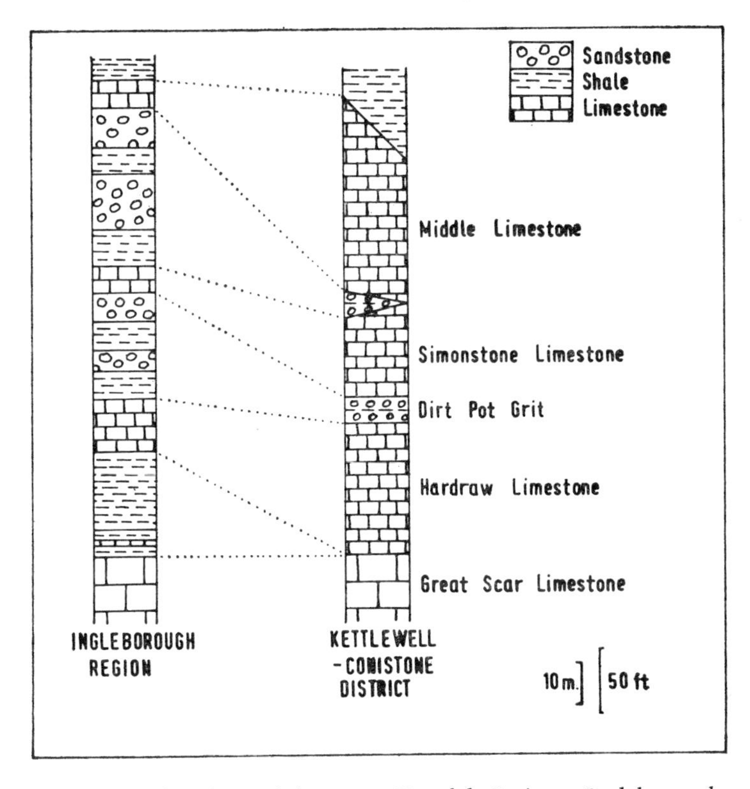

Fig 7 The lithology of the Lower Yoredale Series at Ingleborough and the Kettlewell–Conistone district

for just across the valley Birks Fell Cave starts only at the top of the Great Scar.

The Middle Limestone is the most important of the Yoredale limestones in the number of caves it contains. Goyden Pot, Scrafton Cave, Mossdale Caverns, the lost caverns of the 'Grassington Naturals' and the first 5km of passage in Langcliffe Pot are the best known examples. Underlying the Middle Limestone are a series of non-calcareous beds and these are normally capped by a strong, impervious sandstone bed which is clearly seen where it floors the whole length of Marathon Passage in Mossdale

Cavern. Beneath these shales and sandstones lies the Simonstone Limestone, containing no known major lengths of cave passage, and underlying this is another non-calcareous sequence, including the Dirt Pot Grit, which separates it from the Hardraw and Great Scar Limestones (see Fig 7).

However, all the waters which sink into the Middle Limestone in the Mossdale and Langcliffe areas are known to pass right through this varied sequence before resurging from the Great Scar Limestone at Black Keld. It is only the discovery, in 1970, of the extensions to Langcliffe Pot which has given a partial solution to this hydrological problem. The Langcliffe stream flows for over 3km in the Middle Limestone and then drops abruptly into the Simonstone; this appears to take place where the sandstone is locally very shaly and calcareous, though the initiation of the passage was probably aided by jointing. In the Simonstone Limestone there is only about 100m of passage, and this is floored by the Dirt Pot Grit; then the stream goes through the grit in a boulder choke which obscures all the geological detail. From there the cave continues in the Hardraw and Great Scar Limestones which make an unbroken calcareous sequence.

South of Mossdale, the Yoredale clastic beds may thin out completely, so there appear to be, in all, three ways in which water passes through the grits and shales between the Yoredale and Great Scar Limestones:

(1) Down through faults and major joints;
(2) through locally calcareous facies within the sandstones; and
(3) over the edge of the clastic beds.

Overlying the Yoredale Series is the Millstone Grit which has almost no influence on the karst hydrology. However, the base of the Millstone Grit is strongly unconformable and this is important in the east, around Wharfedale, where the basal Grassington Grits rests directly on successively lower beds to the south, so that around the villages of Grassington and Greenhow some or all of the Yoredale limestones are completely absent (Black,

1950; Dunham & Stubblefield, 1945). At Greenhow the complete thickness of the Yoredales is reduced to 4m.

STRUCTURE OF THE CARBONIFEROUS ROCKS

Situated on the southern margin of the relatively stable Askrigg Block, the Carboniferous rocks of the karst region have a broadly simple structure, dipping a few degrees to the north but sharply truncated to the south by the Craven Faults.

Over most of the area the regional northerly dip is only interrupted by very gentle folds with east–west axes, and Fig 8 shows the distribution of some of these to the west of Ingleborough. Over these folds the dips still rarely exceed 5° though the direction may vary, but even folds of these low amplitudes have significantly influenced cave development where vadose water has tended to collect in the synclines.

Adjacent to the Craven and Dent Faults the Carboniferous rocks have been very much more distorted. The extreme case is in Barbondale where the Great Scar Limestone has been tipped to a vertical position against the Dent Fault. Dips of 45° and more are not unusual in the blocks of limestone between the various elements of the Craven Fault zone. This is clearly seen in the Meal Bank Quarry at Ingleton, and beside the Kingsdale road where the Great Scar is exposed in a series of small but tight folds immediately north of the North Craven Fault. The contorted periclinal anticline through Stump Cross and Greenhow is also worthy of mention, and it is noticeable that all these folds adjacent to the Craven Faults have their axes, where measurable, parallel to the same fault zone.

The Carboniferous rocks were principally affected by Hercynian movements, though important intra-Carboniferous earth movements also took place locally along the Craven Fault zone; these have been responsible for the erosion of the Yoredales and subsequent unconformable succession of the Grassington Grit around the type locality.

Any study of faulting and jointing in the region is naturally dominated by the Craven and Dent Fault systems. The Craven Faults run mainly west-north-west to east-south-east along the southern margin of the main karst area (see Fig 3) and have an unusually long history. They were active in Carboniferous times when they formed the depositional boundary between the limestone facies on the Askrigg Block and the more varied trough facies to the south. However, the fault zone has also been active in post-Carboniferous times and over most of its length has at least two major elements, while a third, the South Craven Fault, diverges to the south near Settle. The North Craven Fault is perhaps the most constant and it is clearly seen on the west bank of the River Doe (Kingsdale Beck) just below Pecca Bridge where it is nearly vertical with a breccia-mineral zone less than 60cm thick. Each of the major faults of this zone downthrows the rocks to the south by an order of 300–600m, though Wager (1931) suggested there has also been some transcurrent movement. Between the Craven Faults the Great Scar Limestone is variously fractured and inclined, but contains few known caves other than those between Settle and Malham.

In contrast the Dent Fault, running almost north–south along the western margin of the karst, forms a somewhat narrower zone. One major fault has thrown down the limestones about 600m against the Ordovician slates of Middleton Fell, though there are also many, very much smaller, parallel faults. Contrasting to the clean-cut nature of the Craven Faults, the limestones have been dragged up into an almost vertical position against the Dent Fault, and a number of small tight folds lie parallel to the main fault.

Besides these two major bounding fault systems, there are numerous other smaller faults throughout the area, many being marked by the development on them of unusually large potholes. Death's Head Hole, Meregill Hole, Juniper Gulf, Hull Pot and Birks Fell Cave are but a few of the fault-guided potholes. The orientation of these faults is variable, though many are subparallel to the Craven Faults, and of particular note are the series

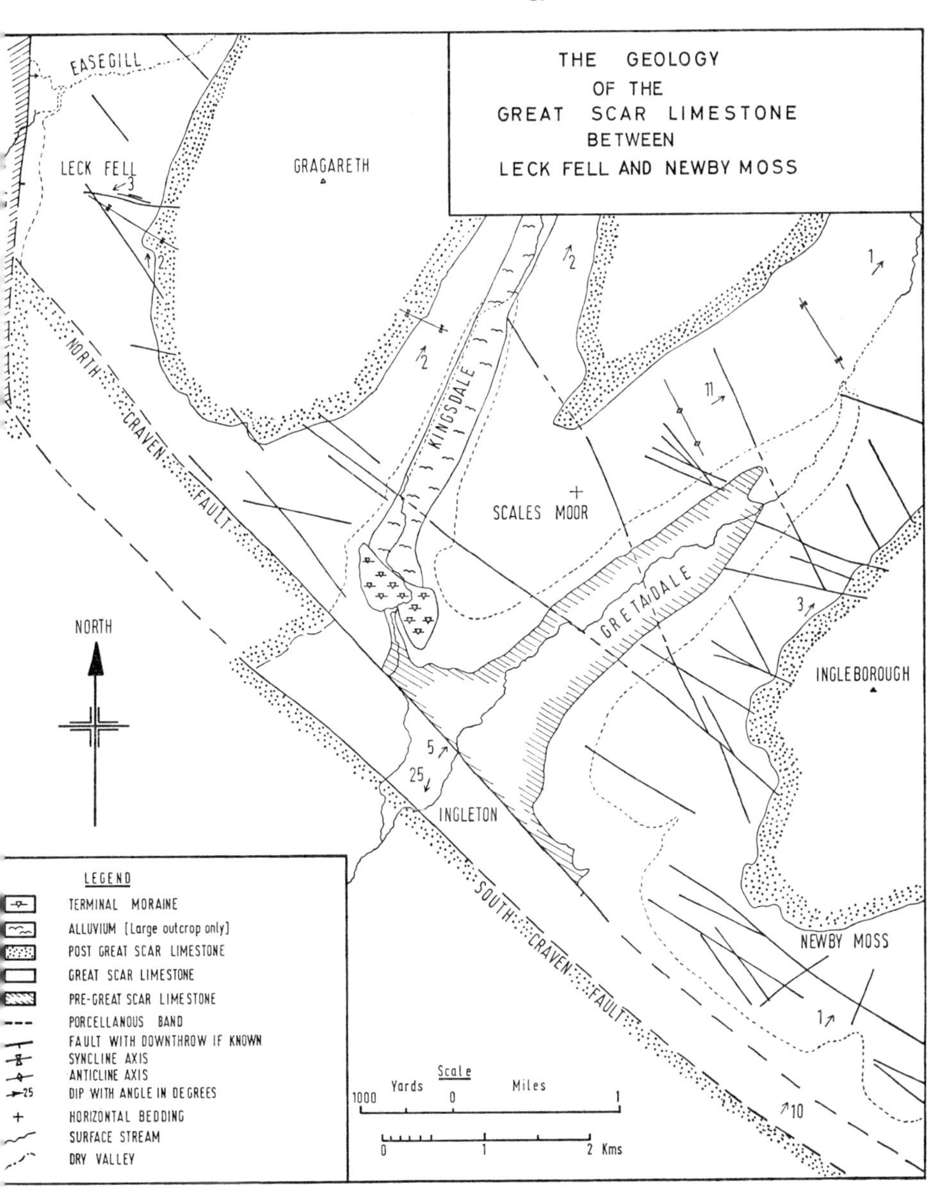

Fig 8 Geological map showing features of the Great Scar Limestone in the area around Ingleton

of stepfaults downthrowing the limestone to the south across Newby Moss, adjacent to the North Craven Fault. Many faults are buried beneath superficial deposits but some of the known ones in the west of the area are shown on Fig 8. An interesting feature of the faults is that they may exhibit two stages of movement, for example, in Rumbling Hole where the displaced shale beds clearly show a vertical movement but the horizontal slickensides indicate a later phase of tear movement.

There is a concentration of faults in the Grassington Moor area, again close to the Craven Faults, and most important is the Bycliffe Fault zone running from the south edge of Conistone Moor, across Hebden Moor to truncate the eastern end of the Stump Cross–Greenhow anticline before meeting the North Craven Fault. This and adjacent faults have probably had a great influence on cave development as they appear to have provided the water in the Yoredale limestones with an open passage through the Dirt Pot Grit into the Great Scar Limestone. Faulting has also been important in influencing the caves of Nidderdale, for it is the erosion of small horst blocks which has exposed the Yoredale limestones, through the cover of Millstone Grit, so that the River Nidd has been diverted underground to form Goyden Pot and the route to Nidd Heads (Ford, 1963).

Most of the faults in the limestone are filled with calcite or dolomite, both white in strong contrast to the dark grey of the limestone. There is usually a zone of brecciation and mineralisation up to 70cm wide, though in some places wider (for example, in the very broad breccia zones running parallel to the North Craven Fault, across Clapham Bottoms, Clapdale Scars and Newby Moss); well-terminated calcite crystals in the Tatham Wife Hole fault also indicate the existence of open cavities. Consequently the faults have been preferentially eroded as they provide open hydrological routes, and this is particularly so in the case of the calcite-filled faults. In contrast, the faults filled with dolomite are less easily eroded, and part of the dolomitised North Craven Fault even forms a topographic prominence. Gouge is rarely seen in the faults as it is so easily eroded, so no estimate of its initial distribution can be made. The extensive sulphide–baryte mineralisation of the Grassington mining district really falls outside the scope of this account (see Dunham & Stubblefield, 1945 and Dunham, 1959), but it is worth noting that traces of galena and barite in particular have been found throughout the region and even mined on the south-west flank of Ingleborough. Secondary calamine in cave infillings has also been mined at Pikedaw, near Malham (Raistrick, 1954).

Joints are a prominent feature of the limestones throughout the region. The majority of the joints fall into two trends which form a conjugate system. These were first shown by Wager (1931) to exhibit a regional swing: the NW–SE set in the west of the area becoming N–S in the east, and the NE–SW set in the west turning to E–W in the east. A third, but minor, system of tension joints bisects the angle between the conjugate sets. Not all joints however fall into these systems and most abundant of the exceptions are those adjacent and subparallel to the many small faults throughout the area. Also within this general pattern there is a slight variation of direction of the main joint sets with stratigraphic level in the Great Scar Limestone. This is very clearly visible on air photographs of the Allotment area on Ingleborough where the joints swing through about 10° over a stratigraphic range of less than 15m.

The extent of the joints is variable and their vertical amplitude ranges from 3 to 100m. They are all close to vertical, with the exception of thinly distributed and relatively minor joints dipping at angles as low as 30°. The term 'master-joint', though frequently used, seems of little significance; Ford (1971 *a*) has suggested an arbitrary definition where a master-joint is one passing through at least two cycles in the limestone—so it is normally more than 20m deep. Undoubtedly there are some joints of unusually great extent, for example, the one in Alum Pot, but beyond their size they seem to have no special characteristics to distinguish them from their smaller neighbours. As would be expected many of the joints have had a considerable influence over cave development by providing relatively open hydrological routes.

An important control of jointing appears to be the proximity of the North Craven Fault. Close to the Fault the joints are frequent and, more important, very deep, for example on Newby Moss, while away from the Fault the joints are perhaps less frequent but also rarely extend through more than one bed of limestone, for example in Penyghent Gill. This variation in joint depth is reflected in the depths of the caves across the region, as, in the majority of cases, cave streams have descended to greater depths by the utilisation of joints. Doughty (1968) also related joint densities to the limestone lithology.

Bedding plane joints were described by Schwarzacher (1958), and it is certainly true that the weathered beds of limestone are frequently separated by stratigraphic joints. However, Schwarzacher's 'master bedding planes' are a very dubious concept, as the great majority of them are openings weathered out along shale beds, the existence of which he did not recognise. The fresh underground exposures reveal a distinct lack of horizontal joints which have a structural, as opposed to weathering or lithological, origin.

SUPERFICIAL DEPOSITS

A considerable proportion of the solid rocks of the area are covered by a variety of generally thin superficial deposits. As these are more pertinently related to the geomorphology of the region, they are only briefly noted in this chapter.

Glacial deposits predominate and most widespread is featureless boulder clay ranging up to about 12m thick. This covers much of the area up to an altitude of about 450m (1,500ft) and has been found on Ingleborough up to 600m (2,000ft) OD (Tiddeman, 1872). There also occur median and lateral moraines (eg in Gretadale and on the east side of Kingsdale), and terminal moraines (eg across the end of Kingsdale). The glacial erratics perched on the limestone pavements of Norber and Winskill are well known, and a very fine drumlin field is a prominent feature of the topography between Horton and Ribblehead.

Fluviatile alluvium floors many of the valleys, and there are also areas of Recent lacustrine sediments, most notably in Kingsdale. Talus banks are a ubiquitous feature of the limestone scars, and a thick wedge of cemented talus is also exposed below the scar of the North Craven Fault in the Ingleton gorge of the River Doe. Larger alluvial cones also occur, for example, the one on which the village of Arncliffe is built. Considering the extent of the limestone, tufa is a relatively rare feature, but the tufa bank of Janet's Foss in Gordale Beck is worthy of mention. Youngest of all the deposits in the area is the thick layer of peat which is so widespread on the higher fells and exerts such an influence on the local drainage.

3

Karst Geomorphology in North-West England

M. M. Sweeting

The two major karst areas of north-west England—the Ingleborough district and the Morecambe Bay area—have distinctly different characteristics. The differences are basically geological; the low isolated limestone blocks of Morecambe Bay contrast with the extensive high karst plateaux of the Ingleborough district. The geomorphological features also show similarly marked contrast, in the scale of relief, disposition of the limestones and their relationship to impermeable rocks. The two areas are therefore described separately in this chapter.

THE INGLEBOROUGH–MALHAM DISTRICT

As described in chapter 2, the Carboniferous Limestones in this area form a unit dipping gently north-east; they are 200m thick and made up of limestones of great vertical variation but little lateral variation (Sweeting & Sweeting, 1969). These limestones lie upon a rigid floor known as the Askrigg Block which is bounded by the Craven Fault system in the south and the Dent Fault system in the west. The Askrigg Block has been uplifted relative to the surrounding regions and it is this perched nature of the limestone series that gives rise to many of its main, characteristic, landforms and karst hydrology. The effects of rejuvenation along the Craven and Dent Faults are most marked in the south-west of the region near to Ingleton; this is where the limestone is most free draining, and most perched above the adjacent lowlands, giving in parts an available relief of 300m. Though differential uplift of the Askrigg Block has been much less in the eastern part of the area towards Wharfedale, the limestone beds reach their highest extent above sea level in the Malham Tarn area, at about 500m (1,600ft). It was in this eastern part of the area that the Carboniferous Limestone was first exposed and this is the part where the oldest karstic relief forms occur. The whole region was also modified during the Pleistocene glaciations. Much of the relief was carved out during this period, and many landforms from the pre-glacial times have become removed or obscured. The three major influences in the relief of the area are lithology, tectonics and the Pleistocene glaciations.

The effects of lithology

The considerable vertical variation in the beds of the Carboniferous Limestone in north-west Yorkshire is a fundamental feature in the development of both the relief and the caves of the region. In parts of the district, seven different types of limestone may occur in a section of only 70m (Sweeting & Sweeting, 1969), and there are also important shale and mudstone layers in the geological succession (Waltham, 1971 *a*). The limestones vary from fine porcellanous varieties, such as the Porcellanous Band, with little or no sparite, to sparry limestones with about 90 per cent sparry calcite. Doughty (1968) has also shown how the nature and intensity of the jointing varies with the type of limestone. Thus each bed reacts to weathering and erosion in its own particular way. In general the more sparry limestones tend to form the massive beds and in these both the frequency of the bedding planes and the frequency of the jointing tend to be reduced. Biomicrites and micrites form the less massively bedded sections in which both bedding planes and joint planes are more prominent. In Crystal Beck and Potts Beck, Littondale, where systematic sampling of every bed has been carried out, the massive canyon-like sections of the becks are cut into the sparry limestones, the gentler and wider sections are cut into biomicrites (Sweeting & Sweeting, 1969). Hodgson (1950) has also shown for the same part of the area how the details of the erosion by the present-day streams have been controlled by the initial depositional structures of the limestones.

The more massive sparry limestones have fewer planes of weaknesses for the penetration of water than less sparry types; they therefore tend to remain in vertical walls. They form the more conspicuous scars, and tend to weather slowly. Less sparry beds with their more frequent bedding planes and joints tend to be undercut and to form shallow rock-shelters. Sparry limestones are also the less susceptible to frost weathering, and this is clearly seen in Twistleton Scars and in Trow Gill.

The Porcellanous Band is the nearest approach to a true micrite in this area (Wood, 1941). It is recognised in the field by its white patina acquired by weathering; its fracture is normally conchoidal and the porcellanous lithology is usually cut by many more planes of weakness than sparry beds. The band therefore does not as a rule form conspicuous scars.

Regular alternation of limestone lithologies occurs all over the Ingleborough and Malham district. Together with the thin, weak shale beds within the limestone succession, it is the reason for the development of the stepped scar-like landscape characteristic of the region. Limestone beds of variable lithology give rise to vertical scars of varying height.

Page 50 shows scars typical of the area. In addition to its determining the variations in the type of scar, each bed produces its own type of scree (Sweeting, 1965) and weathered break-down material. As will be shown below, the reaction of each bed to the effects of glaciation has also varied. Thus the limestone pavements of the area, caused essentially, but not only, by glacial erosion, depend for their morphological detail upon the nature of the lithology of the beds.

In general the limestone beds also show variation in porosity,

Page 49

(*above*) Ingleborough from the south-west, with the snow-covered slopes of the Yoredale Series rising above the limestone bench (*below*) Austwick Beck Head where the water flows out at the base of the Great Scar Limestone

Page 50

(*above*) Twistleton Scars formed by the contrasting lithologies of the different beds of limestone (*below*) Joint and bedding patterns in one of the scars at Twistleton

and this was measured on a suite of samples from Littondale. The sparry beds and the pure micrites have a porosity of 2–3 per cent, much less than the biomicrites with a porosity of 8 per cent; the latter have a more mixed microlithology, consisting of sparry calcite, fossil remains and micritic material. There is also some evidence for differential solubility of the two main types of limestone in Littondale, the sparry limestones being less soluble in a given time than the biomicrites (Sweeting & Sweeting, 1969).

The reaction therefore of the different kinds of limestones to weathering and erosion forms an important study in the karst and cave geomorphology of Ingleborough and Malham. Little attention has so far been given to this aspect of the Yorkshire caves, though one initial study (Waltham, 1971 *b*) has shown the scope and complexity of the problems involved.

The effects of structure and tectonics

One of the main features of the Ingleborough–Malham area is the perching of the limestones above the surrounding lowlands, a feature which is particularly marked in the western sector. This means that most of the limestones are, to a large extent, freely draining. The perching tends to give the area an aspect of high mountain karst despite the relatively low absolute height of the district—around 500m (1,500ft). The main reason for the perched nature of the limestone beds is their situation upon the edge of the Askrigg Block. That differential movement is still taking place along these fault systems is evidenced by the repeated occurrence of earthquakes, albeit small ones, along the edge of the Block. Thus tremors have occurred in 1947, and more recently in August 1970. It is possible that these tremors are but the continuation of movements which have taken place along the Craven and Dent Faults throughout the later Tertiary and Quaternary periods.

The proximity of the limestones to major fault zones is therefore always evident, especially along the southern and western margins. Furthermore, Wager (1931) demonstrated that the directions of the joints were partly related to the major faulting along the Craven Fault system. Doughty (1968) has shown that the frequency of jointing varies within the limestone sequence, some beds (usually the massive sparry beds) having a significantly lower frequency than others (the less massive, more biomicritic ones). The direction, variation and frequency of joints and faults therefore profoundly affect the surface and underground weathering and erosion of the limestones. In the neighbourhood of the main fault systems and minor faults, the joint frequency increases, facilitating the erosion at the surface of small gullies for instance, or underground of cave passages. Penetration of water along joint or fault planes gives rise to solution, and, underground, the breaking away of limestone blocks along lines of joints is widespread. These purely tectonic factors in the development of the caves of the region must not be forgotten.

On a more major scale, rejuvenation of the Askrigg Block along the line of the Craven Fault zone has influenced the relief and the caves even more markedly. This is because in the relatively recent past, probably at the end of the Tertiary and also during the Quaternary and possibly even continuing today, the Askrigg Block has been uplifted relatively to the Craven and Lonsdale lowlands. This rejuvenation has enabled the waters of the block to descend more rapidly into the limestones and to form even deeper underground channels. As a result of the rapid descent of the waters through the limestones, vertical potholes and shafts are common, particularly along the southern margin of the area. This is admirably shown on Newby Moss, where step-faulting associated with the South Craven Fault has assisted in the production of vertical potholes (Waltham, 1970 *b*).

The effects of the Pleistocene glaciations

The tectonic factors have been accentuated by glacial melt waters during the Pleistocene glaciation and particularly during the final stages of the ice melting; during this stage, large quantities of melt water cascaded through the limestone joints and faults and greatly enlarged them. Thus the limestones of the Ingleborough area became even more freely draining.

However, there is evidence to show that the perched drainage of the limestone is not only the result of glaciation, but is also due to rejuvenation. Each river draining off the south-west margin of the Askrigg Block is characterised by rejuvenation gorges, nick points and waterfalls. The valleys have been modified by glacial melt-water channels, and are also affected by marked lithological variation of their bedrock, for example, the alternating grits and slates of the Ingletonian series responsible for the Ingleton waterfalls. But the recession nick points in the different valleys are so regularly disposed and so obviously related to the valley sizes, that regional rejuvenation must have been an important factor in their development. East of Ribblesdale, rejuvenation features do occur in the valleys, but in this part of the area the edge of the Askrigg Block is not so simply defined and is complicated by the resurrected relief along the Middle Craven Fault. Thus the splendid embayments of Malham Cove and Goredale Scar probably have a more complex origin.

Superimposed upon these recession nick points are the effects of glacial melt waters; these are seen in Leck Beck valley and in Trow Gill, Clapdale. In Clapdale, sections of the valleys both above and below Ingleborough Cave have been deepened by melt waters. The nearby dry valley of Clapham Bottoms, however, has not been substantially altered by melt water, and nor has Cote Gill at Newby; indeed the former valley contains deposits of boulder clay. In the Ribble valley between Stainforth and Settle sufficient remains of the old meander scars can still be seen to remind one of the fact that despite glaciation, river erosion and downcutting has still been very important. Stages in the downcutting of the Ribble between Stainforth and Settle are shown in Fig. 9. Thus the relief of the Ingleborough and Malham districts, although much altered by glaciation and glacial melt waters, cannot be regarded entirely as of glacial origin. None the less the influence of glaciation is as important as that of the other factors already discussed.

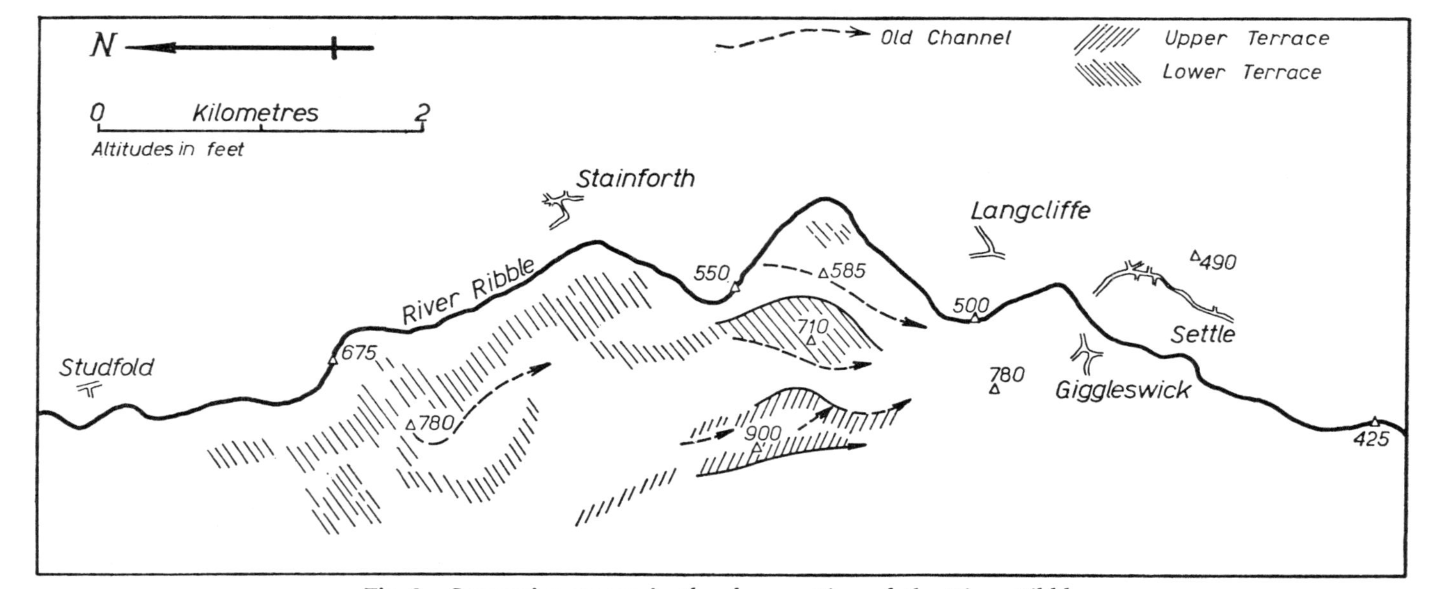

Fig 9 Successive stages in the downcutting of the River Ribble

The main relief forms

As yet the glacial history of this part of northern England is by no means fully known. From rather fragmentary evidence it is assumed that there have been two glaciations in the area, and there are indications of two types of drift. The older is sandier, and found in only a few relatively protected locations such as beneath the terminal moraine at the southern end of Kingsdale. The newer drift is the characteristic drift of this part of the Pennines, very stony and with prominent relief forms, covering almost all the region except where the rock is bare. No intensive study has been made of the drift deposits of this part of northern England since the original geological survey (Dakyns *et al*, 1890), though Raistrick (1930) has described some of the valley moraines.

Despite the large quantities of boulder clay that plaster the limestone hills of the Ingleborough–Malham district, the glaciation in this area was predominantly erosional in its effect. Ice advanced upon the region chiefly from the north and came over the divide between Wensleydale and the southern Dales. In its advance up the southern valley tributaries of the Ure it was under great pressure and several of these valleys show evidence of extreme modification by ice (for example the head of Bishopdale, the head of Snaizeholme, and the head of Deepdale). Clayton (1966) maintains that ice erosion was so great that the 'through' valleys of Cam and of High Green Field at the head of the Ribble and of the Wharfe are the result of glacial erosion and diffluence. He also explains the arc-shaped valleys between the head of the Ribble and the Wharfe as due to glacial erosion; however, the arc-shaped divides between these valleys are so conspicuous and so often repeated features of the relief of the region that they may have a structural rather than a predominantly erosional cause (Hughes, 1901). In addition to the creation of these through valleys, the main valleys themselves—Ribblesdale, Wharfedale, Littondale, Gretadale (Chapel-le-Dale) and Kingsdale—have all been much deepened by ice. Thus in Wharfedale and Littondale at least 60m of overdeepening has taken place. Glacial erosion and excavation is greatest in the central part of the area, in the neighbourhood of the headstreams of the Wharfe and around the north-east and north-west flanks of Ingleborough.

In addition to deepening the valleys, ice scoured and planed the limestones. It is probable that the limestones were subjected to differential weathering during the Tertiary period; this was further accentuated by the scouring action of the ice, which basically created the scars, terraces and limestone pavements.

Goodchild (1875) showed that the scars are best developed where the direction of ice flow was along them, as in most of the main valleys (Chapel-le-Dale, Kingsdale, Littondale); and that they are least developed where the ice flowed across them, as at the head of Snaizeholme Beck where the ice flowed over into the head of Langstrothdale. It is possible that in places the erosive action of the ice was so great as to obliterate the terrace effect of the scars, by a streamlining glacial effect, as on the northern slopes of Penyghent. The main hills such as Ingleborough projected through the ice, so that their southern slopes were protected from the main erosive action of the glaciers. Instead the action of the ice was largely depositional in the lee of the hills, resulting in thick layers of boulder clay and few scars and terraces, for example on Newby Moss and Ireby Fell, respectively south of Ingleborough and Gragareth. The scars also give the impression of being glacially scoured limestones of variable weakness and strength; the variability in strength of the limestone beds and the associated shale and mudstone bands has been picked out by glacial erosion to cause the platform-like nature of the relief. Thus glacial erosion has emphasised the lithological differences already referred to.

The limestone pavements

The effects of glacial scour upon beds of different strength has also helped to produce the limestone pavements of the area. The term 'limestone pavement' is given to the bare rock outcrops occurring in the karst areas, usually, but not essentially, on more or less horizontal beds. Pavements consist of flat-topped rock outcrops known as clints, separated by widened joints or cracks known locally as grikes (Sweeting, 1965). In general the more massive and sparry limestones give rise to more conspicuous pavements and clints which in very sparry limestones may occupy up to $100m^2$. Average clints in sparry limestones are about 2m × 1m in size. The clint surfaces are often dissected by grooves or runnels, formed by solution by acid peaty waters or waters draining from morainic deposits. The less strong limestone beds, the biomicrites, give rise to less extensive clints which have a greater tendency to flake and peel away, especially under the influence of frost. The widened joints or fissures associated with the clints, the grikes, normally vary with the type of limestone. In the sparry beds the grikes tend to be deep, but narrow, in keeping with the general rule that the sparry beds are only soluble along their lines of weakness; grikes in sparry limestones may be from 12 to 25cm wide and from 2 to 3m deep. In the less sparry beds grikes are usually wider but normally much more shallow, from 25cm to 1m wide and 50cm to 1m deep.

Glaciation must have removed the top soil from the limestone and though there is enough evidence to show that during some phases of the post-glacial the limestones have been covered by trees and even by peat, the effect of the glaciations in removing the soil cover must have been such as to make it extremely difficult for the soil cover to be replaced. Thus (as Roglič (1965) has said of the karst of Yugoslavia), the limestone region of Ingleborough and Malham is 'sensitive to cultural influences'; slight changes in both climatic conditions and in conditions of grazing and management give rise to great changes in the vegetation cover. During the post-glacial period the limestone pavements have been subject to many vegetational changes. At the present time it is known that the grazing of sheep in particular, and also of cows, leads to a reduction of vegetation cover. For instance, in the neighbourhood of the Malham Tarn House an area of pavement has been fenced in and grazing restricted; in this area shrubby vegetation is regenerating. The writer herself has also closely watched for about twenty years an area of limestone pavement on Twistleton Scars near Ingleton. This area is at an altitude of 350m (1,100ft) and is an open site; it has been examined yearly and the depth of the soil and vegetation in the grikes and the limit of vegetation on the pavement have been recorded. In general, during the last twenty years the height of the plants and the depth of soil in the grikes has increased, and overall the grikes have become from 15 to 30cm shallower. On the exposed pavement the spread of the limit of the vegetation has been much less, from 4 to 10cm only. On some parts of the pavement, marks made in the preceding ten years are all visible, indicating great stability in the vegetation. On the same area of pavement, small solution hollows have developed during the period of observation that did not exist at the beginning of the experiment. It is clear that such hollows have been aided in their development by the intermittent growth of lichens and mosses directly on the limestone.

Such changes may have altered the original ice-scoured limestone pavements but many of the beds are of sufficient strength and impermeability to resist weathering attack and may not be greatly altered since they were first formed. This is particularly true of those beds which are highly sparry in content or resistant in other ways to chemical- or frost-weathering. Such beds occur on Scar Close on the north-west slopes of Ingleborough and on Scales Moor on the south-east side of Whernside, where unindented and scoured pavements occur. In these beds not only is the limestone very sparry but it also contains quartz grains which may make it more resistant to chemical attack. Pavements formed on beds that are more susceptible to chemical weathering have been lowered by solution since their formation and glacial erratics (often of Yoredale or Millstone Grit) are perched by up to 30cm upon pedestals of limestone protected from rainfall by the erratics. This kind of pavement can be seen on Souther Scales

Scars, south of Scar Close and also on Long Scars near the famous perched boulders at Norber. Pavements formed on weaker beds, that are particularly susceptible to frost action, tend to break down into flakes and are the most rapidly destroyed; they are frequently grassed over, though the pattern of the former pavement can often be seen through the grass cover, as on parts of Thwaite Scars near Clapdale.

Though it is probable that the pavements have been covered by vegetation and re-exposed possibly more than once, they are in general relict landforms; this is proved by the controversial issue raised by their being quarried away in the area under consideration. For many years farmers have been willing to let quarry men remove the top strong beds of the clints in the hope that the exposed rocks will become covered by turf to increase their acreage of grassland. This procedure was successful in the nineteenth and early twentieth centuries in Littondale. Once the stronger beds are removed there is a greater chance of vegetation cover developing by spreading from that already present in the grikes, and the pavements are then destroyed. Clint removal has more lately taken place on the eastern slopes of Whernside, but it is to be hoped that further destruction of the limestone pavements will be controlled.

The natural erosion of the stronger beds is mainly by chemical action. Most clints are scoured by rounded grooves or runnels (rundkarren) which are usually up to 25cm deep and wide, and typically 1–2m long. Such runnels may have been initiated by solution from peaty waters—solution enhanced by the sponge-like effect of damp, matted vegetation. This is suggested by the smooth outlines of the ridges and furrows, which contrast with the crinkled and sharp karren forms developed by solution by direct rainfall (Bögli, 1960). Most of the runnels are now without any vegetation. Thus it is believed that the runnels were formed when the clints carried more peat or light vegetation than they do now. The decay of the vegetation might have been caused by slight change in climate, but might also have been caused by over-grazing or the cutting down of woodland as occurred in the Iron Age and later periods in this part of England.

Solution also takes place along the edges of the clints, in the joints which become enlarged to form grikes. The age and rate of opening of the grikes is debatable. If limestone surfaces below glacial erratic blocks are examined it will frequently be found that the surface is scoured (often with striae in protected places) and that the joints are unopened. This points to the conclusion that the grikes have been opened since the deposition of the erratic blocks; this time is usually assumed to have been at the final phase of the glaciation in England and gives a period of about 12,000 years for the grike formation. This is consistent with other work by Bögli (1961) in the Alps, and by Williams (1966) in Ireland. At the present time many of the grikes are being filled up with debris and vegetation and it is this kind of debris which gives rise to the acidic waters which dissolve the limestone. Williams (1966) is of the opinion that grikes form only with the assistance of soil and vegetation. Moreover, Piggott (1965) considers that the limestone pavements represent a truncated profile —truncated by the ice—so that any grikes or widened joints that existed prior to the last ice in the area were presumably scoured by the ice; this again suggests that the grikes have developed since the final phase of the ice. From measurements of the carbonate content of the waters in the area, there is also no doubt that the solution rate of the limestone is adequate to allow post-glacial widening of the joints to give the modern grikes.

Glacial striae are not usually preserved on limestones, but there are one or two places where they can still be seen. One is the famous locality discussed by Tiddeman (1872) at Long Kin East Cave on the south-eastern slopes of Ingleborough. This site, which has been closely observed by the writer for a number of years, is one where nearly 2m of sandy boulder clay overlies a striated pavement of hard blue limestone. In 1951 some of the boulder clay was removed to expose a fresh piece of pavement. All traces of the striae then revealed disappeared within ten years (Sweeting, 1965). By 1970, the pavement was incised to a depth of 10cm by small streams coming off the boulder clay. This clearly contrasts the active solution on the exposed limestone with the almost complete lack of solution beneath the protecting layer of boulder clay.

The sinkholes and closed depressions

The glaciation clearly removed much material from the limestone surface, especially where the rocks above the limestone were weak and soft. This is important in connection with the Yoredale shales and thin sandstones which lie above the top beds of the Great Scar Limestone. It is difficult to estimate the amount of material above the limestones which has been removed, but some observations may be made. One of the most striking features about many of the larger potholes and caves is that their entrances lie in front of the present boundary between the Yoredales and the Great Scar Limestone. Normally any small streams which reach the limestones from the Yoredales sink into joints and cracks in the limestones immediately. Only some of the larger streams flow for any distance on the limestone, so the majority of the water today sinks well above the larger potholes which are further out on the benches. This is true of Gaping Gill, where Fell Beck now sinks above the main shaft into joints in the bed of the Beck and only under very wet conditions does any large amount of water fall over the lip of the main shaft. Examination of the shale–limestone boundary also shows that some of the largest collapse hollows of the area lie in front of it, for instance Braithwaite Wife Hole, and many of the holes along the edge of the limestone–shale boundary on the south side of Penyghent.

In limestones of the purity of the highest beds in the Great Scar Limestone, water dissolves the rock along the joints rapidly and it is likely that streams would have sunk almost immediately at the Yoredale boundary. Collapse and settlement of the limestones would have taken place at these points where the water enters the rock. Therefore this is where large swallow holes in the limestone should occur, as distinct from small solutional dolines. It is therefore tempting to suggest that the large hollows in the front of the present limestone–shale boundary represent a former boundary; the glaciations have scoured the limestone surface and removed a substantial amount of the shale cover. In places, as on the north-west slopes of Ingleborough, where erosion has been particularly marked, the shale boundary was pushed back by up to 600m. A reconstruction of the pre-last glacial limestone–shale boundary has been drawn in Fig 10. Since the last glaciation the post-glacial streams have not had enough time to form such

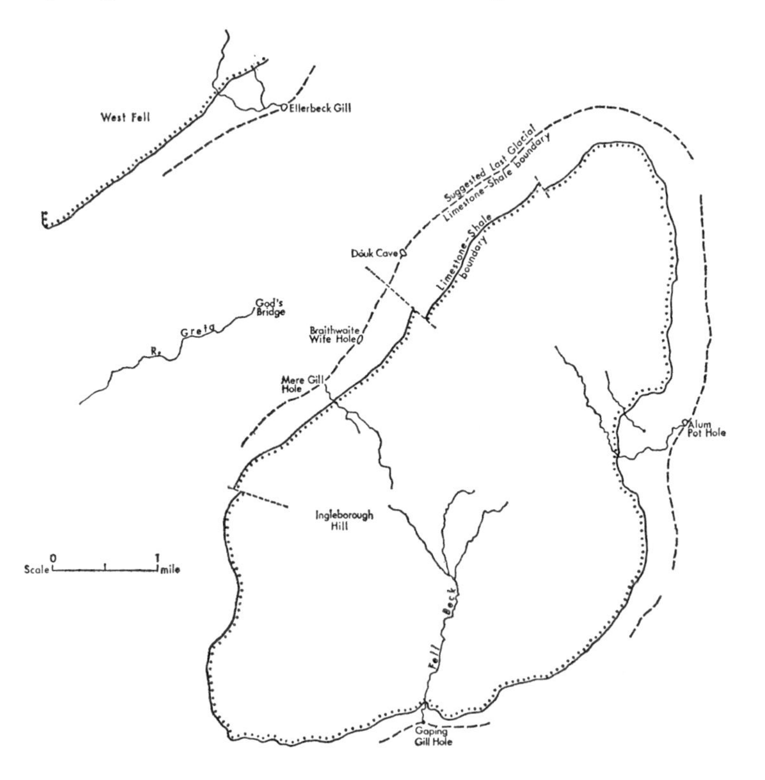

Fig 10 A tentative reconstruction of the shale boundary around Ingleborough previous to the last glaciation

large caves and potholes; thus the streams which enter the limestone south-east of Braithwaite Wife Hole sink into small caves about 1m high. The same argument has been used by Williams (1964) in his study of hollows in the Burren of County Clare, Ireland.

This idea of large closed hollows originating at boundaries between the top beds of the Great Scar Limestone and the overlying Yoredales can possibly be used to explain the occurrence of similar but older hollows in the parts of the area where the Yoredale limestones are now almost eroded away. Large closed depressions—up to 400m in diameter—are found north-east of Malham Tarn around Parson's Pulpit and High Mark, south-west of the Tarn in the Grizedales area, and west of the Ribble valley near the village of Feizor. In all these areas the hollows are in the top beds of the limestone and it is possible that they represent large pothole-type features formerly occurring at the limestone–shale boundary. Their perpetuation (rather than their gradual obliteration by in-filling and collapse) may be due to the patches of shale debris or glacial drift which remain in these hollows causing drainage to collect further in them (Clayton, 1966). Once a wet-soil cover becomes established in such hollows, limestone is dissolved away by peat and humic acids, and, as Clayton says, 'an irreversible process had begun . . . the contrast in soil depth between the incipient ridges and the developing hollows, intensified the contrast in the amount of water present and so contributed to the inexorable development of the hollows and the ever-slower reduction of the interfluves between them'.

Many of the hollows so far described are large in diameter in relation to the depth, and show some signs of slope retreat, if indeed not collapse. However, in many ways most characteristic of the Ingleborough district is the pothole, a narrow, deep shaft at the moor level and usually now with relatively little drainage entering it. It shows no signs of collapse and has fluted solutional walls. This type is characteristic of the Allotment, east of Gaping Gill, and of most of the southern slopes of Ingleborough. Such shafts are not necessarily of the same origin as the sinkholes at the shale–limestone boundary. They are, in many cases, relict features, not being enlarged at the present time. There are two important characteristics of these shafts. First, they nearly all occur in the top, relatively strong, beds of the limestone. Secondly, the locations of most of them are closely related to the occurrence of stony and sandy boulder clay. Few open directly from the limestone pavements and only a small proportion are at the shale margin. On the drift-covered Allotment of Ingleborough, in an area of just over $1km^2$, for example, there are twenty such shafts; further concentrations occur on Newby and Dowlass Mosses, again areas with thick boulder clay. These groups of shafts are therefore related in some way to the glaciation, its drift and its other effects, and are evidence of glacial karst.

Shafts similar to those on Ingleborough have been found adjacent to and actually underneath the glaciers of the Columbia Icefield (D. C. Ford, 1971). It is tempting therefore to consider the potholes in the Ingleborough area as having been created in a similar way, either by being formed under an ice cap or by forming underneath ground-moraine associated with glacial melt water. After the melting away of the glaciers, peaty vegetation growing upon the moraine would provide acidic waters to enlarge the holes further. During the cold of the glacial phases, the potholes and caves were frozen and little erosion and solution took place; but during the retreat phases when there was much stagnant and dead ice, then the limestones would be full of melt water and this could be the times when both the caves and the shaft-like holes were formed.

These arguments reinforce the need for the careful reinvestigation of the boulder clays, moraines and glacial land-forms of the area, an investigation which would throw great light upon the actual disposition of the ice fronts in the last glaciation; they further illustrate the contention that the many Karst landforms and cave systems of the Ingleborough–Malham area cannot be studied in isolation without a thorough understanding of the glacial phases in the area.

The major part of the excavation of the main south draining valleys has clearly been glacial. Littondale and Kingsdale in particular have classic U-shaped profiles diagnostic of valley glacier development. It is perhaps debatable how much pre-glacial and inter-glacial fluvial excavation of the valleys took place. However, the evidence within the cave systems (see chapter 4) does suggest pre-glacial fluvial valleys no more than half their present depth, followed by rapid excavation to today's dimensions. The speed of this second stage of deepening is commensurate with glacial excavation, and not slow fluvial erosion followed by glacial trimming to give the present cross-profiles.

Modification of the valley relief by post-glacial streams has been minor, though glacial melt-water channels are abundant. It is probable that many of the south draining valleys have been modified by melt water flowing from the ice remnants situated on the broad limestone plateaux of hills like Inglcborough, Penyghent and Gragareth. Furthermore, the blocking of older preglacial courses by boulder clay and drumlins has given rise to many new stream courses. This is particularly well seen in Thorns Gill, one of the headstreams of the Ribble, and at Ease Gill at the head of Leck Beck.

The valleys of Leck Beck and Clapdale, in particular, have been overdeepened by melt water, and there is also evidence of this in Cowside Beck, near Arncliffe. Many dry valleys on the limestones have also been regarded as of melt-water origin, notably upper Trow Gill, the Watlowes valley below Malham Tarn, and the lower parts of Conistone Dib, east of Conistone in Wharfedale. It is possible that many others are of melt-water or of periglacial origin, but, as indicated earlier, the regularity in the occurrence of some of the groups of dry valleys suggest they have been formed by factors other than melt-water or periglacial action. Trow Gill has the appearance of a quickly cut gorge formed in association with a melting glacial snout; both Trow Gill and Clapdale are overdeepened compared with Clapham Bottoms.

The solution of limestone under the drift cover causes subsidence of the drift material into holes in the limestone, giving rise to small closed hollows, subsidence cones, known locally as 'shake-holes'. The size of shake-hole is dependent upon the depth of overburden lying upon the limestone, and they are particularly numerous where the thickness is 2–3m. They are rare where the drift cover is over 10m thick; where they do occur in thick drift they are of large diameter—as for instance the group at Notts Pot, on Leck Fell, which are up to 25m in diameter and 10m deep. The sides of shake-holes become stable with a grass cover, if no surface stream enters the hole, at an angle of 30–35°. But where a stream enters the hole the sides steepen and may be 60°. Detailed mapping shows that these subsidence cones are controlled partly by the lines of major jointing and fracturing in the limestones and also by the disposition of the boulder clay and drumlins. Where drainage channels form on the drift hills, such channels terminate at shake-holes at the drift margins. The larger part of the shake-hole depression is formed in the drift material. In most cases even the floor of the depression does not expose limestone, though some reveal clint-like blocks of limestone. In the latter cases the limestone blocks are normally in situ—collapse, within the limestone, has not been an integral feature of shake-hole formation (Clayton, 1966). These depressions have formed by solution of the limestone along joints, and subsequent subsidence of the drift, together with sapping and removal of the finer fraction of the drift deeper into the limestones. In this way some shake-holes are self-sealing though the majority are still drained into the bedrock limestone. Though shake-holes are a feature of the drift deposits, they are also concentrated into spectacular lines along the buried shale–limestone contacts.

Erosional history of the area

Though the Ingleborough area is an example of a glacial karst, the Tertiary period of weathering and erosion, before the glaciation, was none the less important in the evolution of the geomor-

Page 67

(*above*) Clint blocks dissected by karren grooves on Twistleton Scars (*below*) Centripetal karren patterns on clint blocks at Newbiggin Crags, Westmorland

Page 68

Vertical air photograph of Chapel-le-Dale. White limestone pavements down the centre separate the valley floor on the right from the high Yoredale slopes on the left. The main stream on the left sinks into Meregill Hole, and the God's Bridge resurgence is centre right. The large holes north of (*below*) Meregill are Braithwaite Wife Hole and Great Douk Cave. A number of faults cut the limestone in the upper part of the view; Meregill Hole lies on one of these phology. This is because, though the glaciation stripped much of the Yoredale shale cover away, there was a considerable extent of the limestones exposed before the glaciation. This applies particularly to the regions south-west and north-east of Malham Tarn, and to the eastern and south-eastern slopes of Ingleborough between Moughton and Giggleswick Scars. In these areas the oldest relief on the limestones is found. Despite the glacial modification, some, if not a great deal, of the relief of these districts must be of pre-glacial origin.

While it may not be possible to decipher all the stages of the Tertiary dissection of north-west Yorkshire, some facts may be ascertained. There has been much literature on the top summit surface of the Askrigg Block. It was first mentioned by Hughes (1901), later discussed by Trotter (1929) and Hudson (1933), and has recently been statistically analysed (trend-surface analysis) by King (1969). The latter analysis confirmed what had already been established by other methods, that the peneplain surface was probably warped early in its history and that the pattern of the valleys in the area has resulted from that warping. This explains the south-easterly trends of the Wharfe and Skirfare, the southerly trend of the Ribble, and the south-westerly course of the Greta. It also confirms that the minor warpings of an originally eastward-trending erosion surface have played a part in the establishment of the drainage pattern. Some elements in the drainage pattern are shown in Fig 11 and this is based on detailed mapping of the top surface in the area and by a consideration of the valleys and cols; this interpretation is similar to the conclusions reached by trend-surface analysis. It will be seen that in the Wharfe valley the easterly draining valley sections are considered to be the older, and the southerly-trending sections were formed as a result of the subsequent warping.

The closed area of high elevation revealed by trend-surface analysis of the Askrigg Block is matched by the disposition of the strata in the block. As already noted, the top beds of the Carboniferous Limestone are at their highest point above sea level in the

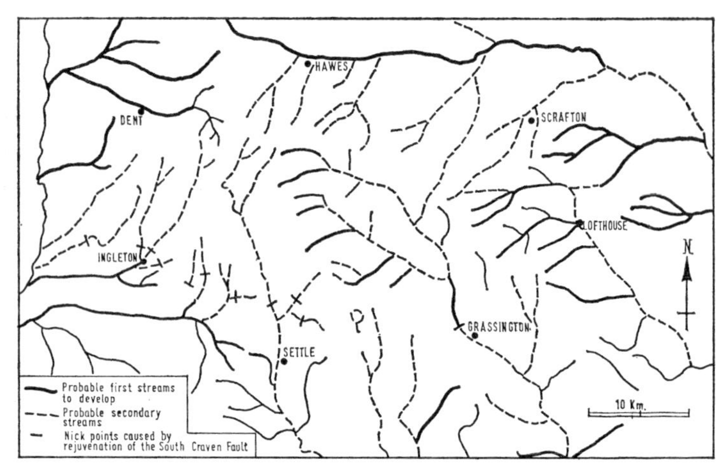

Fig 11 Elements of the drainage patterns between the Rivers Lune and Nidd

neighbourhood of Capon Hall south of Fountains Fell which lies roughly in the centre of the closed area of highest elevation noted by King (1969). At this point, too, the pre-Carboniferous floor also reaches its highest elevation, about 460m (1,450ft). The topmost beds of the Carboniferous Limestone on Malham Moor are at around 550m (1,700ft), whereas on the east in Wharfedale they are between 400 and 430m (1,300 and 1,400ft), and on the west, on Ingleborough and on Whernside, they are also at about 410m (1,350ft). There is, therefore, evidence of a 'structural high' along the line of Fountains Fell and Malham Moor. Thus the initial easterly tilt of this part of the Askrigg Block, followed by a later warping along the line of the Malham Moor structural high might explain the east–west ridges and valleys curving to north–south in the part of the district between the upper Wharfe and the Ribble. Though the divides have probably been overridden by ice and the valleys have become glacial through-valleys, the pattern of curving valleys in the relief is so often repeated that a structural explanation seems more than possible, and is confirmed by the trend-surface analysis.

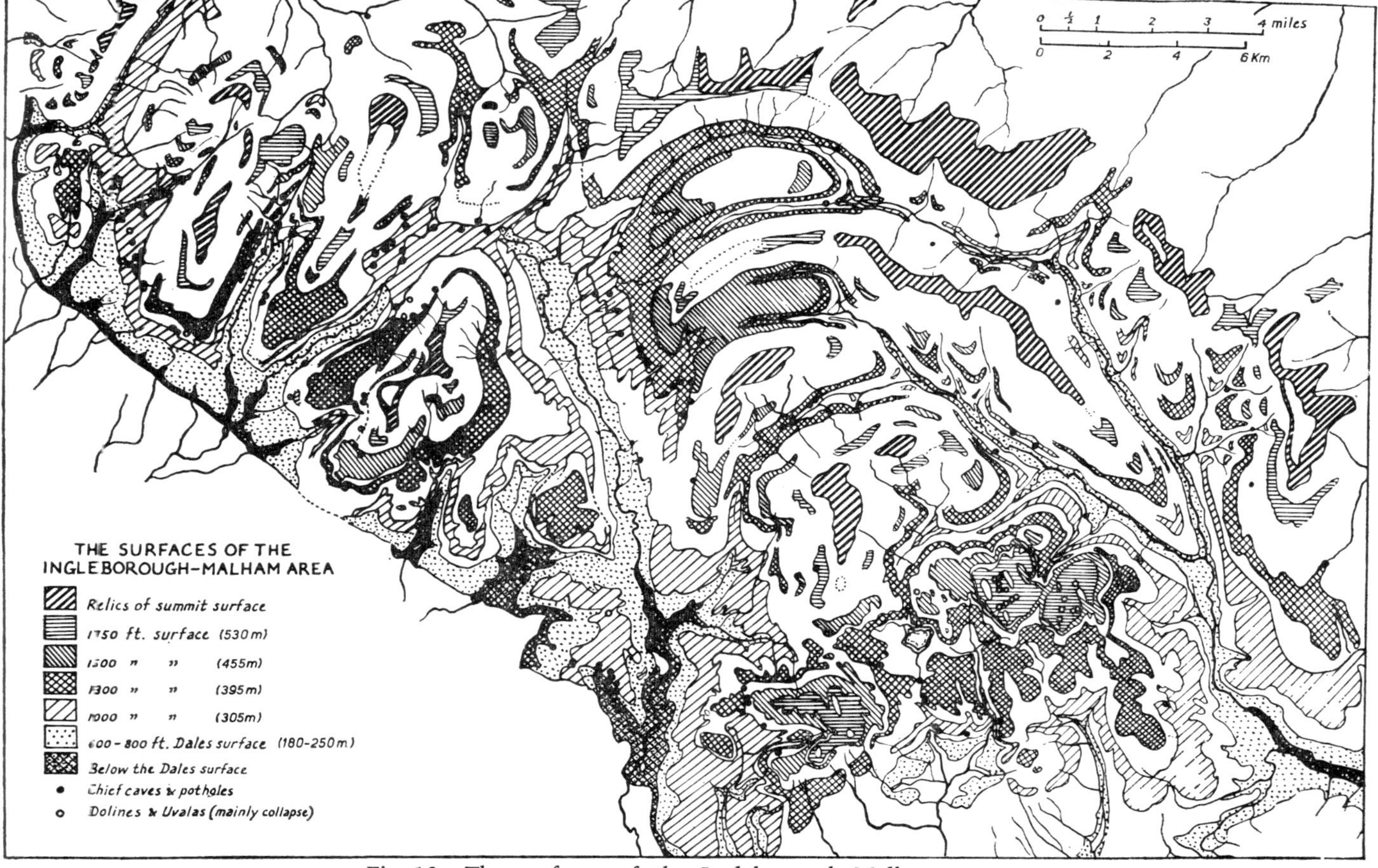

Fig 12 The surfaces of the Ingleborough–Malham area

The dissection of the summit-surface has been accomplished by river and glacial erosion during the Tertiary and Pleistocene. This erosion has been strongly controlled by the generally sub-horizontal structure of the rocks and their contrasting lithologies, and was dependent upon the original and later warpings of the peneplain. Thus many of the present horizontal surfaces are strati-morphs—bedding-plane surfaces stripped by erosion; this, as has been noted by Waltham (1970 *b*), is true of many of the lime-stone surfaces, such as Scales Moor on Whernside, and Moughton and Sulber Scars, on the east side of Ingleborough. However, some erosion surfaces do truncate the bedding of the rocks, for example on Clapdale Scars where the surface cuts across some step-faults, and in the Malham Tarn area where dipping beds are truncated sub-horizontally. Whether any definite erosion stages can be recognised needs much more work in the light of modern geomor-phological thought.

The part of the Malham district between Kilnsey, Arncliffe and Malham Tarn, as pointed out earlier, contains the most evolved karst landforms of the region. This is partly explained by the earlier stripping of the limestones here because of the elevation along the Fountains Fell–Dodd Fell axis; the pre-Carboniferous floor for instance is at over 360m (1,200ft) in the neighbourhood of Malham Tarn, compared with 240m (800ft) at Ingleton.

The regional dip of the limestones is towards the north, but it must initially have been even steeper in the same direction and then tilted back to its present position. This phase of tilting is indicated by the warping of the top erosion surface, when the area was inclined to the south early in its erosion history. Conse-quently all the valleys along the southern margin of the Block drain southwards; the pattern of drainage is clearly visible on Fig 11. Clayton (1966) has pointed out that some of the dry valleys of Malham Moor and Parson's Pulpit may be of periglacial origin following small fault lines. However, such major valleys as that between Knowe Fell and the Highfolds ridge behind the Tarn

House, or that leading up to Middle House, may well be relict valleys of the earlier north–south system. The area around the Tarn itself and Great Close has some of the greatest extent of planed limestone surface in the area, and is much modified by karst processes. However, the large planed area round Malham Tarn and Great Close truncates both the lower beds of the limestone and the pre-Carboniferous non-carbonates. Solution cannot therefore have been the only erosional process, and with the thick cover of glacial drift it is difficult to ascertain how much, if any, of this area may be fairly described as a polje; it is however a randebene (karst margin plain) which is considered to be typical of a karst area (Roglič, 1971).

THE MORECAMBE BAY DISTRICT

The second major limestone area in north-west England is that of the Morecambe Bay district. This extends from near to Kirkby Lonsdale in the Lune valley, round the north shore of Morecambe Bay, into the Furness district (see Fig 31). It is separated geographically from the Ingleborough district by the Lune valley and geologically by the Dent Fault System.

It is an area of contrast with the Askrigg Block. The limestones of the Morecambe Bay area were deposited in a tongue of the main Central Province in which a much greater thickness of Carboniferous rock was laid down; they are 'massif' in type and include thick micrites, sparites and pseudobreccias. They are much more varied laterally than the deposits on the Askrigg Block, and they are (in the present outcrops) without reefs or shoal reefs, except possibly in the south, at Halton Green, near Lancaster (Garwood, 1912). These Carboniferous rocks form much of the northern shore of Morecambe Bay and constitute the southern portion of the ring which borders the Lake District. The limestones are divided by a system of north–south-trending faults into a series of blocks, now separated from each other by estuaries or alluvium-floored depressions. The general dip of the Carboniferous rocks is eastward and the downthrow of each fault is to the west giving a step-like structure to both the geology and the present topography. The general surface of the Silurian, upon which the lowest beds of the Carboniferous were laid down, appears to have been in this area a remarkably even one compared with the rather more uneven relief of the pre-Carboniferous floor in the Askrigg Block. The relief of each block is slightly different, partly because of physiographic factors, but also because the limestone beds and their dip, locally as high as 45°, tend to be different from one block to another, and therefore the weathering is distinctive.

The limestones forming the fault blocks are much lower than those occurring in the Ingleborough district, and the summits range from 100 to 300m (300–1,000ft) above sea level. Their geological history has been more complex than the limestones of Ingleborough and Malham, and the limestones have been exposed for a much longer period of time, starting in the Permo-Triassic periods. In the Ingleborough and Malham area, as has already been shown, the oldest relief is in the south-eastern part where it is believed that the limestones may have been exposed during the later part of the Tertiary era; but much of the relief of the Ingleborough area is relatively new and much of the limestone was exposed for the first time during the Pleistocene period. In the limestone area surrounding Morecambe Bay it is almost certain that the limestones were fully exposed by the end of the Quaternary and that the effect of the glaciation was both to scour the limestones and to deposit a drift cover rather than to strip them of a cover of rocks as has happened around Ingleborough. The existence of haematite in pockets in the limestones in the west of the region, around Ulverston, Millom and Broughton-in-Furness, is believed by some to be due to percolation of iron-rich groundwater from the unconformable Triassic cover, but by others to be a result of upward, hydrothermal mineralisation. It is probable that the limestone has suffered erosion and solution into collapse depressions before the deposition of the Triassic beds; there was thus at least one phase of karsification before the Tertiary and Quaternary periods. The Morecambe Bay karst area is a very old one, but was much modified during the Pleistocene glaciations, and also by the warmer conditions of the Tertiary (Dunham & Rose, 1949). A reassessment of the fossil karst features in the Furness and Ulverston areas is certainly overdue.

The relative antiquity of the limestone relief, and the absence of cover rocks in association with the Carboniferous Limestones, mean that the type of youthful cave and pothole formed at the junction of Yoredale rocks and the Great Scar Limestone in the Ingleborough area is absent; everywhere the limestone forms the highest ground and most caves and potholes that existed at the junction of the limestones with the cover rocks have long since been eroded away, particularly during the glacial period. Thus the full erosional history of the Morecambe Bay area is unknown.

Because the limestones formed the highest hills, Corbel (1957) was of the opinion that the hills surrounding Morecambe Bay represented a relict tropical karst of Tertiary age. This is because under temperate conditions limestone is readily soluble and tends to form low ground, but under tropical conditions limestones form relatively strong rocks and tend to form high ground. Furthermore the isolated blocks of the limestones could be said to resemble isolated cones (or kegeln) typical of a subtropical karst. This thesis is an attractive one, but much work is needed to substantiate it; the isolated hills of limestone may well be expressions of the fault-dissected structure, and many of the other features of the relief equally well relate to the Triassic geomorphology, rather than the tropical climate of the Tertiary.

The limestone blocks are now broken up and are of very limited length and breadth, the largest continuous areas being the Dalton-in-Furness block and the Underbarrow block, south of Kendal. This fact, together with the scour of the limestones by the great glaciers, means that closed hollows in the true karst sense are not very abundant. However, dolines do occur, on Whitbarrow Scar and elsewhere, and hollows a few metres in diameter are found associated with boulder clay. In the neighbourhood of Yealand Redmayne and Beetham there occur larger hollows up to 400m in diameter, and flat-floored, and planed, limestone hollows, which are of a polje type; these have been modified by glacial melt waters and outwash.

The proximity to the coast has complicated the evolution of the Morecambe Bay relief. This is because it is not certain to what extent the late Tertiary and early Pleistocene sea-levels have influenced the relief. Thus a variety of questions remain as yet unanswered. Are the limestone cliffs of the area, as in Silverdale or at Hampsfell, the result of wave erosion, or are they cliffs produced by solution sapping of the limestone in a warmer climate? Are the flat valleys like Silverdale Moss the result of marine planation or of a solutional karst process? And, from the cavers' point of view, are the short caves found in the limestone hills related to former still stands of base-level during the Tertiary and Pleistocene, or are they the result of glacial melt waters over-flowing from stagnant ice, or the result of karst groundwater movements in the limestones long before they became so dis-sected? These and other questions can only be answered by detailed work in the region using modern methods of dating and analysis; some of these discussions in connection with the caves are raised in chapter 11.

The last, and (in terms of the present relief) the most import-ant, event in the region was the last glacial period. As a result of this, the area, despite its low altitude, basically is a glacial karst with many bare rocks, scoured surfaces, and perched erratic blocks. Some of these were mentioned by McKenny Hughes in his article on 'perched blocks' (1886), as for instance those at Cuns-wick Tarn near Kendal and those at Urswick near Ulverston. Some fine glaciated and striated surfaces are to be found on New-biggin Crags, a hill which was in full line of the incoming More-cambe Bay ice; these crags also have limestone perched blocks resting upon other limestone beds. The effect of glacial scour upon the bedded limestones has been slightly different from its effect in the Ingleborough district mostly because of the varying dip of the beds. Dips vary from 5° to 50°, but from 10° to 25° are the most common; the effect of glacial scour upon such dipping beds has been to produce bare exposed dipslopes with limestone pavements alternating with steep cliffs or scars on the up-dip sides. This type of relief is exemplified by Hutton Roof Crag, so named because the succession of stripped beds is said to resemble the slates or flags on a roof; the same kind of relief is seen on Hampsfell near Grange. At the end of the Ice Age these bare limestone hills would have been a stripped (*schichttreppen*) karst, but because of their proximity to sea level and the mild temperatures of the English west coast, they have been more greatly changed by vegetation and solution than the glacial karst of Ingleborough and Malham.

In the massively bedded sections of the limestone, the glacially scoured pavements have become modified by the development of grikes along the joints and the development of runnels and grooves on the sloping limestone surfaces. Because the limestone pavements are inclined, surface and subsurface runnels are much longer than in horizontally bedded rocks and may reach up to 15m as at Hutton Roof. The interrelations of the grikes and the runnels at Hutton Roof have been discussed by Williams (1966). The great majority of the runnels are very well-formed rundkarren—among the best in England. Karren similar to those on Hutton Roof can also be seen as Hampsfell, Farleton Hill, Warton Crag and Whitbarrow Scar. Some of the runnels—the sharp-edged rinnenkarren—have been formed under free atmospheric conditions. However, the majority—the rounded rundkarren—have formed under a cover of vegetation, and the subsurface movement of waters under the vegetation has been controlled by the dip of the limestones (partly because of the thin nature of the soil cover).

The less massively bedded limestones have yielded to frost action, as on Arnside Knott, where scoured limestone pavement has been destroyed and replaced by weathered flaky limestone surfaces; Helsington Burrows, near Kendal, is similar. On Meathop Hill micritic limestones occur and do not support any extensive karren development.

Evidence for climatic change in the Morecambe Bay area has only just begun to be examined but there is no doubt that it is there. Stratified screes are abundant throughout the area and are particularly well developed on the south-east slopes of Whitbarrow Scar and on the south-east side of Arnside Knott. On the south side of Arnside Knott such stratified talus conceals a waterworn and smoothed limestone surface, underneath, with pothole-like features. Thus at this point the limestones have suffered at least two erosion phases. The smoothed and pot-holed surface was formed under a damp, even if not a warm, climate; the layered screes formed under differing climates of severe frost and cold. Evidence for phases such as this should be found in the caves, and the Morecambe Bay area is one which could be a significant one in the history of climatic change in Britain. In particular, a study of the cave deposits, both calcareous and non-calcareous, could be of great value in elucidating their stratigraphy and age.

4

The Geomorphology of the Caves of North-West England

A. C. Waltham

Many hundreds of caves and potholes have been explored in north-west England and their greatest concentration is in the elevated limestone blocks of the Ingleborough district. More than 200km of cave passages are known between Barbondale, in the west, and Nidderdale, in the east. Most caves have been accurately surveyed, and a number have been subject to geomorphological studies. However, the complexity of the morphological detail and the multiphase origins of the caves mean that these studies are not totally conclusive, and there are still many unanswered questions relating to the origin and development of these caves.

The precise mechanism of the opening of the initial fractures in limestone poses a problem which is largely theoretical, and therefore beyond the scope of this account. Useful recent papers on this subject have been published by White & Longyear (1964), Thrailkill (1968) and Atkinson (1968). The initiation of cave development normally takes place with the fractures full of water. These phreatic conditions persist until a time when the fissures are large enough to permit efficient, perhaps downward, drainage of all inflowing precipitation and allogenic stream-water, the latter being precipitation flow which has coalesced in catchment areas on adjacent impermeable rocks. From this time onwards, the sequence and environment of cave development may be inferred from a detailed study of the cave morphology, and this chapter summarises work of this nature in north-west England.

The geomorphic deductions on the controlling factors in development of the explorable caves are relatively easily arrived at; some of these are described below and they have a bearing on the more complex theories of development of the initial 'protocaves'.

PREVIOUS RESEARCH

The many theories of 'cave origin' (or, more accurately, of 'environment of cave development') may be divided broadly into four groups, based upon the division of groundwater environment into the vadose and phreatic zones (Cvijič, 1918).

In 1907, Dwerryhouse suggested a vadose origin for the Yorkshire caves, and this theory was generally popular in other parts of the world at that time (Matson, 1909). Gardner (1935) improved the vadose school of theories by postulating early phreatic opening of the caves by static water, and then the tapping of this water followed by major vadose cave enlargement. A variation on this theme suggested that the main cave erosion was by the invasion of surface streams (Malott, 1937).

Davis (1930) and Bretz (1942) proposed that the major cave development took place in the phreatic zone, but that there were also significant later vadose modifications; they were therefore the main instigators of the 'Two-Cycle Theory'.

The water-table theories of cave formation owe their origins to Swinnerton (1932), though a major contribution to this school of thought was the paper on the Ingleborough caves by Sweeting (1950). It is now widely recognised that a water-table, in the classical sense, does not generally exist in cavernous limestones having an essentially secondary permeability via fractures; local geological features have such a strong influence on the cave patterns that any proposed water-table relating to the main drainage channels becomes an impossibly complex and irregular surface (Zötl, 1959; Drew, 1966). A limestone normally has a dense network of microfissures and the water in most of these may settle to a water-table as in an ordinary uniform aquifer; the main cave conduits will be independent of this water-table—they will be above it or below it, wherever guided by geological structure, and will not relate to it in the way that normal surface rivers relate to a classical water-table. The water in the main conduits accounts, of course, for practically all the flow in cave-bearing limestones, so normally the microfissure water-table can be ignored, except in terms of percolation flow into the main cave passages. However, in the earliest stages of the exposure of a limestone block, before cave development really starts, there are no main conduits, and the microfissure water-table therefore assumes a significant, but short lived, importance. The patterns of future cave development are largely established in these initial stages, so a true water-table may influence cave formation in limestones which, when cavernously eroded, have no overall water-table. A reasonable fracture density may permit caves to develop preferentially in a shallow phreatic environment closely related to a piezometric surface (Ford, 1968). Also, a rest level will exist for the water in any individual passage; this may be perched and local, or it may be at the resurgence level where it will have a wider uniformity. Finally, the main exception to these general rules of cave development is the water-table erosion, with subsequent horizontal cave formation, which is a well-known feature on the margins of poljes in various parts of the world.

A phreatic environment has never been proposed as the sole locus of cave development, though Bretz (1942, 1953) has made a strong case for limestone solution within the phreas. It is perhaps not surprising that there are different caves with origins as proposed in more than one of the above theories. The diagnostic morphological features of limestone caves, as so completely described by Bretz (1942), show that cave systems of both vadose and phreatic origins exist in Yorkshire and north-west England. The relationship between the two main superimposed types is described in a paper by the writer (Waltham, 1970 *b*) which also summarises the modern theories on the region.

When one considers that the limestone topography of north-west England was deeply dissected and rejuvenated by the Pleistocene glaciations, it is only reasonable to expect that the presently active vadose cave systems should be superimposed on a network of abandoned phreatic passages, formed when the limestone was more completely buried and hence containing a more restricted circulation and drainage. A twofold pattern of cave origin is a feature of a great proportion of karst regions, and particularly of those in glaciated regions, such as most of upland Europe. This theory, however, differs from the 'Two-Cycle Theory' of Davis, as it invokes totally independent phreatic and vadose cave systems, in contrast to a series of phreatic passages each modified by vadose erosion, as proposed by Davis (1930).

In north-west England, true water-table caves are only found in the low-lying karst areas around Morecambe Bay. There appear to be no caves of this type in the main upland karst of the Ingleborough region. The concentration, in the latter area, of cave passages at certain levels, originally described as forming at rest levels of a falling water-table (Sweeting, 1950), are now known to be due to geological controls within the confines of the sub-horizontal strata (Waltham, 1970 *b*).

There is today complete agreement among the various students of north-west England karst, that both phreatic and vadose caves are present in large numbers. However, the details of morphology and genesis, including the extent of influence of a water-table or resurgence level, are still debated, and an absolute chronology cannot yet be established to satisfy all cases. This account of the cave geomorphology reviews all schools of thought and emphasises the views which are widely held, or conclusively proved.

THE PHREATIC CAVE SYSTEMS

Most of the larger cave passages in the Ingleborough area are clearly of phreatic origin, as they contain many of the diagnostic features described by Bretz (1942). Roof solution, particularly up

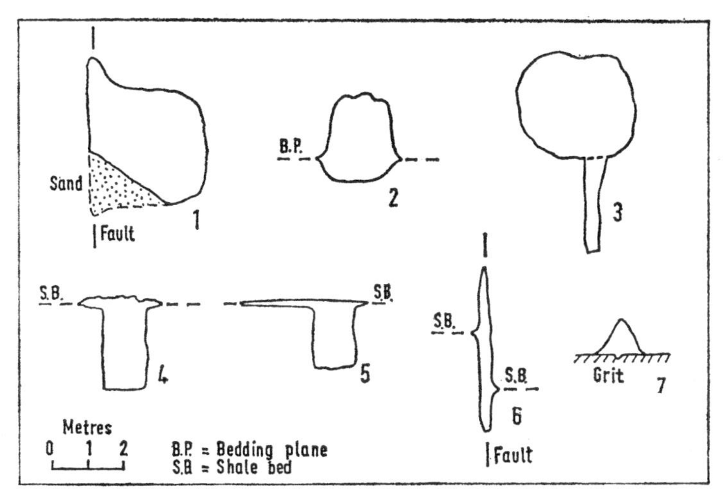

Fig 13 Cross-sections of some characteristic cave passages in the Ingleborough area. (1) Large phreatic tube with some glacial fill and roof solution up a fault plane. Main Passage, Gavel Pot; (2) phreatic passage developed along a bedding plane where the upper bed is considerably more soluble than the lower bed. Death's Head Inlet, Lost Johns system; (3) large phreatic tube with narrow, younger, vadose canyon in floor. Entrance Passage, Notts Pot; (4) vadose canyon incised below bedding plane, which was considerably eroded by a large stream before development of true vadose conditions. Main Passage, Rowten Cave; (5) vadose canyon incised below very wide shale bed opening. Main Streamway, Upper Long Churn Cave; (6) vadose rift passage formed along fault plane. Main Passage, Rumbling Hole; (7) essentially vadose passage with wide floor due to restriction of downward erosion by resistant grit bed; some roof erosion also takes place under temporary phreatic conditions during frequent flooding. Marathon Passage, Mossdale Caverns

cross joints, is usually a prominent feature, and the passages commonly have a sub-circular cross section up to 8m in diameter (see Fig 13) unless influenced by the local geology where they may form high rifts or wide bedding caves. The phreatic systems have a freely three dimensional pattern, though the stratigraphic control results in a dominance of sub-horizontal passages, and in many cases wall scallops demonstrate that the direction of flow was against the gradient.

The best known phreatic caves are the large abandoned systems in the Great Scar Limestone. Much of the Gaping Gill system, the high levels of the Ease Gill Caverns, Duke Street in Ireby Cavern and the Kingsdale Master Cave and Roof Tunnel are good examples. These caves were clearly formed in an older environment, as they are out of phase with the present topography, and the large amounts of boulder clay in them indicate a pre-glacial origin. The same deposits of fill also locally block the passages making exploration of the complete systems difficult. However, each phreatic passage must connect upstream to one or more vadose feeder inlets, and downstream to an old rising, probably vauclusian. For example, Ireby Cavern Duke Street was originally fed by a sink at Marble Steps, but its old rising on Leck Fell is unknown. The entrance passage of Sleets Gill Cave is a fine example of a beheaded and abandoned vauclusian rising where the water originally rose to a resurgence in the pre-glacial floor of Littondale; in the inner reaches of the same cave, a similar inclined phreatic tube, the Ramp, rises over 60m at a steady angle of 60°.

It is very difficult to establish the relationship between main phreatic trunk passages and their resurgence levels. Most of the passages are sub-horizontal, due to geological control, and this merely complicates the problem. Phreatic roof domes along the passages commonly rise to 10m or more above the general roof-level, and a reasonable concordance of apex altitudes of the domes has been taken to indicate the top of the fossil phreas (Brook, 1971). However, such roof features can only conclusively indicate a minimum altitude for the old resurgence level. Absolute indicators are only provided where a vadose streamway develops into a phreatic tunnel at lower altitude; the situation is then complicated by the difficulties of correlating two individual passages across a whole series of complex cave development. Such correla-

Page 85

(*above*) Rundkarren and rinnenkarren on the large steeply inclined slabs of Hutton Roof Crag (*below*) Joint patterns and karren grooves on Newbiggin Crag

(*left*) The enlarged gryke which forms the entrance to Steps Pot on Scales Moor (*below*) Slopes in boulder clay border the vertical shaft where the stream drops into Foss Gill Pot in Wharfedale

tion seems adequately certain in some localities; contemporaneous high-level phreatic passages with vadose inlets in Ireby Cavern and Swinsto Hole indicate that the respective main drainage routes—Duke Street and Roof Tunnel—were formed 60m below the rest level of their water. The inclined tubes of Sleets Gill Cave are conclusive proof of a deep phreas, but the extremely unusual nature of this cave restricts the value of its evidence when applied across a wider area.

Similar large horizontal tubular phreatic passages were also formed in the Yoredale limestones; though fewer such systems are known, the High Level Mud Caverns in Mossdale Cavern are an excellent example.

At the present day phreatic cave passages are still forming, though their depth and scope is restricted by the proximity of the present surface topography base level to the base of the limestone over most of the main karst region. Active phreatic caves have been explored by divers in the Master Caves of the Lost Johns and West Kingsdale systems, and Meregill Skit is an active, vauclusian, resurgence which carries some of the water from Meregill Hole. Further north and east, where the Great Scar Limestone is more deeply buried, the underground water circulation must be entirely phreatic and large water-filled caves have been found by deep boreholes beneath Grassington Moor.

THE VADOSE CAVE SYSTEMS

Perhaps most characteristic of the various types of cave passages in the Ingleborough district is the active streamway—and most of these have been vadose since their initiation. Due to their strict geological control, most of the passages are either sub-horizontal canyons, formed with bedding plane roofs (see Fig 13), or nearly vertical shafts, where the streams drop down joints, though there are some caves which descend the joints obliquely. Except in local fracture zones, the passages are generally smaller than the main phreatic caves. The gently inclined stream passages are rarely more than 1m wide, though they may be up to 30m high and in some cases may be followed for more than 1km; the vertical shafts may be up to 100m deep and are generally 1–10m in diameter.

The vadose origins of these caves is demonstrated by their almost uneroded roofs, continuous downhill gradients and an abundance of abandoned, high-level oxbows; the shafts commonly have stream-cut notches at their tops, and their walls bear vertical fluting and numerous spray corrosion features. Many of the passages still contain the streams which originally formed them, though long abandoned passages also exist; Lost Johns Entrance Series contains a fine series of active and abandoned vadose passages and shafts.

Examples of the vadose caves are numerous, though Swinsto Hole and the Long Churn Caves stand out as classics of the type. Meregill Hole is an unusually large vadose system formed in an easily eroded fault plane, while Rumbling Hole is another fault- and joint-guided system where the main passage drops obliquely across the bedding. Vadose streamways also occur in the Yoredale limestones and Marathon Passage in Mossdale Caverns is among the longest known, though here the passage section is a slightly unusual triangular shape due to the stream flowing on an impervious grit bed (see Fig 13).

In many cases the vadose streams have invaded and modified previously existing phreatic passages in their search for an easier route downwards. Where a relatively small young stream invades an old, large, tubular phreatic passage the result is a clean trench in the floor, such as in the entrance passage of Notts Pot (see Fig 13), and in the Kingsdale Master Cave downstream of the Main Junction. However, where a small phreatic passage is invaded by a large stream it is often difficult to trace the two-phase history; for example, only a few solutionally enlarged roof fissures indicate the short stretch of Swinsto Hole where the present stream has invaded the old phreatic passage from Turbary Pot (see Fig 14).

The term vadose implies an environment of cave formation above the local resurgence level, so that gravity controls the routes of the free-flowing underground streams. However, initial cave development must have taken place when the limestone fissures were full of water (ie phreatic) and it is probable that true

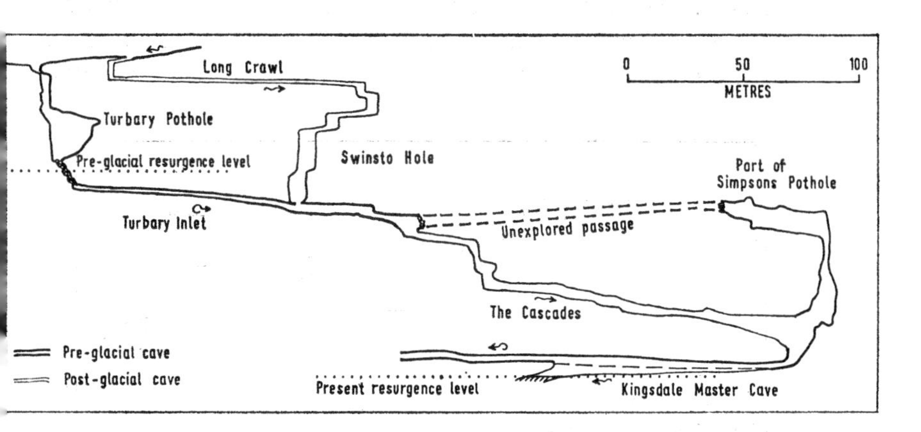

Fig 14 Swinsto Hole, West Kingsdale. A mainly phreatic pre-glacial system has been invaded in two sections by a post-glacial vadose stream

vadose conditions could not develop until the passages were more than about 0·5cm wide (Atkinson, 1968). Furthermore, in some cases, such as the lower parts of Rowten Cave (Brook & Crabtree, 1969), very heavy flows of water resulted in continued phreatic erosion, before the passages were large enough to permit normal vadose development. Also, there may be, in any length of vadose passage, local stretches of phreatic cave; in the Ingleborough district, these are mostly where a vadose passage, strictly controlled by a bedding plane, is locally turned up-dip by the influence of a joint, such as in Heron Pot (Fig 15); these local features may however be tolerated within the concept of a 'vadose system'.

THE WATER-TABLE CAVE SYSTEMS

In the isolated limestone blocks of the Morecambe Bay area horizontal caves have been found which were almost certainly formed at the water-table on the margins of a number of poljes. Networks of passages, mainly small and partly filled by sediment, occur in horizontal and inclined limestones, and in the latter situation the caves cut straight across the bedding. A distinctive feature of these systems is that their known development is restricted to narrow zones adjacent to the polje margins; rarely

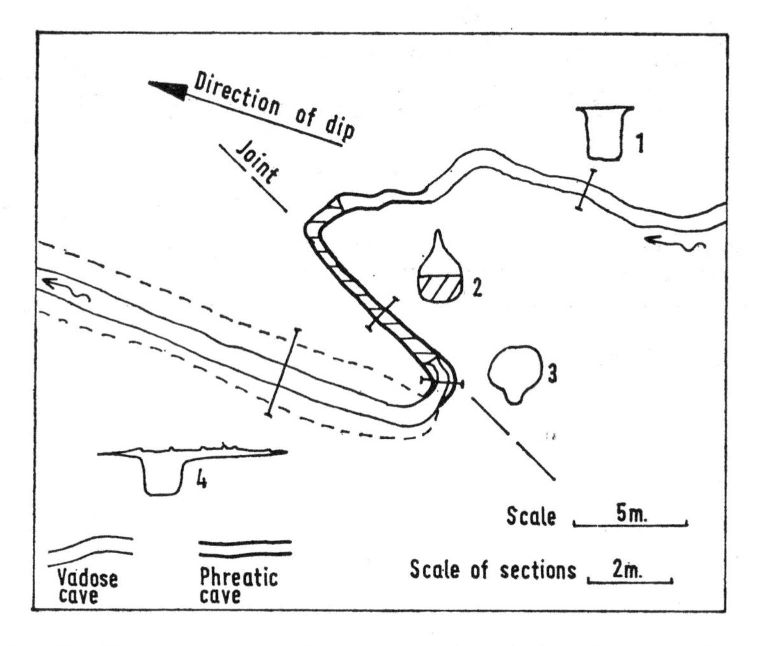

Fig 15 The development of a short section of phreatic passage in a vadose cave, due to the joint-controlled, up-dip deflection of a bedding-controlled streamway. Heron Pot, Kingsdale

do the passages extend more than 30m from the old lake shorelines.

So far, no water-table development of caves has been found in the highland karst of the Ingleborough region, if the possibly shallow nature of some of the phreatic caves (see above) is excluded. The very limited lateral extent of the Morecambe Bay water-table caves does add weight to the theory that the long horizontal caves of the Ingleborough area are not significantly related to a water-table.

SEQUENCE OF CAVE DEVELOPMENT

Though the vadose–phreatic, morphological, classification of caves is perfectly applicable to any single passage, it is artificial when considering a complete karst drainage system. It also cuts across any chronological division of the caves, for vadose and phreatic passages may be developing simultaneously while either vadose or phreatic caves may form during more than one period. The phreatic or vadose nature of a cave passage depends only on the pond-level of the water in its hydrological route, and, excepting local perched pond-levels, this is normally the resurgence level, with perhaps a minor increase of elevation due to hydraulic gradient.

Consequently many of the Yorkshire cave systems have a high-level vadose section feeding a lower phreatic drainage route. The rejuvenated Short Drop–Gavel Pot system on Leck Fell shows this most clearly, for Short Drop Cave has always been a vadose feeder passage, while, downstream of it, the same water formed the deeper phreatic system of Gavel Pot (until the relatively recent rejuvenation and vadose modification). The change in passage morphology, in this case near the Gavel Pot entrance, therefore marks the resurgence level of the system when it was active.

On this basis, a survey of the active and abandoned cave systems in the Ingleborough area shows a concentration of the resurgence levels in two zones about 90m apart, with the lower levels being at or near the present-day active risings. This twofold grouping is clearly demonstrated on Leck Fell by the juxtaposition of the high-level Short Drop system and the lower Lost Johns system (see Fig 16). The Gavel Pot phreatic passages have been abandoned and are clearly out of phase with the present topography, and some of the Lost Johns phreatic passages have also been slightly modified by a small recent lowering of resurgence level. The high-level system with its elevated resurgence level can only have been formed previous to the major excavation of the deep valleys—in this case, Leck Beck; the same applied to most of the other high-level phreatic passages in the Ingleborough district. In contrast, the low-level cave systems such as Lost Johns must be younger, as they are an integral and active part of the present topography.

This main deepening of the valleys was essentially glacial but it is difficult to determine during which stage of the Pleistocene it took place. Surface evidence is confusing, and the underground evidence is mainly relative, and generally ambiguous in terms of absolute chronology. The present state of knowledge on this problem is summarised in chapter 24 of this volume. Most of the evidence appears to place the major excavation of the valleys in an early glaciation—Anglian and/or Wolstonian. Consequently cave passages may fall into broad divisions of 'pre-glacial' and 'post-glacial'—terms which apply to the main erosive glaciation and not to the whole of the Pleistocene.

Further evidence of the pre-glacial origins of the abandoned high-level phreatic passages is provided by the immense amounts of, often unsorted, boulder clay and glaciofluvial sediments found within them. There is locally evidence of multiple pre-glacial and post-glacial phases of development, which further complicate the overall geomorphology and give rise to caves best described as 'interglacial', in chronological terms. However, the most marked contrast is between the 'high-level, pre-main-glaciation' cave systems, and the 'low-level, post-main-glaciation' systems, in terms of detailed morphology, topographic elevation, sedimentation and present activity (see Fig 16).

A chronological division of cave *systems* is, therefore, in reality more significant than a morphological division of cave *passages*, as, in the right geological environment, the latter may show considerable local variation. However, the local details of cave morphology are still relevant, as most systems contain both phreatic and vadose elements, in a ratio broadly determined by age.

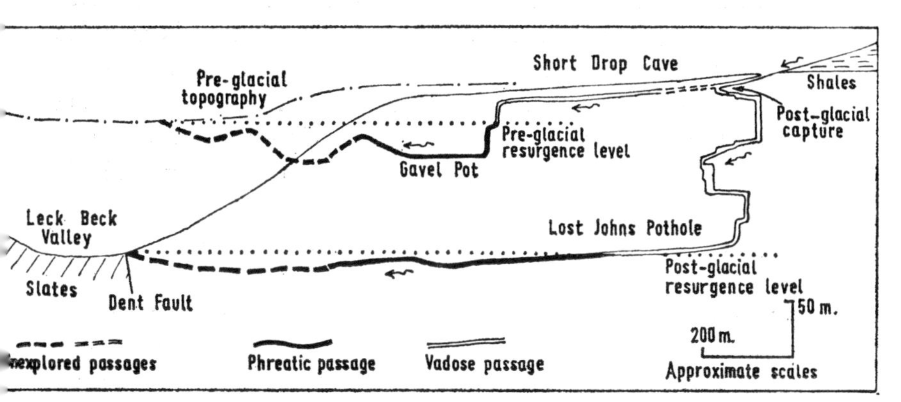

Fig 16 Semi-diagrammatic section across Leck Fell, showing the morphological distinction between the pre-glacial cave system of Short Drop–Gavel Pot and the post-glacial system of Lost Johns Pothole. (*Note*: later post-glacial rejuvenation has drained much of the phreatic passage in Lost Johns Pothole.)

DEVELOPMENT OF CAVES

The precise route of an underground karst drainage channel and the types of factors which control its development vary to a great extent depending on the environment of cave development. The main contrasts are between the vadose and phreatic environments, and these are considered separately below, though there are certain factors which act regardless of environment. Water-table cave development provides a special case where the patterns of cave enlargement are similar to those formed under phreatic condition, except that development is restricted to a sub-horizontal plane. Due to the general lack of water-table caves in the main Ingleborough area karst, this situation is not considered further in this chapter.

As soon as any bed of limestone is exposed to rainfall, its network of fissures and microfractures will fill with water. But for there to be any significant cave development there must be a complete through-system of micro-passages, which can contain an efficient drainage system. Once a flow-through of water is established, continued solution may take place and eventually enterable caves will be formed. Various factors control the routes of the caves, but to a certain extent their location is dependent upon significant supplies of inflowing allogenic water; hence the abundance of entrances adjacent to the outcrop boundary with the shales.

Many cavers have found, to their cost, that, when followed upstream, cave passages frequently have a tendency to decrease in size. This may be due to tributaries joining the lower reaches, but a decrease in size may also be found following up a single junctionless passage. Furthermore there are cases where passages, both phreatic and vadose, can be followed up to impenetrable fissures, where water enters through openings showing almost negligible secondary solutional development. The main inlet in Rumbling Hole is an excellent example of this, for, followed upstream, a small vadose canyon and an older phreatic half-tube in the roof can both be seen to emerge from a bedding plane containing only a very thin elliptical opening. This demonstrates that both types of cave passage, phreatic and vadose, normally develop by headward enlargement, as opposed to downstream enlargement. The upstream development of cave passages is, at least in part, due to the more efficient, and therefore earlier, erosion, where the flows of various tributaries have coalesced in the lower reaches of the stream routes. This effect is of greater overall importance than the tendency for the solutional effort of karst water to be used up in its earlier stages of contact with the limestone. Also, superimposed evenly over this pattern are the more local effects of mixed-water corrosion.

The situation of headward cave enlargement obviates the difficulties of explaining why a cave stream leaves a major geological control, such as a fault, in a downstream direction. The problem now is to explain why a cave stream joins a fault—that is, why the cave left the geological control in its progress of upstream enlargement. In most cases this is due to the tendency for a route to develop with an overall trend towards the greatest available supply of influent water.

Soft but insoluble shale beds are commonly visible in the walls of Yorkshire caves; in some places they have been preferentially eroded to leave deep niches, and elsewhere they project as ledges from the cave walls. Undoubtedly, therefore, both solution and mechanical abrasion have been instrumental in the erosion of the caves, though their relative importance must vary considerably, dependent upon flow-rate and stage (Newson, 1969) (stage being level of flow relative to normal, or degree of flooding).

The importance of collapse in cave formation has frequently been overestimated, as it can only be a secondary process; collapse can only take place into a previously opened cavity. Clearly many of the older caves have been modified by collapse, as their floors are strewn with blocks fallen from a jagged roof, as in parts of the upper passages of the Lancaster–Ease Gill caves. Elsewhere in the same system there is extensive collapse where a large stream passage has developed directly below the old phreatic route and undermined its floor. However, a large proportion of the younger vadose caves bear almost no sign of collapse on any scale; for example there are only two fallen blocks along the 1,500m of cave from entrance to final sump in Lost Johns system. Also, throughout Yorkshire the effects of collapse tend to be very much concentrated in the main fault zones.

CONTROLLING FACTORS OF PHREATIC CAVE DEVELOPMENT

Within the phreas, karst water may circulate on a freely three-dimensional pattern, and the hydraulic gradients along different flow paths must be an important factor in subsequent cave development. During pre-glacial times, phreatic drainage was dominant, and the main routes which developed into large caves were those with the maximum hydraulic gradient between the principal sinks and the cols along the outcrop of the North Craven Fault. Cols on the Fault outcrop developed where the main pre-glacial valleys crossed this line, and the significance of these was that they were the lowest points in the barriers of impermeable rocks surrounding the saturated block of Great Scar Limestone; consequently the contained karst waters in the limestone overflowed at these points, having risen to the surface through deep resurgences.

Superimposed on this main hydraulic control of cave development was the influence of various geological factors; major joints and faults, and limestone beds containing a dense joint pattern (Doughty, 1968) or a more soluble variety of limestone all provided easy hydraulic routes, while the drainage was stratigraphically constricted by the shales, and concentrated along zones rich in shale horizons.

These factors therefore combined to develop deep phreatic systems connecting vadose inlets near major sinks, to vauclusian risings adjacent to the North Craven Fault. The depth of the main phreatic passages was probably controlled lithologically, but varied considerably; the main trunk route in Ireby Cavern was 60m below the resurgence level while that in Gavel Pot was at a depth of only 15m.

The main drainage routes converged on the risings to give a great variation in their orientation. The passage from Marble Steps sink ran parallel to the fault, north-westwards through Duke Street in Ireby Cavern to emerge at a rising (now completely obscured by drift) in the side of the old Leck Beck valley. In contrast, the water sinking in pre-glacial times at Rowten Pot flowed straight towards the fault, down the Kingsdale Master Cave and Roof Tunnel, to emerge at a vauclusian spring in the floor of the old Kingsdale, since removed by the Kingsdale glacier.

CONTROLLING FACTORS OF VADOSE CAVE DEVELOPMENT

In contrast to the above, hydraulic factors have negligible influence over cave formation under vadose conditions, once the initial openings of the cavities have developed adequately to permit true vadose drainage. Clearly in the very early stages of karstic development, the fissures must be so small that they are full of water and therefore enlarged under phreatic conditions.

As soon as strict vadose conditions emerge, the underground streams have a free unconfined flow and therefore take the easiest downhill course, within the confines of the geological structures. This last proviso is most important as the limestone is not a homogeneous porous mass, but instead offers only a variable boxwork of initial fractures which may be utilised for cave formation.

There are five main geological factors relevant to vadose cave formation. Shale beds occur at over twenty horizons in the Great Scar Limestone and provide almost completely impervious sub-horizontal barriers of very great extent (Waltham, 1971 *a*). The downward flow of vadose karst water is commonly interrupted by the shale beds, and consequently the flow, and subsequent cave development, tends to be along the top of the shale beds. However, the shales are mechanically weak, so a stream flowing on a shale bed soon cuts down through it, then to drop sharply through the limestone to the next shale band. Furthermore the stratigraphic distribution of the shales is distinctly irregular, and there is normally an abundance of cave passages developed in the parts of the limestone succession rich in shale bands; the most marked grouping of shale beds is at the very top of the limestone, hence the abundance of shallow sub-horizontal caves throughout the Yorkshire Dales. Lower in the succession, two further zones rich in shale beds have resulted in many of the cave passages again being concentrated in narrow bands. The stratigraphic control of cave development is thus of primary importance, although these cave horizons were originally interpreted as being due to formation of caves at stand-levels of a falling water-table (Sweeting, 1950); more recent cave surveys and further study, however, have demonstrated the over-riding importance of geological control (Waltham, 1970 *b*).

The varying lithology of the limestone beds must have an influence on cave development, though this is not very spectacular. Certain more soluble beds, or those with a high density of micro-joints, are preferentially eroded (Waltham, 1971 *b*). Some caves have also developed along bedding planes containing no shale at all, and this is due to the lower bed at the contact being rather less soluble than the upper, hence restricting vertical development.

Gentle folds are found in the limestone throughout the Ingleborough district, generally having limbs dipping at only a few degrees. Most of the sub-horizontal vadose caves observe strictly stratigraphic control and therefore must, by definition, run roughly down-dip; consequently major vadose cave streams tend to develop flowing down the plunge of the synclines.

Joints are an important feature of the Great Scar Limestone and occur in a number of different orientations. They obviously provide an easy downward route for karst waters, and indeed the majority of vadose shafts in the Ingleborough district have origins traceable to joints. Also many cave passages have their plans influenced by joints even when developed in the confines of a single stratigraphic horizon; these factors are clearly visible on cave surveys (Waltham, 1970 *a*).

Faults are more thinly distributed in the Ingleborough district limestones, but are easily eroded and have an influence on cave development similar to that exercised by joints, though locally on a larger scale.

A study of the detailed morphology of almost any major vadose cave system in north-west England provides many examples and abundant evidence that the above factors are the dominant controls of the cave development. In terms of regional hydrology the synclinal concentration of cave streams is of paramount importance, but for a single cave passage the influences of shale beds and joints are together most significant. A 'model vadose cave' in the Ingleborough district therefore tends to be formed where an allogenic supply of water enters the limestone, most frequently at the shale boundary. The flow is then down a joint into the limestone, as far as the first shale band, where flow is diverted along this in a down-dip direction, eventually to drop down another joint to another shale bed, etc. Some streams may flow obliquely down joints, but even then the resultant caves commonly have a slightly stepped nature due to the influence of shales and bedding.

In the shale-bed controlled caves, the streams initially flowed above the shales, hence the very slight arching and erosion of all the passage roofs. However, as soon as the stream flowed fast enough to abrade its bed mechanically, it rapidly removed the soft shale, over a considerable width, and then flowed in the limestone bed below. The greater overall resistance of the limestone, compared to the shale, led to a narrow canyon being cut in it, resulting in the classical 'T-shaped' passage (Fig 13).

The Lost Johns Master Cave is probably the longest in the region that is developed from a shale-less bedding plane. It was in fact formed by water initially flowing along the top of a more resistant limestone bed, and this is demonstrated by the Death's Head Inlet, formed on the same bedding plane but under phreatic conditions, whose cross-section (Fig 13) reflects the relative solubility of the two beds.

The detailed morphology of the Lost Johns Master Cave clearly demonstrates that its upstream section originated under vadose conditions while downstream it was initiated within the phreas. Furthermore it has a distinct gradient to it (about 1 in 150) and cannot have formed 'at or just beneath' any water-table (Warwick, 1953); the same applies to all the other known master caves in north-west England. Consequently the term 'master cave' should cease to have any genetic significance, though it may well still be used to describe a main cave streamway having numerous tributaries.

Leck Fell also provides a most spectacular example of synclinal concentration of vadose drainage. A single shallow syncline, plunging north-west, contains, along its axis, the main vadose streamways of both the Short Drop and Lost Johns cave systems, both with numerous converging tributaries, the former system being 120m above the latter (see Fig 58).

The overall rate of descent of the vadose caves must depend on the relative influences of the shale beds and joints, as these are the major controlling factors. Across the region shales are almost constant in every way, but the joint types and density vary considerably. Whether or not a single cave stream should descend a certain joint through a certain shale bed is an unmeasurable and unpredictable factor.

However, there is a measurable variation in the depths of individual joints in the Ingleborough district and this has a similar effect. The major factor controlling the joints appears to be the proximity to the North Craven Fault; close to the fault, for example on Newby Moss, the joints generally pass through many beds for tens of metres in vertical extent, while further north, for example around Penyghent Gill, examination of the outcrops shows that most of the joints do not pass through more than one bed, generally less than 4m thick. (This absolute magnitude of individual joints is completely separate from the varying joint densities in the limestone, ascribed by Doughty (1968) to lithological variation.)

Consequently on Newby Moss there is an abundance of deep potholes with almost no horizontal development along shale beds. Similarly, in Penyghent Gill the caves are mainly of the horizontal type with only a few short vertical shafts, and the influence of the shales is so great that most cave streams resurge perched up the valley sides instead of attaining any great depth underground. For the same reason, the Ribblehead area contains only very shallow caves; had the area been more heavily jointed, one would expect there to be numerous potholes up to 75m deep, feeding perhaps to a combined resurgence at God's Bridge.

Superimposed on this general pattern of shallower caves further away from the North Craven Fault, are a number of deep caves formed on isolated faults. Meregill Hole and Birks Fell Cave are two fine examples of fault-guided systems deeper than their neighbours.

The almost complete lack of known cave systems in the heavily jointed limestones between Settle and Malham emphasises one important factor in vadose cave development. There is undoubtedly a very efficient karst-drainage system in the area, for there is a lack of surface drainage, but the reason that there are

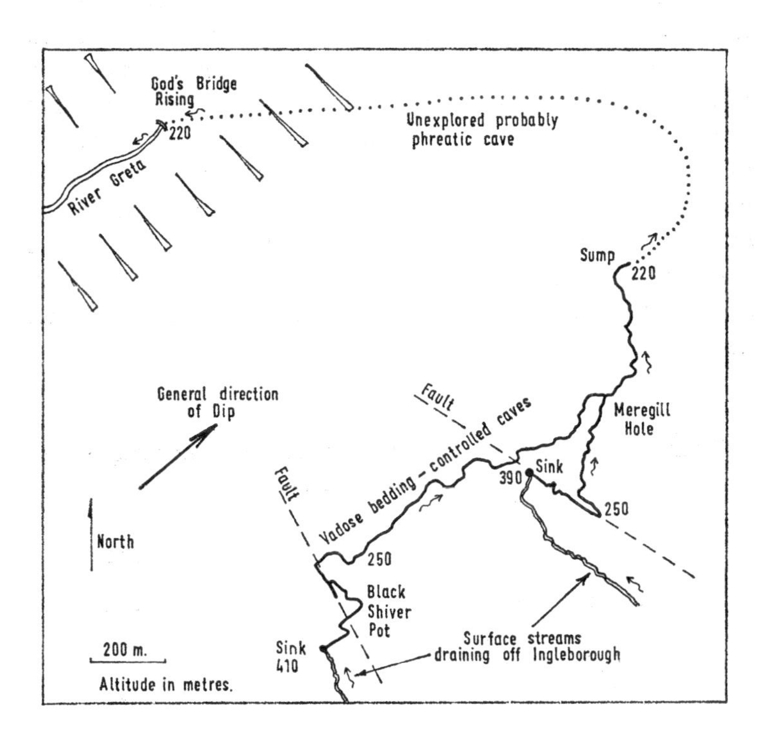

Fig 17 Plan of the main stream caves of the Meregill cave system, showing short, deep, fault-guided potholes and down-dip flow of the vadose drainage to enter a sump level with the resurgence. Within the phreas the water must reverse its direction of flow towards the main surface valley

almost no enterable caves is that there are no sinking streams provided by the coalescing of surface drainage on an adjacent shale outcrop. A large, passable cave can only be formed by a large stream, but in this area all the near-surface drainage is in the form of small trickles of water in comparably small micro-caves.

One final point, relative to the caves of the Ingleborough area, which needs explaining is the scarcity of deep, active through-cave systems, compared to other areas in the world. This is due to the fact that the main underground vadose drainage is down-dip, towards the north, while the main resurgences are in the lowest exposed beds of limestone, in the lower parts of the valleys, which drain to the south. Consequently most cave streams flow north only to enter a sump from where they flow back against the dip, within the phreas, to the resurgences; Meregill Hole is a fine example of this (Fig 17). The only place, therefore, where one would expect to find a complete major vadose cave system is where the limestone dips towards a deep surface valley; this very nearly occurs at Langstroth Cave, which has only two short sumps, and occurs in just such a location.

TABLE 3 *Cave systems in north-west England containing more than 5km of passages*

	(km)
Lancaster Hole–Ease Gill Caverns	27·0
Gaping Gill system	10·5
Mossdale Caverns	9·5
Langcliffe Pot	9·5
Pippikin Hole	6·5
West Kingsdale system	6·5
White Scar Caves	6·0
Lost Johns system	5·0

Page 103

(*above*) The extensive bare glacially scoured limestone pavements of Scales Moor
(*below*) Glaciofluvial debris accumulated in Stream Passage Chamber in the Gaping Gill system

(*left*) The flat-roofed vadose canyon passage above the New Roof Traverse in Lost Johns system (*below*) The almost perfect phreatic tube of Silverdale Shore Cave revealed in an old sea cliff bordering Morecambe Bay

TABLE 4 *Caves and potholes in north-west England deeper than 150m*

	(m)
Meregill Hole	170
Gingling Hole	170
Penyghent Pot	160
Black Shiver Pot	155
Gaping Gill system	155
Long Kin West Pot	155
Tatham Wife Hole	155

5

Introduction to the Karst Hydrology of North-West Yorkshire

R. A. Halliwell, J. L. Ternan and A. F. Pitty

Four aspects of the karst hydrology of north-west Yorkshire might be recognised. First, there are the sink-resurgence drainage systems, with early tracings in the Malham area by members of the Yorkshire Geological & Polytechnic Society (Howarth *et al*, 1900), followed by their more extensive work beginning in 1899 (Carter *et al*, 1904). Not only were many sink-resurgence links then established, and their discrete nature identified, but also their crossings underground at Malham and at Turn Dub were demonstrated. The general pattern of surface drainage sinking at the outer edge of the impermeable higher Carboniferous strata is well known, and current tracer work, taking advantage of refinements in technique (for example, Ashton, 1967; Drew & Smith, 1969; Crabtree, 1971; Glover, 1972), continues to add to the broad general picture outlined by the classic tracings.

Secondly, volumes of flow in resurgences and surface streams, together with their seasonal fluctuations, are important, particularly for comparative purposes with other limestone areas. Little information on discharge volumes is as yet available to cover the area as a whole, although Ternan's recent measurements are an example of the scale and range of flow for one well-defined part of the area. Six resurgences or risings were studied on the periphery of the High Mark upland, a limestone tract some 45km^2 in area, lying at about 450m (1,500ft) OD and rising to 540m (1,765ft) at Parson's Pulpit. Here the dip of the limestone strata, although locally variable, is regionally towards the north-east and east. Although the dip is usually less than 5°, it may influence the direction of water movement, as the majority of risings emerge from the north-east and east margins of the upland. A large proportion of the total flow is discharged into lower Littondale and into Wharfedale. At Moss Beck Rising in lower Littondale a mean discharge of 192l/sec was recorded, compared with 230l/sec at Reynard's Close Rising (Table 5). The two largest risings on the north-west and south-west margins of the upland, one rising near the source of the Skirfare's tributary, Cowside Beck, and the other near Malham Tarn, are estimated to have a mean discharge of about 30l/sec only. Catchments for such small flows need be no greater than 1km^2, and are of the same order of flow as those measured by Halliwell on the south-east side of Chapel-le-Dale (Table 5). The mean figures in Table 5 are affected by a few high flood discharges. Thus, at Moss Beck and Robin Hood's Well, discharges greater than the mean were recorded on only 20 per cent of the recording times, and at Chapel House Well and Reynard's Close, 25 per cent and 34 per cent respectively of the discharge observations exceeded the mean values. The minimum discharge, at the four risings observed by Ternan, was in early September 1969, after a month in which only 28mm of rainfall was recorded at Malham Tarn Field Centre. The discharge at these four sites responds rapidly to rainfall. At Robin Hood's Well, as for Chapel House Well, correlation analysis suggested that the rainfall 24–96hr prior to recording time has the greatest influence on the measured discharges, although the response may be more rapid after the occasional spell of heavy rainfall. At both Reynard's Close and Moss Beck, discharge appears to be, on average, most closely related to the rainfall during the period 24–48hr prior to each observation time. As these are the largest of the risings considered, it could be that these systems are better integrated and more efficient in

TABLE 5 *Examples of some discharge observations at some north-west Yorkshire risings*

	Discharges (in l/sec)				
Rising*	Mean	Maximum	Minimum	Source	Sampling period
Robin Hood's Well (Wharfedale)	28	131	5	J. L. Ternan	May 1969–May 1970 (fortnightly)
Chapel House Well (Wharfedale)	13	62	3	,,	,,
Reynard's Close Rising (Wharfedale)	230	673	29	,,	,,
Moss Beck Rising (Littondale)	192	692	21	,,	,,
Skirwith risings (Chapel-le-Dale)	23	122	dry	R. A. Halliwell	May 1971–May 1972 (fortnightly)
Engine Shed risings (Chapel-le-Dale)	48	392	6	,,	,,
Light Water Spring (Chapel-le-Dale)	28	364	dry	,,	,,
Norber (Crummockdale)	—	486	2	Craven Water Board	July 1969–Sept 1971 (occasional, at low flows only)
Southerscale Fell Spring (Chapel-le-Dale)	—	—	2	,,	,,
Turbary Springs (above Yordas, Kingsdale)	—	—	7	,,	,,

*All springs rise from the Great Scar Limestone except the last two listed which rise from Yoredale limestones.

transmitting rainfall pulses through to the risings.

A third feature of the hydrology, discontinuous surface drainage, typifies many limestone areas. These drainage lines, by following a valley course when in flow, differ from sink-resurgence links which are usually unrelated to valley forms. In north-west Yorkshire continuous surface flow may in part be maintained by the impermeability of glacial deposits, by the frequency of shale horizons, particularly in the Yoredale strata, by the proximity or outcrop of the pre-Carboniferous basement, and by heavy rainfall gathered in catchment headwaters on the non-carbonate outliers. The beds of the Dee in Dentdale and the Skirfare in Littondale are often merely dry limestone slabs in low-flow conditions. On 40 per cent of sampling visits Halliwell has observed the Kingsdale Valley Beck to be flowing the full length of the valley. On 29 per cent of such visits the stream bed was dry for almost the whole length of the valley. The Beck floods, as do several Yorkshire karst streams, with an actual flood pulse. This pulse was observed on 17 October 1971, when an almost vertical wall of water, about 0·5m high, advanced down the river bed at about walking pace. The wave is caused by the flooding of the sinks in the river bed, approximately half-way along the length of the valley. There have also been reports, perhaps doubtful, that a similar 7cm high wave might be seen in Kingsdale Master Cave approximately 45min after the surface pulse.

In Kingsdale the highest water conditions observed in a year's fortnightly sampling were on Saturday 29 August 1971. After fine weather for much of the previous week, there was continuous drizzle on this day and river levels began to rise by midday. Several intermittent springs started to flow during the afternoon. The rainfall ceased about 6pm, but restarted about 11pm and continued throughout most of the night. The Greta, as well as Kingsdale Beck, experienced flood peaks at about 5am on Sunday 30 August. In Kingsdale a second flood peak occurred at about 2pm flooding the road to Braida Garth farm to a depth of 0·6m, and most of the valley below Keld Head, where a large temporary lake was formed. At Keld Head the water was approximately 2m above average level with a pressure dome 0·4m high. At Yordas Cave, the underground system was unable to cope with the flood waters, and the natural rock archway in the cave entrance was scarcely visible above the height of the water flowing out of the cave to join a second stream flowing over the surface from the normal-flow sink. The combined streams had a flow of approximately 130l/sec. Similarly, the major Swinsto Flood sink, at Turbary road, was completely flooded, with an input of 415l/sec and a surface output to the second flood sinks of approximately 70l/sec. On the same day the Greta, in Chapel-le-Dale, removed tarmac from the road by the church, yet the stream bed was dry by 10am. At approximately the same time, however, the Greta at Beezley Farm stepping stones was still approximately 1m above the normal water level. The multiple flood pulses are due not only to successive periods of rainfall, but also to the same rainfall pulses being transmitted by different routes, often different caves with contrasting characteristics.

For tributary risings entering main valleys, as in Littondale and Wharfedale, the length of intermittent flow is usually short, with discharge additional to normal flows occurring from intermittent risings just upvalley or uphill. This contrasts with some of the examples described so far where the filling of a normal sink to capacity leads to the advance downvalley of the intermittent flow. At Reynard's Close, for instance, water was observed to emerge on 82 per cent of sampling visits at a semi-permanent rising 10m to the north-east of the main outlet. At Robin Hood's Well it seems that at least 15mm of rainfall, over a three-day span, is required before discharge occurs at the flood rising. When the flow at Robin Hood's Well was less than 30 l/sec, no discharge occurred at the flood rising. Flow at the flood rising was recorded with combined discharges above 47l/sec. A threshold value therefore exists between 30 and 47l/sec at which water is unable to discharge completely at the present rising and backs up within the system to flow out of the flood rising.

A fourth aspect of the hydrology of north-west Yorkshire relates to quality of the water. Gradually, it has been appreciated that the study of solutes in limestone streams and caves can support an increasingly wide range of inferences about the nature and origin of underground water, particularly from hydrological zones that are not directly observable. One reason for the utility of such studies is that flood waters, or water rising from non-calcareous rocks, drift, or peat, have little calcium carbonate in solution. Richardson (1968) shows that for Yoredale risings in the Alum Pot area, a mean calcium carbonate value of the water is 50ppm (Table 9). This compares with 48ppm observed by Pitty as a mean value for Fell Beck before it enters Gaping Gill. The solute concentrations in short-term flood pulses may be much lower. For instance, whilst the flood wall of the Kingsdale pulse, observed by Halliwell, was 43ppm calcium carbonate, 1½hr later the concentration had dropped to 16ppm, and the temporary lake observed on 29 August 1971 contained only 12ppm calcium carbonate. By contrast, a second feature of karst hydrochemistry is the percolation water. This represents precipitation which, instead of concentrating in stream channels, has moved through narrow, limestone joints only, and is typified by higher concentrations of calcium carbonate in solution. This type of water, observed in many cave seepages and springs, often contains more than 140 ppm calcium carbonate. Concentrations rise the longer such water has been in contact with the rock, apparently at a rate of 76–8ppm per 100 days (Pitty, 1971). Mean values alone, of calcium carbonate and other chemical characteristics, therefore, can give a clear indication of the origin of rising waters, and, as Richardson shows in chapter 8, detailed interpretations can be developed. After the mixing of these two types of water, their chemistry allows the proportion of flow from cave seepages, itself scarcely measurable by physical means, to be inferred (Pitty, 1966 *b*). Thus, from a series of fortnightly samples, Halliwell is able to suggest how amounts of percolation water entering the main stream in Swinsto Crawl varied with changes in hydrological

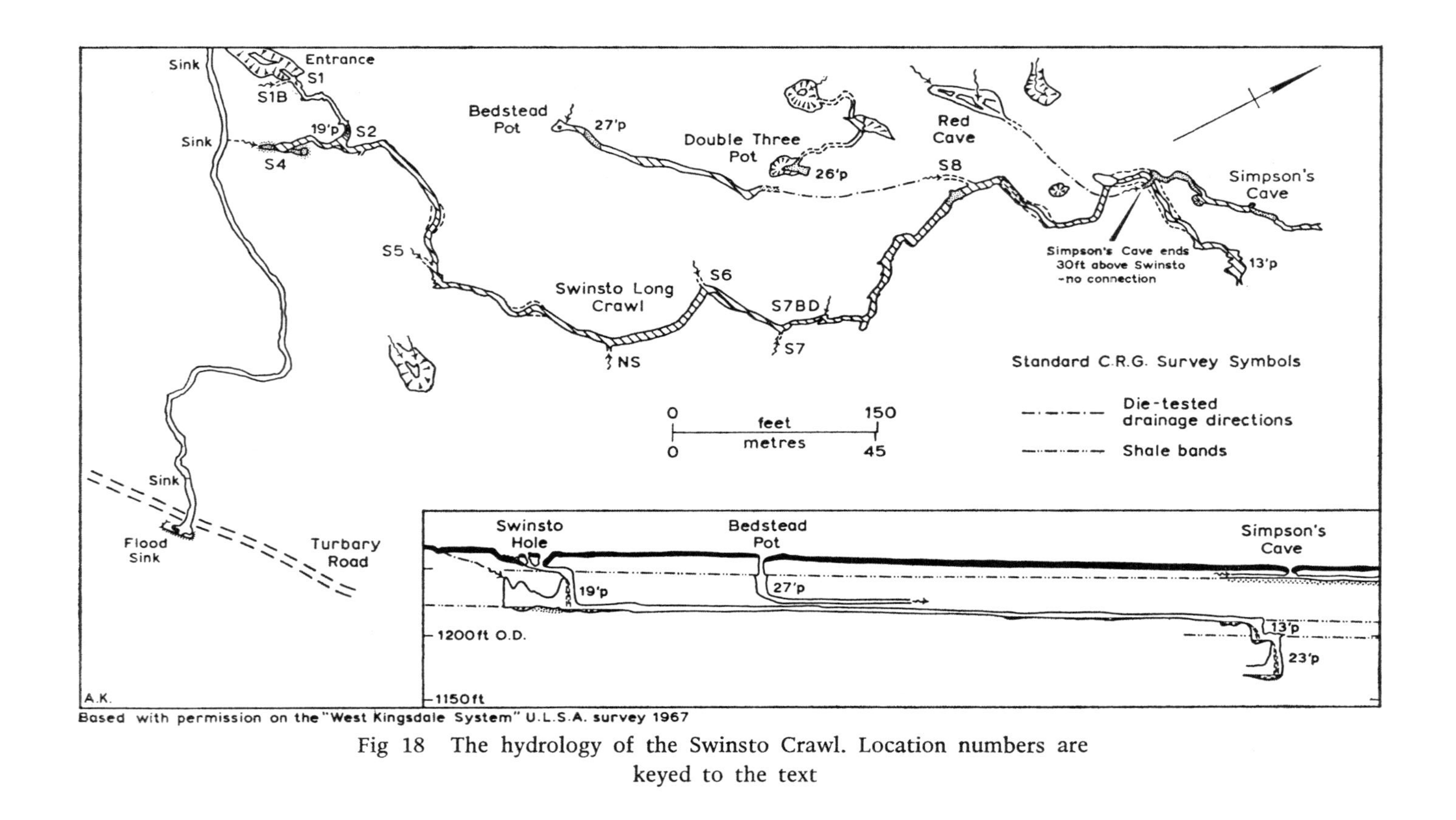

Fig 18 The hydrology of the Swinsto Crawl. Location numbers are keyed to the text

conditions. The discharge of water entering the cave (sampling point S1 on Fig 18) is about 5·5l/sec, held approximately constant by a passage constriction between the surface stream-bed sink and its point of re-emergence in the cave proper. According to the increase in calcium carbonate concentration in the main stream, down-cave from a series of inlets, contributions to the volume of the main cave stream can be inferred (Table 6). For instance, the waterfall inlet, S4, is linked with the surface-stream bed, just downstream from Swinsto Entrance. Unlike the entrance stream, the increased flow in flood conditions (Table 6) suggests that there is no constriction on this inlet. Inlet S5, in contrast to points fed from the surface stream, has the chemical characteristics of percolation water yet responds rapidly to heavy rainfall, its trickling flow becoming a jet shooting across the width of the passage. Inlet S6 is also distinctive, with generally the chemical characteristics of percolation water, but showing some dilution after precipitation. From Table 6 it can be seen that the hydrology of this cave system may alter substantially with changing hydrological conditions. Passages not only fill with water, back up, and overflow, but inlets usually with minor flows may rapidly discharge major volumes of water into the main cave stream.

A final chemical characteristic of hydrological significance is the seasonal change in solute concentration, which adds to the distinctiveness of the hard, percolation water. This characteristic is described and discussed by Pitty (chapter 7) based on observations in Ingleborough Cavern.

The three separate studies which follow this general introduction all deal with hydrological inferences derived from studies of karst water chemistry. Therefore, in this introduction there has been some emphasis on the illustration, by specific examples, of some physical characteristics of the hydrology of north-west Yorkshire. As yet, however, too little is known and recorded of the general physical and chemical characteristics of the karst hydrology of this area for a comprehensive, integrated summary to be attempted. North-west Yorkshire is well known for the

TABLE 6 *Estimates of discharge volumes from inlets to Swinsto Crawl*

Sampling site	Mean flow		Low flow	Flood	
S2	5·5		5·5	5·5	
S1	5·5		5·5	5·5	
S4	6·5	(59%)	dry	12·0	(68%)
S2B	12·0		5·0	17·5	
S5	1·0	(7%)	trace	17·5	(50%)
S5B	13·0		5·0	35·0	
S6	0·1	(1%)	dry	1·5	(4%)
S6B	13·0		5·0	36·5	
S8	2·0	(13%)	—	?	?
S8B	15·0		5·0	?	?

(*Note*: The volumes are in l/sec. Water samples were collected from January 1971–January 1972. The identification of the sampling sites is given in Fig 18. The suffix 'B' indicates a point about 4m downstream from the point indicated by the equivalent code without the suffix. The figures in brackets indicate the approximate volume of flow from an inlet as a percentage of the main stream flow, downstream from their junction. Inlet S8, connected to Bedstead Pot, could not be reached during flood conditions.)

educational and recreational opportunities that it offers. It is therefore with surprise, mingled with regret and apology, that it must be admitted that, since the classic work of seventy years ago defined the broad outlines of the sink-resurgence hydrology, too little has been done to do adequate justice to this classic area.

6

Some Chemical and Physical Characteristics of Five Resurgences on Darnbrook Fell

J. L. Ternan

Measurement of basic processes operating in limestone areas has increasingly become part of research on cave development and limestone hydrology. This type of observation work, which includes measurements of denudational losses by solution from limestone catchments and caves (for example, Smith, High & Nicholson, 1969) and the identification of factors instrumental in causing variations in solutional losses (Pitty, 1966 *b*), may lead to a greater understanding of cave origin and development in any particular area than can be gained from a study of cave morphology alone. The aim of this account is to present some data recorded at five resurgences on Darnbrook Fell. Water samples from these sites were collected at approximately two-weekly intervals throughout the period February 1968–February 1969, and were analysed for dissolved calcium and magnesium carbonate content. Water temperatures were also recorded. From such measurements inferences may be made concerning the source of water issuing from the cave systems behind the risings and the time taken for water to pass through the limestone. The scale of short-term fluctuation in dissolved calcium and magnesium carbonate and in water temperature provides a useful index of the relative contribution of swallet and percolation water to the system. Conclusions regarding the nature of underground flow may also be made from comparisons of risings in close proximity to each other.

DESCRIPTION OF SAMPLING AREA AND THE LOCATION OF SAMPLING POINTS

The geology of this area is typical of the north-west Yorkshire limestone region. A hundred metres of pure massive Great Scar Limestone are overlain by the alternating shales, sandstones and limestones of the Yoredale Series. The regional dip of the beds of the Great Scar Limestone and the Yoredale Series is approximately 5° to the north-east but locally variations occur in both direction and degree.

Much of the solid geology of Darnbrook Fell is masked by a mantle of solifluction and till deposits, often covered by blanket bog peat. These drift deposits, in conjunction with the impermeable beds of the Yoredale and Millstone Grit Series, provide a catchment for numerous small streams which sink on reaching the limestone. A few large streams such as Darnbrook Beck are, except in drought periods, able to maintain surface flow across the limestone. The five sampled resurgences are located around the lower flanks of Darnbrook Fell and in all cases the water emerges from the D_2 beds of the Great Scar Limestone (see Fig 19).

Hesleden Rising is a right-bank tributary spring to Hesleden Beck, located just above Lower Hesleden Farm in Littondale. There is no known cave system behind the spring. Similarly the source of the water is not certainly known, but most of it is probably derived from small sinks straight up the fell behind the spring and just less than 100m higher in altitude.

Fosse Beck Rising (Litton Foss) lies at the foot of a mass of limestone blocks above East Garth Farm near Litton village. Behind the low bedding plane resurgence a short but large stream passage is choked by blockfall 30m from daylight. Most of the

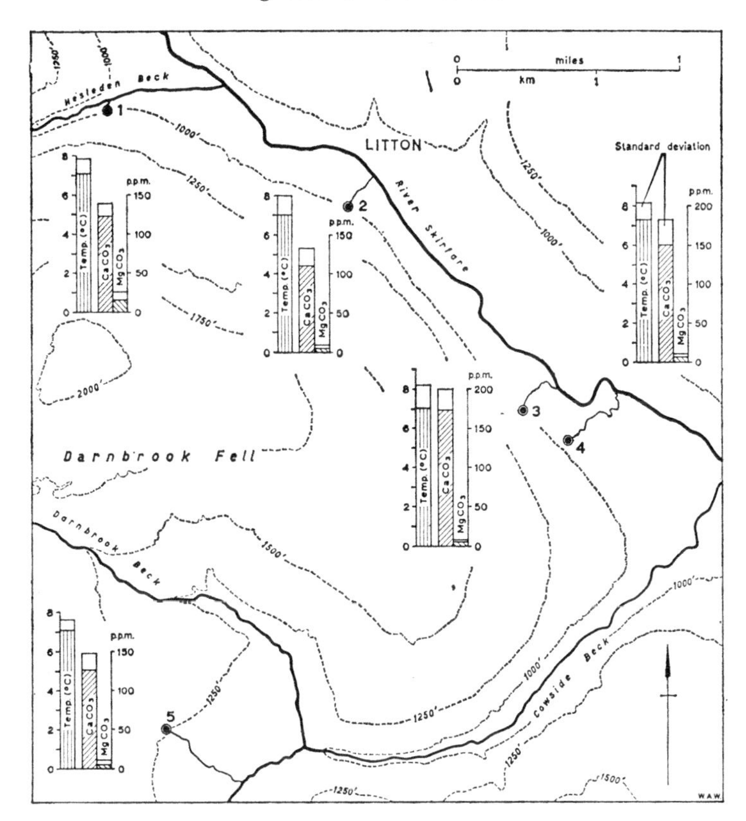

Fig 19 Location map of sampled risings on Darnbrook Fell and their mean characteristics. 1=Hesleden Beck Rising; 2=Fosse Beck Rising; 3=Scoska Cave Rising; 4=Bown Scar Cave Rising; 5=Thoragill Cave Rising

water comes from a group of sinks including Litton Pots on the fell behind the cave at an altitude of 410m.

Scoska Cave Rising is marked by a large cave opening 1·5km west of Arncliffe village. The small stream flowing from the cave mouth may be followed in for nearly 300m to where the passage is blocked by flowstone which the water is at present corroding.

Bown Scar Cave Rising, 400m south-east of Scoska Cave, is a

complex group of resurgences. The main permanent discharge is from a thin bedding plane low down while just above and to the north another bedding cave produces vast quantities of water during flood only. These two outlets are united inside the cave and over 1,000m of active passage is known up the main streamway and its various tributaries. It has been proved that a number of sinks on the fell feed the cave but whether all of this water flows via the inner reaches of Scoska Cave before entering the Bown Scar system is uncertain. A third spring in the Bown Scar Cave Rising group lies just to the south of the others, level with the flood exit. It has a completely different source under low-stage conditions, but under even moderately high flow it is contaminated by water crossing from the main stream just inside the cave. The data discussed below relate to measurements on the combined flow of the main spring and its flood outlet only.

Thoragill Cave Rising is situated just above Thoragill Beck House, overlooking Cowside Beck. Behind the resurgence the stream can be followed in its cave for 60m to where it emerges from low bedding planes. Most of the water appears to be derived from a stream sink no more than 500m to the west up the dry valley above the rising; the proportion of water derived from sinks higher up the fell is, as yet, unknown.

Discharge at four of the risings responds rapidly to rainfall and considerable variation was noted over the year. Discharge variations at Scoska Cave were, however, generally small and, except in very wet conditions, little response to rainfall was observed.

CALCIUM AND MAGNESIUM CARBONATE VARIATIONS AT THE RISINGS

The dissolved calcium carbonate concentration in the water at the five resurgences exhibits a broad seasonal variation which reaches a peak in the period July–September on all five resurgences. At Fosse Beck, Scoska Cave and Bown Scar the highest recorded amounts occurred on 3–4 August whereas at Hesleden Rising and at Thoragill this peak occurred two weeks later, on

17 August. The amount of dissolved calcium carbonate in solution reached its lowest point on all sites in the late March–early April period. Pitty (1966 *b*) considered that increased CO_2 production in the plant-root zone of the soil associated with higher temperatures and greater insolation will produce a seasonal variation in calcium carbonate solution that may be recognised at a rising after a period equivalent to the time taken to pass through the limestone. The near coincidence of the seasonal air temperature wave and the seasonal wave in dissolved calcium carbonate concentration in the resurgent waters is, therefore, indicative of the rapid flow-through times known to occur on such swallet-resurgence systems.

Superimposed upon this seasonal trend, short-term fluctuations in the amount of dissolved calcium carbonate may be attributed to the influence of dilution following periods of heavy rainfall. This dilution effect is an important characteristic and is particularly marked where a large swallet system, or where a large number of swallets contributing to a system, lead substantial volumes of allogenic water, low in dissolved carbonates, underground. Such a marked short-term effect occurred at all risings on 2–3 October 1966 following a week in which nearly 10cm of rainfall was recorded at Malham Tarn.

The contrast between short-term variations in dissolved calcium carbonate recorded at Scoska Cave and those at Bown Scar Cave Rising is striking. A number of short-term dilution effects such as those occurring on 15 January, 11 May and 30 June at Bown Scar resurgence were not recorded at Scoska Cave Rising. During flood conditions on 2 October a strong dilution pulse was, however, recorded at both risings. At Scoska Cave the dissolved calcium carbonate concentration in the water preceding this pulse on 18 September was 185ppm. During the passage of the pulse on 2 October the concentration dropped to 113ppm and by 16 October it had recovered to 171ppm. This dilution pulse at Scoska Cave was accompanied by a considerable increase in discharge, the highest flow over the sampling period being recorded

TABLE 7 *Some physical and chemical attributes of the sampled resurgences on Darnbrook Fell*

Site	Altitude (m)	No of Samples	Mean $CaCO_3$ (ppm)	SD* $CaCO_3$ (ppm)	Max recorded $CaCO_3$ (ppm)	Min recorded $CaCO_3$ (ppm)	Mean $MgCO_3$ (ppm)	SD* $MgCO_3$ (ppm)	Max recorded $MgCO_3$ (ppm)	Min recorded $MgCO_3$ (ppm)	Mean Temp (°C)	SD* Temp (°C)	Max recorded Temp (°C)	Min recorded Temp (°C)	Ca:Mg Ratio
Hesleden Rising	300	22	123	15	148	90	15	7	30	6	7·08	0·78	8·62	5·89	6·9
Fosse Beck Rising	290	24	111	23	157	88	7	3	12	3	7·03	0·96	9·06	5·70	13·4
Scoska Cave Rising	305	24	174	27	217	113	6	2	10	1	7·03	1·17	8·83	5·05	24·4
Bown Scar Rising (south side)	280	24	151	31	207	105	7	3	16	3	7·34	0·78	9·12	5·95	18·2
Bown Scar Rising (main spring)	280	16	152	33	180	94	7	1	9	4	7·41	0·67	8·58	6·50	18·3
Thoragill Cave Rising	380	18	127	20	159	94	6	1	22	3	7·14	0·48	8·10	6·10	17·8

*SD = standard deviation

on 2 October. Owing to the proximity of the Scoska Cave and the Bown Scar systems, the comparative data for Bown Scar Rising on these dates are of some interest. On 16 September the dissolved calcium carbonate concentration at Bown Scar was 173ppm. This dropped to 107ppm on 2 October during the passage of the flood pulse and by 16 October, with a fall in discharge, had recovered to 129ppm. The recorded water temperature at Bown Scar Rising was 9·12°C on 2 October but only 8·81°C at Scoska Cave. From both the dissolved calcium carbonate and the water temperature data flood pulses at Scoska Cave Rising appear to be less intense. Low discharge variability, the generally high dissolved calcium carbonate concentrations, the occurrence of rare but pronounced dilution pulses followed by rapid recovery in the dissolved calcium carbonate concentration are indicative of a system essentially supplied from percolation sources although still subject to sudden influxes of swallet water during high flood conditions. Wide fluctuations in discharge and frequent dilution pulses at Bown Scar point, however, to an essentially swallet-resurgence system. The existence even in flood conditions of contrasts between the two systems, in dissolved calcium carbonate concentrations and water temperature, suggests different sources for their waters rather than the possibility of Scoska Cave providing a simple flood route for Bown Scar water at high flow.

Short-term dilution effects are less extreme on the southern upper bedding plane rising of the Bown Scar group than on the main spring. On 2 October water emerging at the southern spring contained 172ppm of dissolved calcium carbonate although 1m away water emerging at the upper bedding plane rising on the north side (part of the main spring flow) contained only 107ppm. Two quite different flows of water can therefore be identified at Bown Scar Rising.

All five resurgences show a pattern of dissolved magnesium carbonate fluctuation which tends to follow that of the calcium carbonate. At Hesleden Rising the amount of magnesium carbonate in solution reaches a peak in the mid-June to mid-September

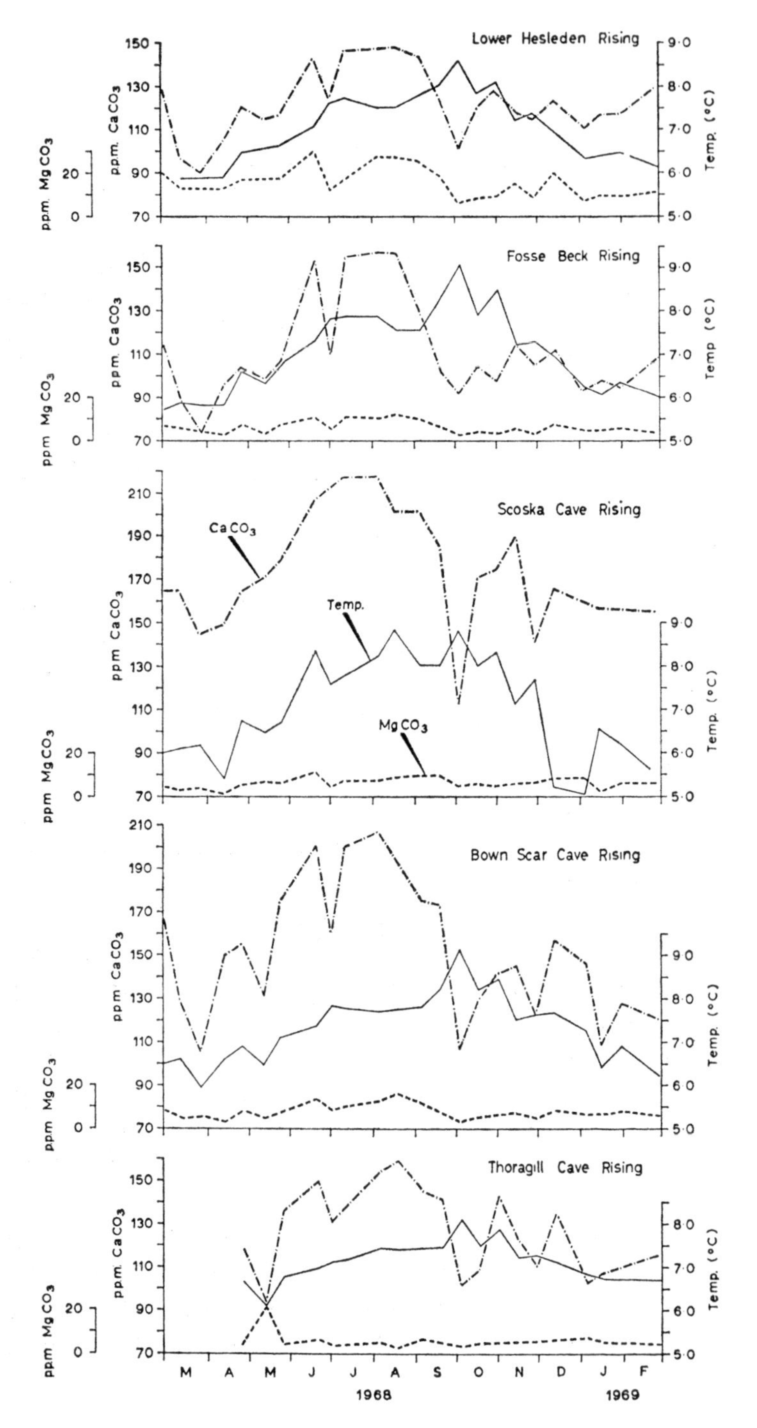

Fig 20 Calcium carbonate, magnesium carbonate and temperature variations at five resurgences sampled on Darnbrook Fell

period (Fig 20), this period being approximately coincident with the peak period of calcium carbonate solution. Although best developed at Hesleden this seasonal variation in magnesium carbonate in solution is also detectable at Fosse Beck and Bown Scar Risings, but is scarcely distinguishable at Scoska Cave Rising. At Thoragill Cave Rising no seasonal trend is apparent. However, the accuracy of the analytic technique is in the order of $\pm$ 2ppm and with very low amounts of dissolved magnesium carbonate this margin of error may be sufficient to mask any real variations. Hesleden Rising has the highest mean dissolved magnesium carbonate content (15ppm) whereas Scoska Cave and Thoragill have the lowest (6ppm).

TEMPERATURE VARIATIONS AT THE RISINGS

Temperature changes at the five sites recorded the familiar pattern of seasonal variation, an annual range of between 2·00°C (Thoragill Cave) and 3·80°C (Bown Scar Cave) being observed. On four resurgences water temperature reached its recorded maximum on 23 October 1968, Scoska Cave being the only system which attained a fractionally higher temperature on 16 August of 8·83°C as opposed to 8·81°C on 2 October.

It is suggested that this apparent lack of coincidence between the solution peaks and the temperature peak may be due simply to sudden influxes of warm flood water, probably from swallet sources. In October 9·7cm of rainfall was recorded in the seven days prior to 2 October, 6·3cm prior to 16 October and 6·1cm prior to 30 October. The earlier much higher midsummer air temperatures, however, occurred generally in conditions of lower rainfall and high evapotranspiration when percolation or seepage water probably provided the main source of water supply to the systems. This percolation water, being in more intimate contact with bedrock than swallet water and moving at a much slower rate (Pitty, 1966 *b*), might, in terms of temperature, be expected to approximate more to the temperature of the bedrock than to the temperature extremes experienced at the ground surface. The short-term temperature fluctuations are therefore considered to be analogous to the short-term fluctuations in dissolved calcium and magnesium carbonate.

One fundamental difference, however, exists between short-term dissolved calcium and magnesium carbonate pulses and water temperature pulses. Being generally low in dissolved solids concentration, flood water entering an underground system through open joints or swallets normally has a negative or dilution influence on the dissolved solids concentration of the water already in the system. Consequently the passage of a flood pulse at a rising is usually characterised by a definite fall in the amount of limestone in solution. The flood water may however be either warmer or cooler than the ground water and the passage of the flood pulse may additionally be characterised by either a rise in the water temperature at the resurgence or a fall. An examination of the dissolved calcium carbonate and water temperature patterns of the five resurgences (Fig 20) shows that the passage of any flood pulse between March and May 1968 and in January and February 1969 was marked by a fall in water temperature. By contrast flood pulses occurring between July and October were characterised by a rise in water temperature. At Thoragill Cave Rising the fall in dissolved calcium carbonate recorded on 13 May was accompanied by a drop in water temperature from 7·15°C on 26 April to 6·25°C on 13 May. By the following recording date (25 May) the water temperature had climbed back to 7·20°C. On 2 October another strong dilution pulse was recorded on all sites. As mentioned above the transmission of the pulse on this occasion was accompanied by a marked rise in water temperature at all resurgences.

When these short-term temperature effects are eliminated, it is evident that the temperature of the base-flow component of the discharge at the risings shows a marked seasonal trend. Although much of this water may be derived from percolation or seepage sources its passage into and through the system is sufficiently swift to preserve the broad seasonal temperature wave. On

Darnbrook Fell rapid percolation through numerous swallets is probably an important factor in this process. Only on systems with a very slow rate of water circulation is this seasonal wave likely to be eliminated.

THE CALCIUM: MAGNESIUM RATIO

The concentration of magnesium in karst waters has been used by Richardson (1968) as a guide to the origin of such waters. Hem (1959), however, recommends the use of the ratio of calcium to magnesium (computed from equivalents per million) in this context and Meisler & Becher (1967) found a close relationship between this ratio in ground water and the composition of the source rock. Low ratios (around 1·0) were indicative of dolomitic rocks and higher ratios (greater than 5·0) indicated a source in relatively pure limestone. Of the five resurgences discussed in this account the lowest Ca: Mg ratios were at Hesleden Rising (6·9) and Fosse Beck (13·4). Dolomitic limestone has been recorded by Moore (1958) at the top of the Hardraw Limestone and at the base of the Undersett Limestone in Wensleydale. Although it is not known if dolomite occurs in the equivalent parts of the Yoredale succession on Darnbrook Fell water emerging at Hesleden Rising is probably derived from the Lower *Lonsdaleia* Series which includes the Hardraw Limestone on Darnbrook Fell. During a wider survey of karst waters in the Fountains Fell–Darnbrook Fell region very low Ca: Mg ratios were recorded on springs in the Black Hill area in the vicinity of the North Craven Fault. Here values of between 1·3 and 2·3 were found and provide definite evidence of the presence of dolomite. The source of resurgent waters and the presence of dolomite is of particular interest in that recent research by Rauch & White (1970) has shown that cave development is minimal in strata with a high proportion of dolomite.

CONCLUSION

Although data from five resurgences only have been presented in this account some general conclusions may be made. The low calcium carbonate concentrations at four of the sites (Scoska Cave excluded) is due to the high swallet water dilution, and their mean value (128ppm) is only slightly higher than the mean recorded by Pitty (1972) for surface streams in north-west Yorkshire (123·5ppm). Scoska Cave with a higher mean calcium carbonate content appears to be essentially supplied from percolation water although flood pulses presumably from swallet sources were occasionally recorded. Differentiation between swallet and percolation sources on Darnbrook Fell is however difficult with much of the percolation component rapidly entering the subterranean drainage systems through open pipes in the floors of the numerous depressions on the limestone.

The degree of development of short-term fluctuation of calcium carbonate in solution provides an index of the importance of sinking stream supply to the systems, and comparison between systems may readily be made. Because no laboratory analysis is necessary water-temperature measurements are of particular value and the passage of water from swallet sources may be identified from either a marked rise or fall in water temperature. Comparison of the physical and chemical characteristics at the two risings at Bown Scar shows the presence of two distinct water flows, rising about 1m apart. This is evidently a multiple rising similar to those described in East Mendip by Drew (1970) and provides further evidence of the discrete nature of water movement in this area.

7

Karst Water Studies in and around Ingleborough Cavern

A. F. Pitty

A decade ago it was decided to test a simple hypothesis about water seeping, drop by drop, into caves. It was supposed that the chemical characteristics of such waters, and their changes in flow rate, might demonstrate distinctive patterns that could throw light on their mode of movement through the limestone above a cave. The more detailed argument behind this supposition, and the description and interpretation of an initial study in Poole's Cavern, Derbyshire, have already been outlined (Pitty, 1966 *b*). In the extension of this work to a second year's sampling in Poole's Cavern, and adding a year's observations from Peak Cavern (Pitty, 1971), the study also of Ingleborough Cavern seemed important for comparative purposes. Apart from the prerequisite of accessibility for sampling, and the celebrity and scale of the cave systems to which it is linked, Ingleborough Cavern appeared to be sufficiently representative of the near-horizontal, bedding-joint cave which typifies many Yorkshire cave systems, for any results to have some broader significance.

Ingleborough Cavern joins Clapham Dale about 1·5km north-north-east of Clapham village (see Fig 73). The adjacent resurgence at Clapham Beck Head is in part the waters that fall down Gaping Gill (see chapter 18). As an area of exciting pioneer and contemporary exploration, the cavern and its surroundings need little introduction, and the uninitiated reader can turn to the introductory guide (T. D. Ford, 1971 *a*).

LOCATION OF SAMPLING POINTS

For the purposes of sample collection and for measuring seepage rates, nine sampling points were located in Ingleborough Cavern at approximately 40m intervals. The first seepage, labelled S1, is the south-westerly one of a pair of seepages falling 30cm apart in the centre of the sand-covered slope at the inner end of the show section of the cave, 3m beyond the 'Pool of Reflections'. S2 falls on to a rounded boulder down-cave from the Pool of Reflections, where the path crosses from one side of the Long Gallery to the other. S3 occurs in a cross-rift, along the line of which the main cave is offset, near the 'Peal of Bells' stalactites. It drops on to a low rock bench on the south-west side of the passage. The rock surface has a stalagmitic veneer, the drops from S3 creating that one of several punctures encountered in this veneer which is nearest to the main cave. In the low but broad continuation of the main cave the bedding-plane roof rises gradually and S4 descends as a shower from a narrow cross-rift. Volume measurements were impracticable at this point. Down-cave, S5 falls 5m before the stream crosses from one cave wall to the other, opposite to a curtain which is symmetrically shaped like half of an old-fashioned beehive. S5 falls on to a stalagmite-covered ledge on the west side of the passage, about 30cm off the path and about 60cm above it. S6 is the seepage falling into the small depression in the crown of the large 'Jockey Cap' stalagmite which commands the down-cave end to the 'Pillar Hall'. Unlike the other seepages, which were selected as the first drops caught in the light when the headlamp was switched on, S6 was deliberately selected because of the distinctiveness of the formation with which it is associated. S7 falls on the north-east side of the cave 3m down-cave from a post near the 'Beehive' curtain. Drops at S7 fall from a 7cm long, hook-shaped stalactite on to a narrow platform 15cm below. The platform is 1m from the ground. The site of S8 is similar to that of S7, in that the point at which the water is sampled would have been beneath the level at which formations ceased to grow as determined by a former lake level. The lake was drained artificially in 1837. S8 is situated on the north-east side of the cave, 3m beyond the lower end of a downward slope in the path, just before the passage becomes more constricted. It falls from the overhang on to a low stalagmitic mound, shaped like a meat chop. Finally, S9 falls at a point 2m before the path straightens as it approaches the cave entrance, some 20m from the open-air. S9 is situated on the south-west side of the cave and falls only 15cm from the bedding-plane roof on to a shelf fringed with irregular stalagmitic material a few centimetres in height.

In addition to the nine seepages, water samples were collected from the stream in the cave and also from three surface streams in the vicinity of the cave. First, high on the plateau 1,500m from the cave entrance, Fell Beck was sampled just before the stream falls over 100m down the shaft of Gaping Gill. Fell Beck, with some tributaries rising at altitudes of more than 600m (2,000ft) on Simon Fell, begins to sink at an altitude of 410m (1,350ft) where the limestone crops out from beneath the overlying sandstones and shales, upstream from the point at which the remaining surface flow plunges down Gaping Gill. Immediately below the entrance to Ingleborough Cavern, is Clapham Beck Head, the resurgence for the Gaping Gill water and all the water draining out of Ingleborough Cavern. This stream was sampled just in daylight. In addition, the main cave stream in Ingleborough Cavern was sampled close to the point at which it disappears into the 'Abyss', which is between seepage sampling points S5 and S6. A rising just upvalley from Clapham Beck Head, draining down the lowest 6m of the south-east valley wall, was also sampled, at the level of the valley floor. This tufa-depositing rising was included as Spring 13 in the Yorkshire Geological Society's survey at the turn of the last century (Carter *et al*, 1904).

CALCIUM CARBONATE ANALYSES

Plastic containers were placed at the sample points in the cavern and, whilst these filled up, the surface stream samples were collected. The water samples were analysed in the laboratory within two days of collection. Occasionally the duplicate determinations differed by more than 2ppm and a third determination was made to ascertain which of the first two was nearer the probable true value, before a mean figure was calculated. Annual mean values of $CaCO_3$ for the twelve months from August 1964 to July 1965 ranged from 156ppm at S2, to 227ppm at S9. The mean values for the cave stream, 197ppm, and Spring 13, 165ppm, fell within this range, but both Gaping Gill (48ppm) and Clapham Beck Head were decisively lower. The range of variations was of the same order for all the seepages, ranging between coefficients of variation of 9·9 for S8 and 17·0 for S3. The spread of three of the four streams was beyond this range (Table 8).

When the month-by-month observations are plotted (Figs 21 and 22) a seasonal pattern emerges, particularly in the $CaCO_3$ content of the seepage water. When these figures are correlated with mean air temperatures for selected intervals prior to each sampling day, many of the sampling sites, particularly the seepages, show a high degree of correlation with antecedent temperatures as recorded at Malham Tarn House. For instance, the positive association between air temperature and $CaCO_3$ for S2 is closest for the mean temperature figures for the interval 21–41 days prior to each sampling day ($r=0{\cdot}89$). The cause, indicated by the temperatures, is believed to be far more complex than merely a direct influence of temperature alone. After an extensive review of literature on the subject it was concluded that the process involved is seasonal changes in soil carbon dioxide due to the stimulus imparted to soil microbes by plant root activity, reaching its maximum in the long hours of daylight and during the higher temperatures of midsummer (Pitty, 1966 *b*).

The time lag before the chemically altered water reaches the cave appeared to be a useful guide to the time taken for the

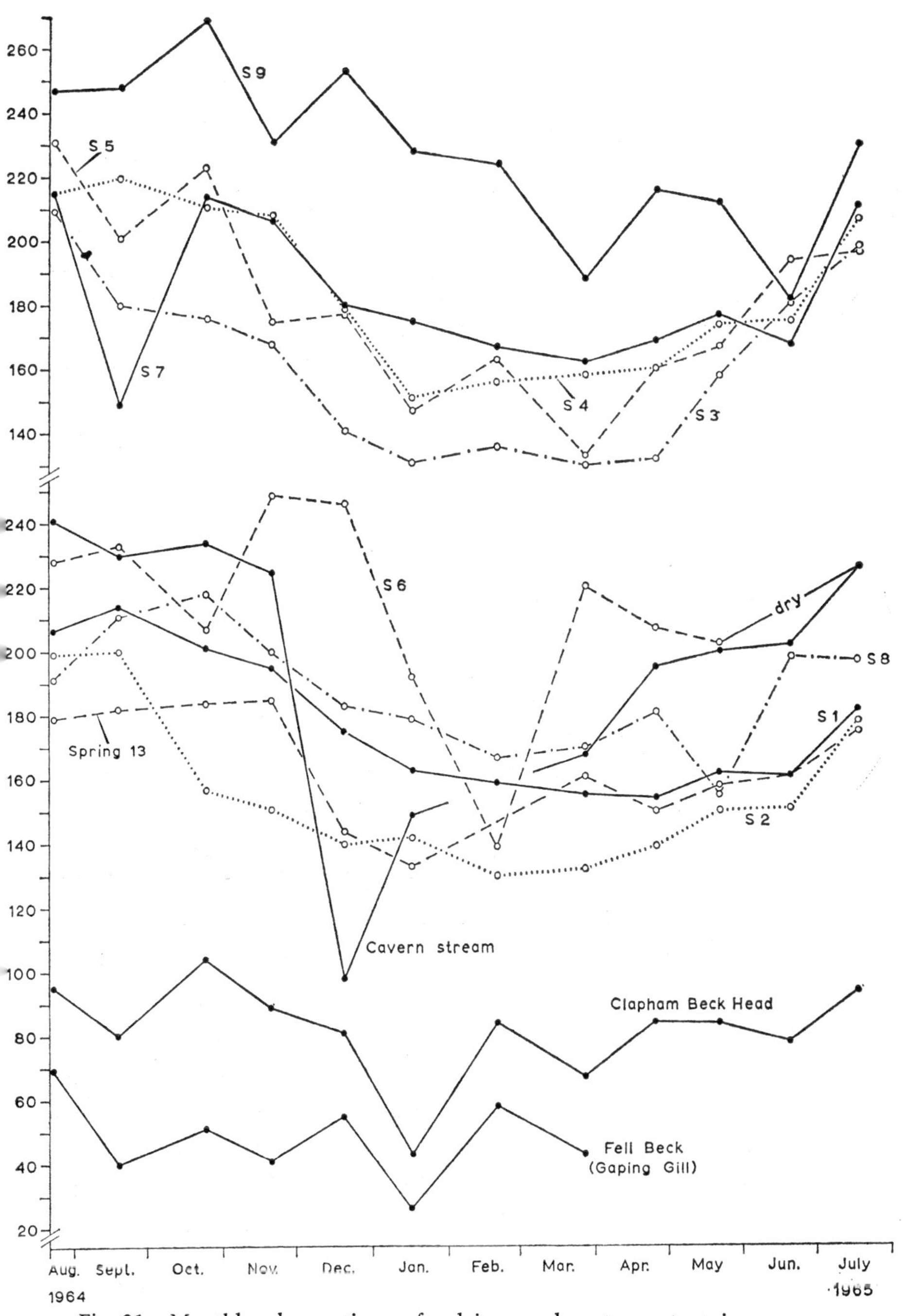

Fig 21 Monthly observations of calcium carbonate content in cave and surface waters at sampling sites in and around Ingleborough Cavern

water to travel through the limestone strata overlying the cave. This can be shown more clearly by estimating the time within the 3-week intervals for which the correlation might be closest. For instance, the correlation coefficients for S1 are $r = 0{\cdot}76$ for the 21–41-day period, $r = 0{\cdot}94$ for the 42–62-day period and $r = 0{\cdot}91$ for the 63–83-day period. In this case it seems that the correlation is closest for a longer time lag than that indicated by the mid-point for the 42–62-day period, and that this can be estimated by interpolation based on the degree of correlation shown by the 3-week intervals on each side of the interval showing the highest correlation. When the mean annual $CaCO_3$ content of the seepages is plotted against the estimated flow-through time, a clear relationship emerges, with the long flows having progressively higher $CaCO_3$ values (Fig 22). A striking addition to Fig 21 compared with an earlier diagram (Pitty, 1968 *a*) is the inclusion of S6. As this point was dry on the sampling day in June 1965 and since the original computer program was too inflexible to deal with the problem of 'dry' days, S6 was omitted from the original calculations. The present, revised program is more general and calculates correlations at any given site between climatic data and karst water data for only those days on which the latter was obtained, and results show that S6 has a mean $CaCO_3$ content of 214ppm and an estimated flow-through time of 63 days, and conforms with the previously described trend (Fig 21).

When the $CaCO_3$ observations are plotted against the appropriate antecedent mean temperatures, the pattern in detail of the departures from the close positive association between antecedent temperatures and $CaCO_3$ content shows clearly defined contrasts between sampling stations. Whereas the autumn fall in $CaCO_3$ for S1 is higher on average than for corresponding temperatures in the spring rise, the rise and fall of observations for S2 and S3 cross and recross. Whereas for S2 the observations resulting from soil conditions in the alternate months June, August, October and December are lower compared with the trend of adjacent months, the same months are relatively high points on

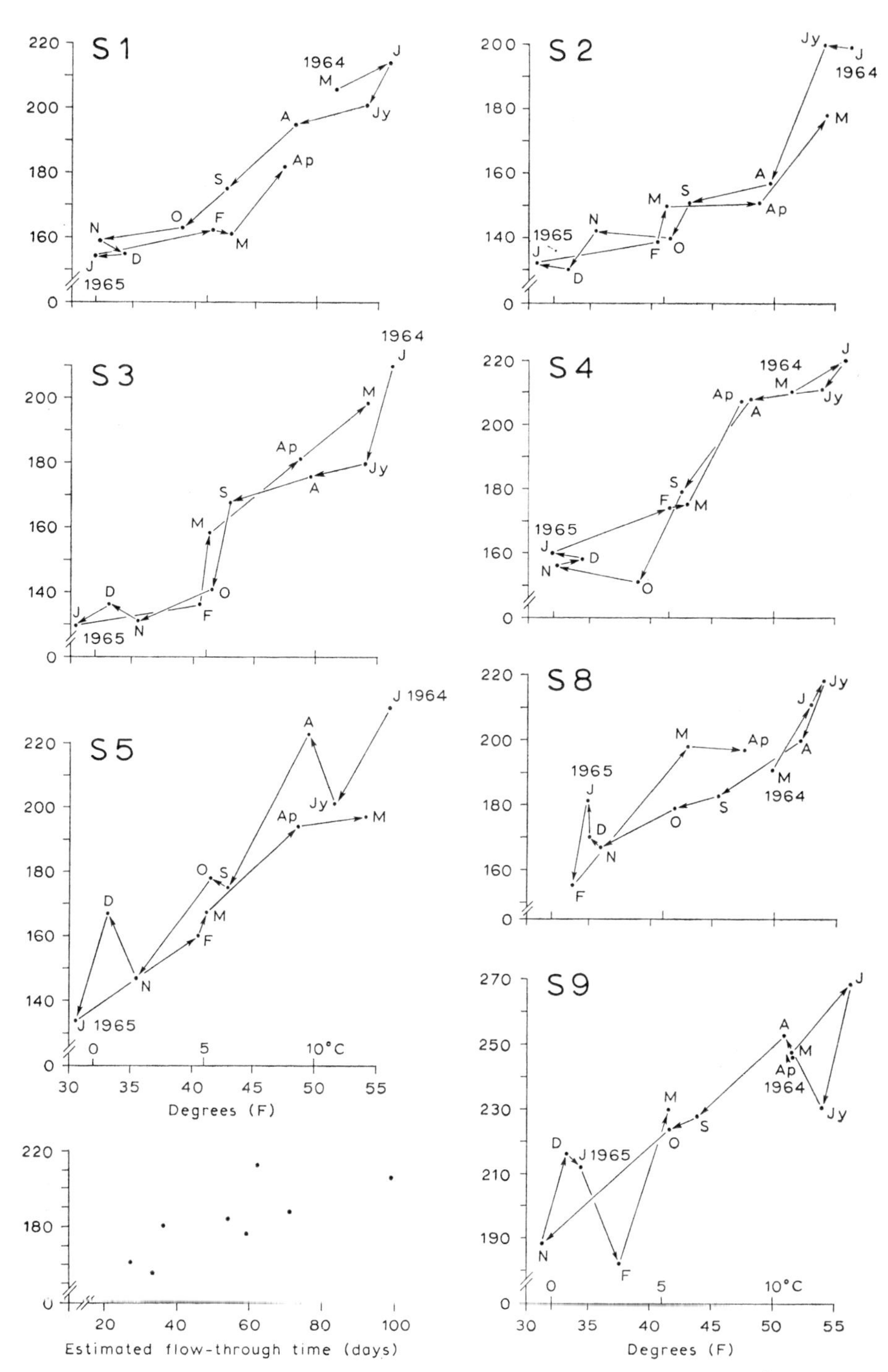

Fig 22 Patterns of calcium carbonate content of seepages in Ingleborough Cavern in relation to antecedent temperature and to (bottom-left frame) estimated flow-through time

the S5 graph. S5 also provides the contrast between an erratic autumn decline in $CaCO_3$ and an apparently regular late-winter/spring rise. S4, S8 and S9 each show an observation somewhat isolated below the general trend, being the water emerging in January and related to the November temperatures at S4, that of the March-influenced water emerging in May at S8, and that of the February-influenced water emerging at S9 in June. Clearly these contrasts in detail indicate that distinctive, if minor, influences are at work as well as the plant root activity initiating changes in the soil carbon dioxide output. For the surface streams, dilution after precipitation might be anticipated from inspection of the plunges shown in the graphs of seasonal changes in $CaCO_3$ (Figs 21 and 22). None the less all the streams show some indication of a positive association with temperature, particularly Spring 13 and the cave stream.

The distinctive statistical nature of precipitation data makes it difficult to correlate accurately precipitation with $CaCO_3$ content in cave waters. Results give a guide to the time intervals most likely to link a precipitation effect to a response in the amount of $CaCO_3$ rather than a reliable measure of the closeness of this relationship. They are adequate to bring out clearly the main contrast between surface water and cave seepage water, the former revealing how a pulse of water, which has had little time to increase its solute load, dilutes the amount of $CaCO_3$ in the stream. Thus the amount of precipitation falling over the two days preceding sampling has a controlling influence on the degree of dilution of the $CaCO_3$ content of Fell Beck as it enters Gaping Gill. This dilution effect is transmitted rapidly to Clapham Beck Head as again the 0–2-day interval is important. The dilution effect does not reach the Ingleborough Cave stream immediately, and it is precipitation over the longer 2–13-day period which is required for dilution to be most effective. For Spring 13 the pattern may be influenced by dilution from precipitation in the immediate area around the rising rather than from the flow-through of a sink-resurgence system, as $CaCO_3$ concentrations, in so far as they are slightly correlated with precipitation, follow most closely precipitation amounts on sampling day. S1–5 cave seepages may be similarly affected to a slight degree, and for S6, where the positive influence of antecedent temperature is least marked, the possible indication of slight dilution from precipitation is most clearly marked for the 0–2-day period prior to each sampling day. Similar indications of a slight dilution of $CaCO_3$, in cave seepage water which is largely governed by antecedent temperatures, by precipitation falling shortly before sampling days has been illustrated for a seepage in Poole's Cavern, Derbyshire (Pitty, 1968 *b*; Fig 3). It seems that in addition to the dominant pattern of water in discrete microfissures which are partly or largely filled with water, moving at rates of 30–60cm/day, there may also be the occasional larger fissure which is usually dry. After rain, small, and short-lived, pulses of dilute water may thread their way down quickly to the cave roof level. However, on present evidence this is a possibility only for S7, S5, S3 and perhaps S4 in the Ingleborough Cave; by comparison only one of the five seepages in Poole's Cavern appears to be slightly diluted in this way by precipitation in the two days prior to each sampling day.

The Jockey Cap (seepage) S6 shows a distinctive response to precipitation with the amount of $CaCO_3$ tending to increase with precipitation in the fortnight preceding sampling. A trace of dilution immediately after precipitation may also occur, as it is the 3–13-day period which gives the highest positive correlation. This result may be significant in the understanding of the genesis and growth of cave formations as this distinctive positive correlation with precipitation of the Jockey Cap seepage resembles the similarly distinctive pattern of water seeping from Mary Queen of Scots Pillar, one of the largest active formations in Poole's Cavern.

S9 may show a third type of response in $CaCO_3$ content to precipitation; here the $CaCO_3$ amounts tend to correlate positively with precipitation in the five weeks prior to sampling days. At a point only 20m from the cavern entrance, evaporation may well be a factor, and as seepage rates are correlated most closely with a similar interval of antecedent precipitation at S9 (see below), the positive correlation between precipitation and $CaCO_3$ content may indicate the reduced effectiveness of evaporation on seepage water when it is falling more quickly.

SEEPAGE RATES

Mean seepage rates range from 14 to 120 drops/min. However, calculations for S7 in Table 8 omit the March 1965 sample when water at the sampling point was a continuous stream, and with streaming occurring on four out of the six winter months at S8 and on the September 1964 sampling day, the calculations for non-streaming days at S8 is also unrepresentative of the streaming seepage type of flow. The eight mean summer rates vary from 12 to 119 drops/min. Excluding the streaming seepage, S8, the summer seepage rates at the remaining seven stations where seepage was counted are on average 74 per cent of the winter seepage rates. This reduction gives an idea of the amount of evapotranspiration loss in summer and compares with a mean value of 81 per cent for three seepages in Poole's Cavern (Pitty, 1969).

In correlating the monthly observations of seepage rates with antecedent precipitation, again the purpose is to search for the most effective time intervals involved rather than to measure the degree of association with precipitation. S1–3, S6 and S9 show a pattern of closest correlation with a longer time interval in summer, and a response in winter linked with a shorter, or much shorter, period of antecedent precipitation: S1 from 0–13 days to day 0, S2 from 0–41 days to 0–1 days, S3 from 0–13 days probably to day 1 rather than to the 7–13-day period as this is the pattern for S6, and for S9 from 0–41 days to 0–20 days. S7 probably also follows this pattern as it was the streaming volume in March, 2 days after 1·2cm of rain fell, that made it statistically impossible to calculate correlations for the winter half of the year as a whole for this seepage.

S9, distinctive in showing no difference between mean rates for winter and summer seepages, is also distinctive in the relatively

TABLE 8 *Chemical and physical characteristics of Ingleborough Cavern karst waters*

Sampling	Calcium carbonate content (in ppm)		Seepage rate (in drops/min)						Summer/Winter
			Annual		Summer		Winter		
	Mean	Coefficient of variation	Mean	C of V	Mean	C of V	Mean	C of V	%
S1	177	12·2	66	31·6	60	27·9	72	34·0	83·9
S2	156	15·4	14	29·5	12	29·4	15	26·6	78·3
S3	162	17·0	27	60·9	19	62·4	35	48·2	50·6
S4	184	13·8	—	—	—	—	—	—	—
S5	181	16·1	120	2·9	119	2·4	121	3·1	97·7
S6	214	16·7	93	65·1	76	61·9	110	65·4	69·3
S7	183	12·5	121	64·9	83	69·0	166	48·5	34·3
S8	188	9·9	37	97·3	38	107·0	—	—	—
S9	227	11·3	50	44·3	51	45·1	50	47·8	100·6
Fell Beck	48	28·0	—	—	—	—	—	—	—
Cavern stream	197	22·1	—	—	—	—	—	—	—
Clapham Beck Head	82	19·0	—	—	—	—	—	—	—
Spring 13	165	10·7	—	—	—	—	—	—	—

(*Note*: Values are based on monthly observations from August 1964 to July 1965. Those for S6 are based on eleven observations and for Fell Beck on eight only. Seepage calculations for S7 and S8 omit days when streaming occurred; they are included here to show that even the dropwise falls at these points show greater variability than observed in the more regular seepages where streaming was not observed.)

long time interval over which the effect of winter precipitation builds up. For S5, precipitation over 2–6 days prior to sampling produces a similar effect, but in winter no clearly indicated time interval emerges. Perhaps the constant head reservoir which seems a likely type of source for this very regular seepage is always at least brim full in winter. It is noteworthy that S8, for which winter volume observations were often impossible due to the transformation of the discrete seepage into a stream of spray, also shows the highest correlation with the comparatively short 3–6-day period, even in summer.

CONCLUSIONS

The observation of $CaCO_3$ content of cave waters in Ingleborough Cavern and of the water in nearby surface streams, and also the counting of seepage rates in the cavern, have added evidence to knowledge about the general characteristics of karst water, and have specified also some of the individual characteristics of processes in a well-known limestone area.

The degree to which each seepage operates independently of adjacent ones supports the suggestion that seepage water in massive limestone tends to move in isolated networks (Pitty, 1966 *b*). Secondly, the over-riding positive correlation of $CaCO_3$ with antecedent temperature has been demonstrated again most clearly for cave seepage water. Thirdly, the similarity of processes recorded for the Jockey Cap in Ingleborough Cavern, and for Mary Queen of Scots Pillar in Poole's Cavern, may suggest a working hypothesis on the creation and growth of large formations in caves. Fourthly, in addition to the antecedent-temperature-dominated cave seepage and the dilution-affected surface flow, a third type of karst water is perhaps recognisable at Spring 13. Judging from the relief in the area it seems unlikely that a sink hole could feed this issue of water which, in its chemical variations, resembles the cave seepages closely, particularly the streaming S4 water; nor was Spring 13 observed to dry up. One possibility is that the drainage line feeding Spring 13 was formerly a down-cave continuation of a system on the Ingleborough Cavern side of the valley before valley incision severed it from its former headstream areas. The water now draining out at Spring 13 could therefore be the seepage water dropping into this abandoned section and, with flow now reversed, draining back towards the present valley wall. A final general point is that, despite the complications of streaming seepages and the difficulties of handling precipitation data statistically, the consistency of results between Ingleborough Cavern, Yorkshire, and Poole's Cavern, Derbyshire, is striking.

In north-west Yorkshire where the discrete courses of underground streams, at Turn Dub in Ribblesdale, and in the sources of the Aire near Malham, have been known for more than two generations, it is perhaps not surprising to find that seepage lines in microfissures are apparently also discrete. It is possibly also useful to have established that the amount of $CaCO_3$ in the water about to descend Britain's most spectacular vertical shaft at Gaping Gill is virtually an inverse function of precipitation. The rapidity with which the pulse of dilute water passes through to Clapham Beck Head is also significant and reflects at once both the effect and the cause of a large cave system. Thirdly, the way that the Ingleborough Cavern stream normally drains out seepage water only but may still occasionally experience substantial dilution (225ppm, November 1964, to 98ppm, December 1964) demonstrates the way in which this passage, now normally abandoned as an outlet in favour of a lower course, may still function during high water. Finally, a distinctive feature of Ingleborough Cavern is that the faster flow-through times and the correspondingly lowered $CaCO_3$ values are distinctly different from observations in Poole's Cavern and other areas in Derbyshire. The lower values of $CaCO_3$ in some parts of north-west Yorkshire, when compared with other British limestone areas, in addition to the diluting effect in streams of allogenic water from non-calcareous outcrops, is, it seems, due to the relatively rapid flow-through time of the seepage water.

8

Karst Waters of the Alum Pot Area

D. T. Richardson

The streams of the Alum Pot network drain the eastern slopes of Simon Fell, on the north-east side of the Ingleborough block. On the steep slopes of the Yoredale rocks are many small surface catchments and springs whose combined flows cross on to the Great Scar Limestone at an altitude of 370m (1,200ft) (see Fig 23). Nearly all the water disappears underground at the edge of the limestone outcrop, and between there and the open shaft of Alum Pot over 3km of cave passage are known.

The majority of the caves are vadose canyons formed below very gently dipping bedding planes and shale beds. Most passages are 1–3m high and wide, and transmit the water rapidly and efficiently eastwards. Nearly all the cave passage known is in one integrated system which is broken up by very short lengths of surface stream course; the waters of Borrins Moor Cave flow right through the Long Churn caves into the Diccan Pot–Alum Pot system. Most of the passages are active, and have a constant, but gentle, gradient, dropping only 30m from the highest sinks to the lips of the shafts on the Alum Pot joint. Here the water falls nearly 100m to sumps level with the resurgence at Turn Dub, 2·5km further east on the far bank of the River Ribble (see Fig 80).

The thin bedded nature of the upper beds of the Great Scar Limestone and the general lack of deep joints have together been responsible for the extensive shallow karst drainage of the Alum

Pot area. Beside the main system, focused on the Long Churn caves, there are a number of other small underground drainage systems which resurge on the fell itself, the water then flowing over the surface to fall into Alum Pot or sink back into the limestone. Eventually all would seem to resurge at either Turn Dub or other small springs near to the floor of Ribblesdale.

In the Long Churn Cave system the normal flow of the main stream is just under 0·1cumec. All the drainage systems independent of this one, with the exception of the Fell Close–Gillgarth system on the southern margin of the area, have flows of less than one-tenth of that in Long Churn.

Vegetation on the higher slopes consist of grass, sedge and peat bogs, overlying boulder clay which in turn blankets the Yoredale Series and masks nearly all topographic expression of this Series' lithological variation. Some of the Great Scar Limestone is exposed as bare pavement, though most is covered by thick grass grown on a thin cover of boulder clay.

Over a period of seven years more than 200 water samples have been analysed from sixty-eight sites in the area. Stage conditions have a very considerable influence on nearly all parameters of the water chemistry, so this account mainly refers to suites of samples collected on just two dates. However, the conclusions are also based on data over a broader spectrum of time, stage and place, which all fit into the recognisable patterns presented by the summary analyses with this account.

THE WATERS OF THE UPPER CATCHMENT AREA

On the slopes above the Great Scar Limestone, the permanent sources of water fall into two distinct groups—the springs and seepages from the peat bogs, and springs from the thin limestone horizons within the Yoredale Series. The water is chemically very different from the two types of source, though periods of heavy rainfall increase the direct run off and dilute the permanent flows of stored water.

Waters derived from the peat bogs (Table 9) have very low

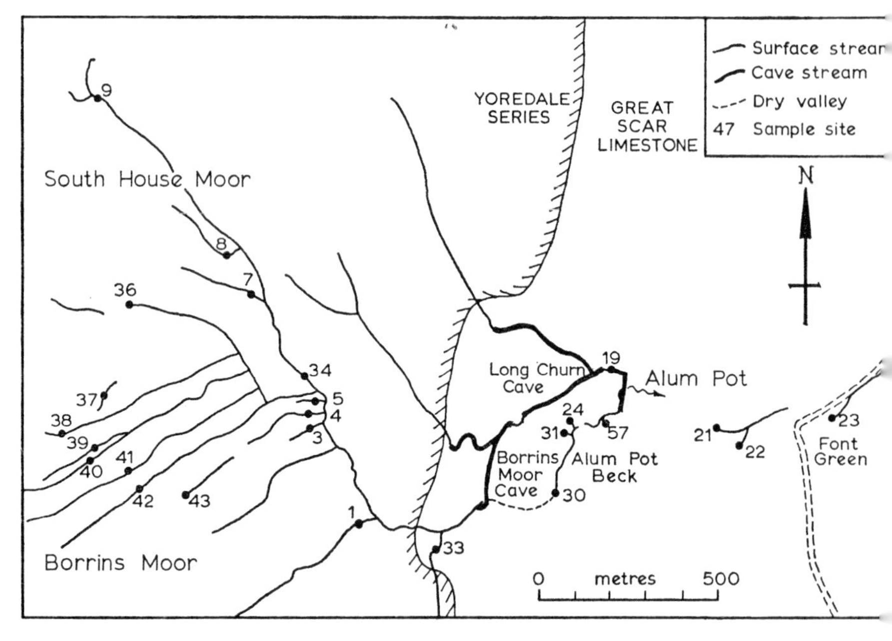

Fig 23 Hydrology of the Alum Pot area

pH and are strongly aggressive towards $CaCO_3$. They have a definite solute load ranging 8–18ppm, but the amount of dissolved carbonate and bicarbonate is extremely low. Instead, sulphate and chloride contents are high, and consequently the Mg/Ca ratio is mostly greater than unity—in complete contrast to the carbonate-rich, low-magnesium, waters of the limestones. The sulphate is almost certainly derived from the oxidation of pyrite, small quantities of which are very widespread in the shales.

Water rising from springs in the Yoredale Series is typical of limestone karst. High pH values and high carbonate content, with low Mg/Ca ratio, reflect long contact with the limestone. The analyses are in fact very similar to analyses of water from large resurgences at the base of the Great Scar Limestone. This may be taken to indicate quite extensive karstic erosion within the very thin Yoredale limestones, though at present no caves are known in these beds on this part of Ingleborough. Spring waters at low altitudes in the Yoredales are noticeably harder than those at

TABLE 9 *Analyses, meaned by groups, of karst waters sampled on 13 August 1970, in the Alum Pot area*

		Yoredale springs		Great Scar springs		
Water type	Peat water	High	Low	Seepage	Caves	Long Churn Cave stream
Sample sites (see Fig 1)	7, 8, 9, 34, 37	36, 38, 39, 40, 41, 42, 43	1, 3, 4, 5, 33	30, 23, 21	31, 24, 22, 57	19
Altitude (m)	455	510	400	320	345	350
Altitude (ft)	1,495	1,665	1,310	1,050	1,125	1,145
Total hardness (ppm $CaCO_3$)	11	50	94	227	89	65
Alkaline hardness (ppm $CaCO_3$) (essentially bicarbonate)	1	42	80	207	76	51
pH	4·9	7·5	8·1	7·8	7·8	8·1
Temperature (°C)	13·6	10·9	11·8	11·4	12·2	13·0

higher altitudes in similar geological situations (see Table 9). This must reflect longer contact of the lower-level water with the limestone, but this does not necessarily infer integration of the Yoredale underground drainage. The water flowing down the fell sinks at the top of each Yoredale limestone bed, and reappears a few metres lower down at its base; water from the lowest springs is merely returning to the surface for perhaps the third or fourth time on its journey down the fell.

The high temperature of the peat water (Table 9) reflects the warm summer day on which the samples were collected. The lower Yoredale water temperatures are due to the longer periods of underground flow, and the difference between the two Yoredale figures is probably only an effect of altitude. Variations of flow rates have little effect on the quality of the peat waters, though the total hardness of water from the Yoredale limestones is considerably reduced during periods of high stage. The samples analysed for the compilation of Table 9 were collected during a relatively dry summer period; the annual mean hardnesses of the Yoredale waters are therefore lower (by about 25–50 per cent) than the tabulated data.

HYDROCHEMISTRY OF THE ALUM POT CAVE SYSTEMS

The cave systems of the Great Scar Limestone have three types of water flowing into them—relatively soft, aggressive drainage from the peatbogs; harder water derived from the Yoredale limestones; and direct rainfall which lands on the Great Scar Limestone itself. The Yoredale water is quantitatively dominant with respect to the Alum Pot caves.

Through the known cave systems, from Borrins Moor to Alum Pot, the overall pickup of dissolved carbonate is very low. Fig 24 shows the analysed total hardnesses at a series of points down the cave stream on two dates. Downstream variation in solute load is influenced by four factors: (1) addition of Yoredale-derived water; (2) addition of aggressive peat water; (3) addition of percolation water; and (4) solutional pickup from the cave walls. No significantly aggressive peat-water streams have been recorded as tributary to the main Alum Pot drainage; all the measured input waters are dominantly of Yoredale origins.

Much of the percolation water, which joins the stream as a series of small inlets, is very high in dissolved salts. Analysed total hardnesses of a number of small trickles and seepages of percolation water, inside and outside the cave, range between 130 and 250ppm, with a mean of 180ppm. Continued pickup of such percolation water must increase the hardness of the stream water. However, these cave systems are very shallow—mostly with less than 10m of rock over their ceilings—and, in wet

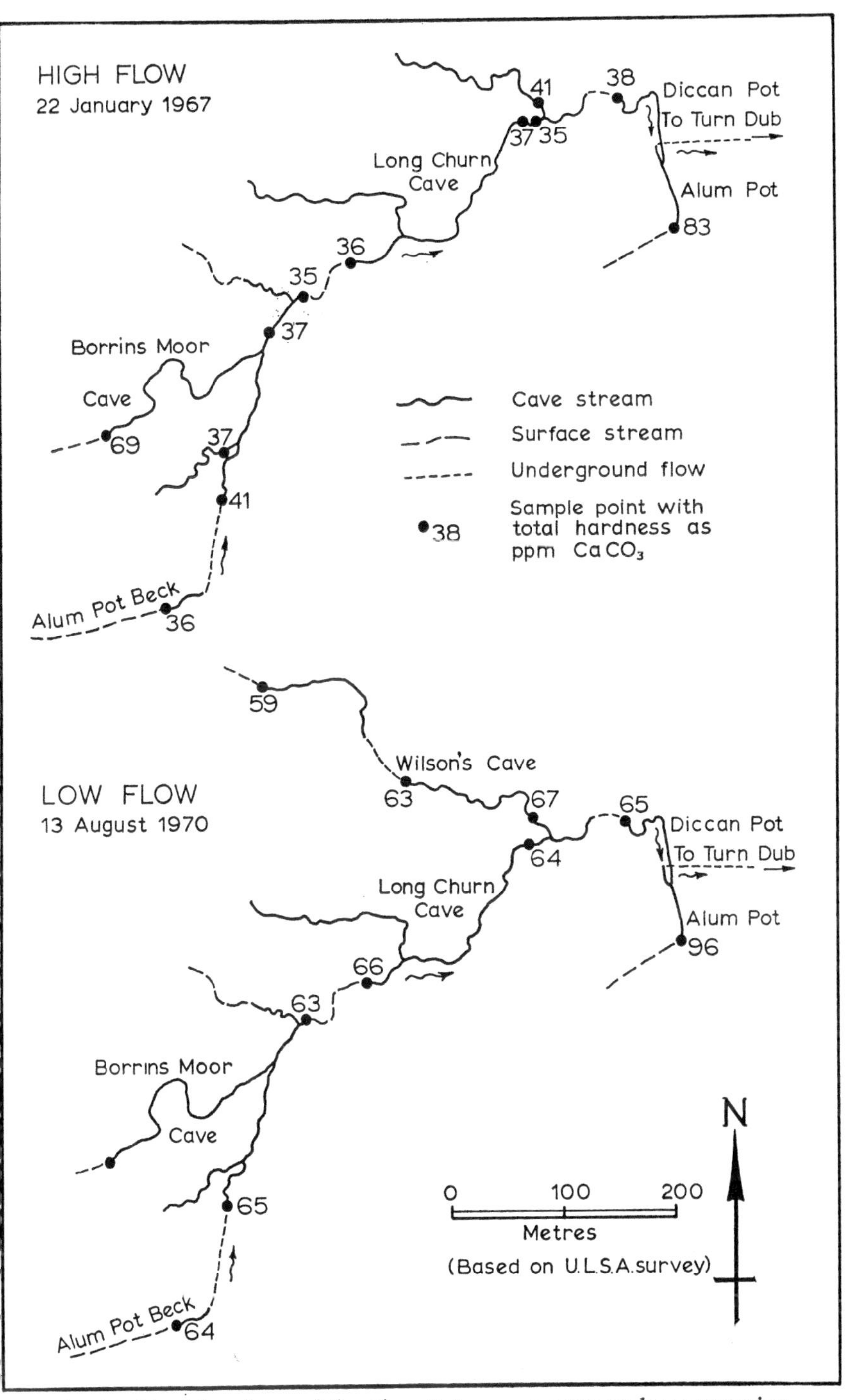

Fig 24 Total hardness of the Alum Pot cave streams under contrasting flow conditions

weather at least, much water must pass very rapidly through the fissured limestone; with consequent minimal solutional pickup this must have a contrasting diluting effect on the hardness of the main stream.

Analyses have not yet been comprehensive enough to quantify the effects of the combining of the chemically different stream water. Consequently it is impossible, so far, to estimate either the effect of the mixing of percolation and stream waters, or the amount of carbonate dissolved from the passage walls.

After dropping down the shafts of Alum Pot, the water passes through a long phreatic zone to resurge from Turn Dub. Total hardness of the rising water has been analysed at 125ppm, showing that the pattern of low overall solutional pickup in the shallow vadose caves is continued through the phreas.

The effect of rainfall on the hardness of the Alum Pot waters is just one of overall dilution. Fig 24 compares total hardness data collected in conditions of contrasting stage. There had been no rain for a considerable period before 13 August 1970, whereas there were several days of very heavy rain on and before 22 January 1967. All the waters show considerable reduction of solute proportion with increase of stage. However, the total solute load removed is greater in flood conditions than in relative drought; total hardnesses drop by a factor of about 2, to match flow rates increased by factors of 3–5.

SPRING WATERS FROM THE GREAT SCAR LIMESTONE

Downstream of the sink to Borrins Moor Cave, the course of Alum Pot Beck is abandoned where it cuts across the benches of Great Scar Limestone. In all but very dry weather, a number of small springs and seepages produce an underfit stream which falls over the lip of Alum Pot. Below the open shaft, another dry valley continues down the fell.

Hydrologically, this dry valley system is very similar to an unroofed cave system, in that the springs which occur in its banks and floor fall into two distinct groups (see Table 9). Most of the very small seepages, from boulder piles or joints in the bedrock, produce water of very high total hardness—characteristic of its percolation origins. In contrast, the larger springs, mostly issuing from very small cave openings, are of much softer water—chemically comparable to the water in the main Long Churn Cave stream. These latter waters have all passed through the limestone in open, discrete passages similar to inlet streamways in the cave system; typical is the spring just below Alum Pot (site 22, on Fig 23) which has nearly all its water derived from the long and complex, shallow cave systems of Gillgarth and Fell Close, further to the south. The only reason that these waters reappear on the surface above Turn Dub (to which they eventually flow after sinking again) is the strong stratigraphic control which keeps much of the drainage perched well above the regional base-level.

The close juxtaposition of some of the contrasting types of springs on the Alum Pot Fell demonstrates the generally poor degree of integration of the Great Scar Limestone karst drainage.

SUMMARY

Widespread, less systematic sampling of waters across the whole of north-west Yorkshire has shown that the Alum Pot area is hydrologically representative of the karst region. The general situation in the Great Scar Limestone is that water draining into the caves already has a significant solutional load derived from the overlying Yoredale limestones, and resurges from the lower levels of the Great Scar with only slightly increased dissolved carbonate content.

The Mg/Ca ratio of the dissolved salts is nearly everywhere low—in the order of 0·05—which reflects the low dolomite content of the source limestones. The main exception is the magnesium-rich waters of the peatbogs where the dominant anion is sulphate instead of carbonate.

Perhaps the most distinctive feature of the north-west England karst hydrology is the very low total hardness of the resurgent water, especially when compared with that in Mendip (Smith & Mead, 1962). This contrast is also reflected by the rates of increase of the solutional load through the known caves in the different regions. One of the main factors responsible for these differences must be the much larger quantities of water flowing through the caves of north-west England. Only a small proportion of the total karst water is responsible for nearly all the solutional effort of cave enlargement in this region. In contrast, the majority of the cave streams of the Mendip Hills have small flows and gain a considerable proportion of percolation water within the known lengths of passage. The low total hardnesses of the Yorkshire resurgences further reflect the very high ratio of swallet water to percolation water in the largest karst drainage systems, such as Borrins Moor–Alum Pot–Turn Dub.

9

Biospeleology in North-West England

J. M. Dixon

In common with many of the speleological sciences, the surface of biospeleology in north-west England has barely been scratched. Most collecting has been sporadic and isolated; a few projects have been more systematic but no really comprehensive studies have been undertaken.

The *Biological Records* of the Cave Research Group have listed details of fauna collected in the caves of north-west England since 1938 (Glennie, 1955, 1956; Hazelton, 1958–1971). However, the maximum number of records per year—fifty in 1968—indicates the sparsity of available data. The few systematic projects are summarised below, and these are followed by a discussion of their significance, and appendices of the fauna and flora so far recorded. Scattered records of finds which were not parts of projects are listed only in the appendices.

FAUNA OF MORECAMBE BAY LIMESTONES

Moseley (1969) investigated the fauna of the Hale Moss Caves at Beetham.

He commented on the fact that since these caves run close to the surface, comparatively large amounts of food can be washed down by rainfall; also the temperature will be more variable than in most caves and there is a continual interchange between surface and cave terrestrial fauna. An accompanying faunal list shows that the terrestrial specimens are very similar to those of the endogean domain. (Specimens recorded are included in the Appendix to this chapter.)

Aquatic species in these caves are found in isolated pools with no direct connection with surface populations. In these conditions numerous specimens of a few species are expected. This was borne out by one pool examined which supported a flourishing colony of *Rivulogammarus pulex* (L.) and a colony of *Pisidium personatum* Malm and very little else. Both the aquatic and terrestrial fauna are helped in survival by the presence of a plentiful food supply. In the case of terrestrial fauna this is primarily organic matter carried down through the various openings by seepage water. The pools are fed by seepage water and lined with rich deep mud which contains a high proportion of organic matter. This is broken down by bacteria and provides food for protozoa. Numerous ciliates were revealed by microscopic examination of a sample of the water. These are presumably eaten by the *Rivulogammarus pulex* and the *Pisidium personatum*. *Rivulogammarus* also eats vegetable detritus.

The fauna of the caves and mines in the Morecambe Bay area was described in some detail by Moseley (1970). Physical data such as temperatures are given for some locations and the total number of species is recorded for each different biotope in the mine or cave. A brief note dealing with food sources, and one with the distribution of fauna in the Warton mines are included. There is also a short discussion on the history of the hypogean fauna of the area. Moseley found that each biotope shows a characteristic fauna. These are summarised below.

Dark zone terrestrial species occur in mine timbers, a very rich habitat; the fauna consists mostly of lumbricid worms, isopods, diplopods, and Collembola. Vegetable detritus supports isopods (*Andronicus*), diplopods (*Polymicrodon* and *Blaniulus*) and Coleoptera (*Quedius*). Rock surfaces are dominated by adult and larval Diptera; isopods and diplopods sometimes occur. The surface film of water is mainly occupied by Collembola, but mites and Symphyla are also present.

No direct evidence is available regarding interstitial fauna, but the isopod *Trichoniscoides saeroeensis* Lohmander may be primarily interstitial.

Dark zone aquatic fauna found in pools are mainly planarians, copepods and sometimes *Rivulogammarus pulex*. Flowing water has not been investigated. Seepage water and springs within the caves and mines did not yield any fauna, but a single specimen of *Hydroporus obsoletus* Aube may indicate a phreatic fauna.

Threshold region terrestrial fauna found on rock surfaces are typically various species of Diptera and, in winter, Lepidoptera, together with accidental fauna. The surface film of water supports Hemiptera (*Velia caprai* Tamanini) and various Collembola. *Rivulogammarus pulex* is common in the threshold region aquatic zone.

The work is concluded with a comparison of the areas and suggestions for future work.

FAUNA OF WHITE SCAR CAVE

Fauna studies in White Scar Cave were carried out by Schofield (1964). He reported on the different species and relative sizes of populations from the show portion of the cave and from the rarely visited further reaches. He recorded the surprising fact that the larger and more numerous colonies, with the exception of Diptera species, appear to be present in the unvisited section rather than the commercialised section with its organic contamination.

Interesting observations were made on the effects of flood conditions on the fauna and the length of time taken for re-establishment after severe floods; also on the effects of drought conditions.

D. Platt carried out a study of Collembola in Stump Cross Cavern during 1961–2. One location was taken in the show cave portion and one location in a seldom visited area. Contrary to Schofield's findings in White Scar Cave, Platt found greater concentrations of Collembola in the show section than in the unvisited section. He also noted that the springtails tended to congregate at the ends of the pools nearest a light point in the show section.

In 1960, a visit was organised to White Scar Cave on behalf of the Freshwater Biological Association to try and determine if the lakes in this cave supported fish. No fish were caught but the visit yielded specimens of the crustacean *Bathynella natans* Vejdovski, the first time this had been recorded from Yorkshire. Following this discovery, Gledhill & Driver (1964) also found a single specimen in the silt of a wall pool in May 1960, and in May 1961 eleven specimens were collected from stream gravels and wall pools. In September 1961 a phytoplankton net suspended for about three hours below a waterfall in the cave yielded two specimens; on the same occasion, samples were taken from the stream gravels of Great Douk Cave and *Bathynella natans* was again found.

According to Gledhill & Driver (1964), Chappuis (1943) showed that the true habitat of *Bathynella* was in the interstitial spaces of the permanent water table where conditions were suitable, its appearance in wells, caves, etc, being accidental. If the appearance of *Bathynella natans* in White Scar Cave is accidental then its true habitat must be the microfissures of the Great Scar Limestone.

Further investigations, by Gledhill & Serban (1965) in Great Douk Cave, yielded forty specimens all referable to the subspecies *Bathynella natans stammeri*. A study of this population together with a population found in Romania (Pestera Lazubiu, Gorjolteria) showed that *Bathynella natans stammeri* was common to both countries.

FAUNA OF SCOSKA CAVE

C. Caywood and pupils of Victoria Secondary School, Dewsbury, carried out a comprehensive survey of Scoska Cave, Littondale. The cave was divided into zones and each zone carefully examined for specimens. These are included in Appendix 1. Three bats were observed during the course of the survey (1965–7) and information obtained locally implies that a reasonable colony of bats inhabited Scoska until 1964. Evidence of digging and burrowing were found on a mudbank in the zone 70–95m from the entrance, and hazel nuts were also found in this region on several occasions. This suggests the presence of long-tailed field mice which construct burrows underground and store-up quantities of nuts, etc, for winter use. This area of the cave is relatively dry and not liable to flooding.

Moth counts were made during the winter months, and the threshold flora was recorded, including several species of mosses and liverworts.

FAUNA OF SPRINGS WOOD LEVEL

For the first biological survey of Springs Wood Level (Dixon, 1965 and 1966) the mine was divided into zones of 15m (50ft) commencing at the entrance and working inwards. A short description of each zone and the fauna recorded there was given, together with a list of temperatures, percentage relative humidity and pH of waters present.

It was found that Diptera and Araneae were predominant in the deep threshold zone up to 60m. Beyond this spiders were rare, but flies were still common up to 90m and occasionally still present at 180m. Tissue Moths (*Triphosa dubitata*) which hibernate in the mine occurred between 75m and 135m from the entrance. Beyond 180m only a few *Sciara* larvae were found. The stream samples revealed *Rivulogammarus pulex*, cyclopoid copepods, harpacticoid copepods and lumbricid and enchytraeid worms. Specimens collected are listed in Appendix 1. The survey concludes with a table of counts of hibernating moths and a paragraph describing sampling of the stream.

In 1967, J. Dixon and D. T. Richardson of the Northern Cavern and Mine Research Society, together with C. A. Willoughby of Bradford University, obtained the lease of Springs Wood Level. Subsequently an underground research station has been established in the Level, and present studies are mainly oriented towards biology and hydrology.

A survey is being carried out into the habits of the Tissue Moth (*Triphosa dubitata*) which hibernates in the Level. Date of arrival and departure, preference for any particular substrate and distance from the entrance are factors under investigation. An interesting observation was made by E. & J. Dixon in October 1969 when seven pairs of moths were seen *in copula*. Many of the moths not copulating were present in pairs on the rock surface. The number of moths present in the Level varies from 50 to 300 during midwinter. A small number of Herald Moths (*Scoliopteryx libatrix*) are also present in the mine during the winter months.

The aquatic fauna of the Level is investigated by placing fine mesh sieves under various water inlets along the length of the mine. These are washed clean into bottles and taken away for microscopic examination. To date these have yielded ostracods, cyclopoid copepods, harpactid copepods, ciliate protozoa, enchytraeid worms, free-living nematodes, small specimens of *Rivulogammarus pulex* and small planaria (probably *Crenobia alpina*). A coarse mesh sieve placed under the main water inlet has held caddis larvae on all occasions of inspection and in July and August of 1968 and 1969 a number of empty cases. Adult caddis flies are present during the summer months in numbers averaging thirty. These are usually the species *Plectocnemia conspersa* (Curtis).

The Diptera have been observed and *Speolepta leptogaster* Winn. has been seen *in copula* on several occasions. A survey, conducted by Richardson, showed that *Helomyza serrata* (L.) tends to occur mainly in the 30–75m zone. This is clearly shown in the histogram (Figure 26). The counts of flies were carried out over a number of months.

An abundant population of the planarian *Crenobia alpina* is present in the mine at all times and appears to wander at will through the Level, specimens having been found in different locations on different occasions. A small colony of *Dendrocoelum*

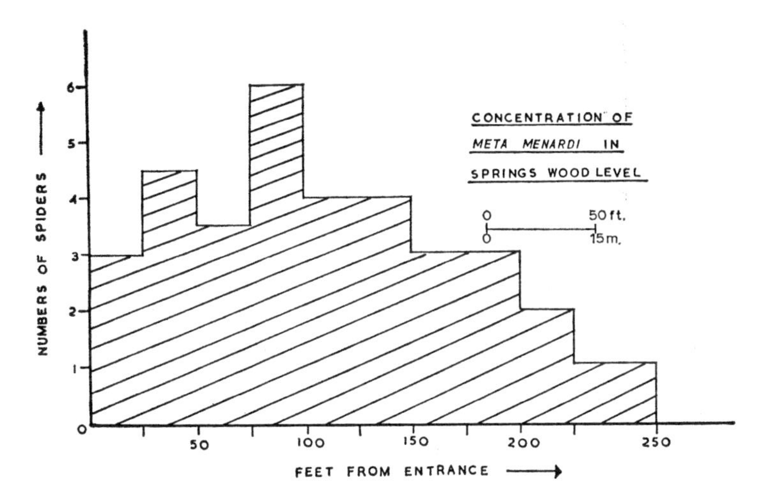

Fig 25 Concentration of *Meta menardi* in Springs Wood Level

lacteum (Mueller O. F.) has been found but only in a single location and not in conjunction with *Crenobia.*

Several specimens of the spiders *Meta merianae* (Scop.) and *Meta menardi* (Latr.) have been found on well-developed webs and some specimens have been seen feeding on various Diptera. The histogram (Fig 25) shows the distribution of *Meta menardi* in the Level, determined by Richardson over a period of several months.

In 1966, the writer started a project to try to determine whether *Rivulogammarus pulex* is a troglophile or merely an accidental trogloxene carried in by streams and water inlets. The shrimps were first bred on the surface, and a second generation was then transferred to Springs Wood Level, where they were kept and fed under controlled conditions. So far, results have been rather negative; the colonies have shown little signs of breeding and have completely died out on more than one occasion, whereas control colonies, kept in similar captivity under normal lighting conditions, have survived and reproduced normally over a period of four years.

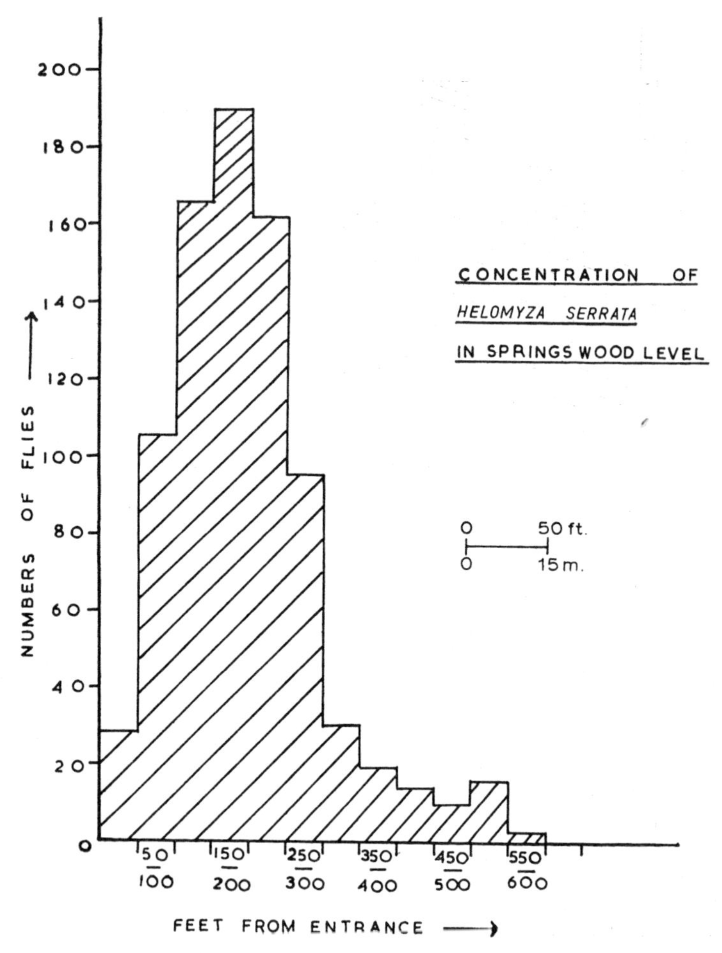

Fig 26 Concentration of *Helomyza serrata* in Springs Wood Level

CAVE VERTEBRATES

Reports of fish being seen in northern caves are numerous. Schofield (1964) reports that any literature dealing with the possibility of fish being present in British caves usually states that, due to the comparatively low temperatures, only a limited fauna can be supported and this will not provide a sufficiently regular food supply to support a vertebrate. However, some evidence has been found to suggest that this objection is not necessarily valid. In February 1959 a group of cave divers investigated Keld Head Rising and brought out a 6 pint sample of mud and gravel from the floor of a submerged passage. This yielded sixty lumbricid worms (*Stylodrilus heringianus*) and twenty larvae of three species of Chironomidae. Similar samples collected in late February and early March of the same year produced similar results. This was a surprising yield considering the unfavourable time of the year.

In May 1960 two Northern Speleological Group cavers collected an 18cm long white fish from Upper Long Churn Cave. This fish was examined by a member of the Freshwater Biological Association and was identified as a six-year-old female brown trout. Scale examination showed that it had spent four years underground. It had recently spawned and if the eggs had been fertilised and hatched a trout indigenous to underground conditions would have been produced (Schofield, 1964).

Fish have been reported in Great Douk Cave, Swinsto Cave, several times in the sump of Alum Pot, and in Ingleborough Cave. In the latter, divers have observed numerous fish, and in a submerged portion R. D. Leakey noted several almost transparent trout, only the eyes appearing pigmented (Schofield, 1964).

Members of the Northern Cavern & Mine Research Society have seen and tried to capture small fish from Borrins Moor Cave, so far without success, due to the constricted nature of the passages. Similar investigations in Hurtle Pot, Chapel-le-Dale, have yielded six specimens, five brown trout and a salmon parr. These all showed some loss of pigmentation but appeared to have fully developed eyes.

It is impossible to say whether any of these fish have become adapted to hypogean conditions until more specimens have been caught and studied.

Bats do not seem to be common in northern caves and very few reports have been made over the years. Single specimens have been seen in Gaping Gill, Long Kin East Pot, Buckden Gavel Mine, Scoska Cave, Stump Cross Cavern, Springs Wood Level, and Fairy Cave and Aragonite Band Mine in the Furness area. None of these has been conclusively identified. A specimen from Springs Wood Level was photographed and tentatively identified from the photograph as a Daubenton's Bat (*Myotis daubentoni*). Other odd reports of bats seen in caves and mines filter through from time to time, but no systematic studies have been pursued.

CAVE FLORA

The flora of northern caves and mines has been neglected even more than the fauna. In 1965, K. Wright of Bradford University conducted a preliminary investigation into the microbiology of Borrins Moor Cave. He isolated and identified a number of species of bacteria and fungi (see Appendix 3). Unfortunately Dr Wright left the district shortly afterwards and so no further progress was made.

Cubbon (1970) collected and identified several species of fungi from Springs Wood Level, and recently (1970) members of the NCMRS collected four species of fungi from Firestone Mine, Middleton-in-Teesdale.

Some of the threshold flora of Scoska Cave, Long Churn and Borrins Moor Cave have also been recorded. The above appear to be the total records for cave flora for north-west England.

PRESENT RESEARCH PROGRESS

The paucity of the above records (flora and fauna) may seem to indicate that north-west England has little of interest to offer. This is very far from the truth, the area has rich potential as the following examples show.

The small blind isopod *Trischonoides saeroeensis* was found in two mines in Lancashire by Moseley. This was the first record for the country and the first time anywhere under hypogean conditions. It is described by Sheppard (1968). The opilionid *Mitopus morio* was recorded by the Cave Research Group for the first time in Scoska Cave. Dixon found a specimen of a very rare millipede *Cylindroiulus (Aneuloboiulus) parisiorum* in Borrins Moor Cave, and Schofield produced the first record of the collembolan *Onychiurus schoetti*, from White Scar Cave. A mite found by Moseley in Moss House Mine, *Microtrombidium sucidum*, is a very rare species and the first record for mines. A further mite from the same mine, *Rhagidia diversicolor*, is a first British record. This mite is known from caves in the French Jura and has also been found in felled logs. In Connonley Lead Mine Dixon found an interesting colony of the mite *Veigaia transisale* which contained abundant mature specimens, some with eggs, which obviously indicated that the colony was breeding there. Finally a planarian collected by Moseley in the Furness area showed variation from normal form with regard to the eyes.

Moseley is continuing his research in the Furness area. He is particularly interested in determining whether *Niphargus* is present in the area. Glennie (1968) stated that, according to Ruffo (1953), *Niphargus* was not found north of the limits of the Anglian glaciation. However, discoveries of *Niphargus aquilex aquilex* in gravels of the River Teme near Gwerneirin, Radnorshire, *Niphargus aquilex aquilex* Schiodte in Holwell Cave high in the Quantock Hills, and *Niphargus kochianus kochianus* Bate near Bury, Lancashire, suggest either that *Niphargus* survived well within the limits of the glaciation or that it has migrated to these positions since glaciation. Glennie explained that from Gwerneirin the River Teme meanders through flat water meadows, down the valley, with a gradient of less than 1 in 150, and hence there would be no obstacle to the migration of the shrimps in the submerged gravels. Beyond Gwerneirin, however, the gradient increases rapidly and further travel upstream by the shrimps is unlikely. This process of migration could not be used to explain the presence of *Niphargus aquilex aquilex* in the Quantock Hills nor *Niphargus kochianus kochianus* at Bury. Thus there seems to be good evidence of hypogean fauna having survived the Anglian, and subsequent, glaciations. There is some evidence for a glacial refuge in the Irish Sea basin during the last glaciation, and so the Morecambe Bay area seems a favourable region in which to search for *Niphargus* or other aquatic troglobites.

Members of the NCMRS have been making comprehensive biological collections in the caves of the Alum Pot system. The work is still continuing, so no summary account has yet been prepared, but the individual findings have been recorded by the Cave Research Group (Hazelton, 1967, 1968, 1971) and in the Appendix 2 to this account. Four members of the Society, including the writer, are conducting an investigation to try to determine whether there is any difference in the type and/or numbers of organisms found at high and low altitudes. Sites over 500m and under 60m above sea level are being examined to give as wide a difference as possible over the area accessible to the people concerned. Factors such as barometric pressure, temperature and relative humidity are being measured and a systematic collection of fauna in each location is being carried out. Seasonal variations are also being noted.

DISCUSSION

Many of the species listed in Appendices 2 and 3 are merely chance visitors to caves and mines carried in by floods or wandering into the threshold zone accidentally, and known technically as accidental trogloxenes. Other species spend part but not all of their life cycle there, and are called habitual trogloxenes. Some species can live permanently and reproduce successfully in the dark zone, but are also found living permanently and equally successfully in the endogean or epigean domains. These are known as trogophiles. (The epigean region is the surface of the earth; the endogean is the soil immediately below and includes the zone penetrated by tree and plant roots and the burrowing of surface animals. The hypogean is the region below the endogean and includes most caves and mines beyond the threshold zone.) Finally there are those species which are the true inhabitants of the hypogean domain and which are not found in the epigean or endogean domains. These are the troglobites. Being the true cave fauna these species are generally of greater interest to biospeleologists. It is interesting to look at the troglobite fauna of the North of England and compare it with that of the South.

From a total of almost 240 species recorded in Appendices 2 and 3 only about ten can be classed as troglobites. These include three species of Collembola or Springtails: *Onychiurus schoetti*, *O.* cf *dunarius* and *O. schaeferia* cf *emucronata*. Collembola need a constant high humidity and they feed on fungi and scraps of organic debris which can be carried down by a mere film of seepage. Springtails are often found on sticky mud deposits, which are found in many caves and which have been found to contain organic matter up to 0·66 per cent dry weight of clay. Thus Collembola are well adapted to hypogean life.

As far as the Diptera are concerned it is possible that some of the *Sciara* sp whose almost transparent larvae feed on fungi (Actinomycetes?) could be classed as troglobites but more work on this group is needed. *Speolepta leptogaster* (Mycetophilidae) and *Triphleba antricola* (Phoridae) were stated by Glennie & Hazelton (1962) to be troglobites. However, Vandel (1965) stated that neither species should be regarded as true troglobites. Vandel listed *Speolepta leptogaster* as a guanobius species. This species is found very frequently, and often in abundant numbers, and yet the guano-producing fauna of caves, notably bats, is very sparse in northern caves. The food of *Speolepta leptogaster* is unknown.

Triphleba antricola is often found running over the rock face and floor, seemingly reluctant to fly, although the wings appear normal.

The only other species of Diptera, which may be a troglobite in this country, is *Limosina racovitzi* (Sphaeroceridae).

Of the Coleoptera only one species, the dytiscid *Hydroporus obsoletus*, has been suggested (Glennie & Hazelton, 1962) as a possible troglobite.

None of the Araneae can be classed as troglobites and to date only one species of Acari, *Rhagidia logipes* (Rhagidae), has been found. This is a totally depigmented form with fragile elongated legs. This mite may well feed on the dead bodies of insects (Collembola?), or it may be coprophagous.

Finally, there is the solitary crustacean (excluding the one record of *Niphargus kochianus kochianus* from Bury), *Bathynella natans*, an interstitial species.

The southern troglobite fauna comprise those species already mentioned, and, in addition, the cave beetle *Trechoblemus micros* (which has been found in Derbyshire, so could possibly be present in Yorkshire and Lancashire), the spider *Porrhomma rosenhaueri* and a mite, *Eugamasus anglocavernarum*.

With regard to the Crustacea, however, three species of *Niphargus*—*Niphargellus glennei*, *Crangonyx subterraneous* and *Asellus (Proasellus) cavaticus*—have all been recorded from the South of England and in some cases South Wales.

The only crustacean troglobite found in the North so far is *Bathynella natans*. Of the crustaceans mentioned *Bathynella natans*, *Crangonyx subterraneous*, *Niphargus kochianus kochianus*, *Niphargus aquilex aquilex* and *Niphargellus glennei* are interstitial fauna. These are generally considered to be eurythermal, as opposed to cavernicoles which are rather more stenothermal. It is possible, therefore, that, during the periods of glaciation when the North of England was covered by ice and temperatures were at their lowest, the interstitial fauna stood a better chance of survival than the stenothermal cavernicoles inhabiting the same region.

It is only comparatively recently that *Bathynella* was found in Yorkshire and its size makes collection rather a specialised task. It is possible therefore that other interstitial fauna could be present and the record of *Niphargus kochianus kochianus* from a well at Bury substantiates this. It is interesting to note that when *Bathynella* was scientifically searched for and collected a comparatively large number of specimens was found, eg eleven specimens from White Scar Cave on one occasion, and forty specimens from Great Douk Cave on another. Until, therefore, a more systematic and thorough search of silt from cave pools, wells and springs in the North of England has been made, it is not reasonable to state categorically that the interstitial crustacea of the South are not present in the North.

Regarding *Niphargus fontanus* and *Asellus* (*Proasellus*) *cavaticus* which appear to be more cavernicolous than interstitial, it is probable that the range of temperature within which they can tolerate and breed successfully is less than for the interstitial fauna, and they are unlikely to have survived glaciation, and so are unlikely to be found in the North. As mentioned previously, the Morecambe Bay area was possibly a glacial refuge and this area probably offers the greatest possibility of finding any of the above-mentioned species.

In conclusion it is worth noting that, although there are nearly a thousand caves and potholes in the area concerned, recorded collections have only been made from some ninety localities. This figure also includes some of the numerous mine levels in the area, and for many sites there are only one or two records. Clearly much more collecting is required, and, with identification by the biological recorder of the Cave Research Group, this can certainly provide useful and interesting results.

Appendices

Appendix 1 List of caves and mines in the North of England from which fauna has been recorded

Yorkshire		NGR	Entrance altitude (m)
1.	Ashberry Windy Pits, Ryedale	SE 571.850	150
2.	Attermire Cave, Settle	SD 841.641	430
3.	Beck Head Stream Cave, Clapham	SD 755.712	250
4.	Birkwith Cave, Ribblesdale	SD 803.769	340
5.	Black Reef Cave, Ribblehead	SD 774.795	275
6.	Blackside Pots, Dentdale	SD 792.885	580
7.	Blind Gill Level, Gunnerside	SD 935.018	450
8.	Borrins Moor Cave, Ribblesdale	SD 771.754	380
9.	Browgill Cave, Ribblesdale	SD 748.798	320
10.	Cappy Gill 2, Ribblesdale	SD 803.773	350
11.	Combs Cave, Helwith Bridge	SD 802.702	365
12.	Connonley Lead Mine, Connonley	SD 995.458	200
13.	Cuddy Gill Cave, Ribblehead	SD 766.799	325
14.	Dale Barn Cave, Chapel-le-Dale	SD 712.756	230
15.	Dow Cave, Kettlewell	SD 985.743	340
16.	Eglins Hole, Nidderdale	SE 093.735	190
17.	Elbolton Cave, Thorpe	SE 008.616	330
18.	Gaping Gill, Ingleborough	SD 751.726	400
19.	Gillfields Level, Grassington Moor	SE 116.649	290
20.	Gill Rigg 1, Ribblesdale	SD 801.782	330
21.	Gill Rigg 2, Ribblesdale	SD 799.787	330
22.	Gorton Level, Gunnerside	SD 939.014	450
23.	Goyden Pot, Nidderdale	SE 099.761	220

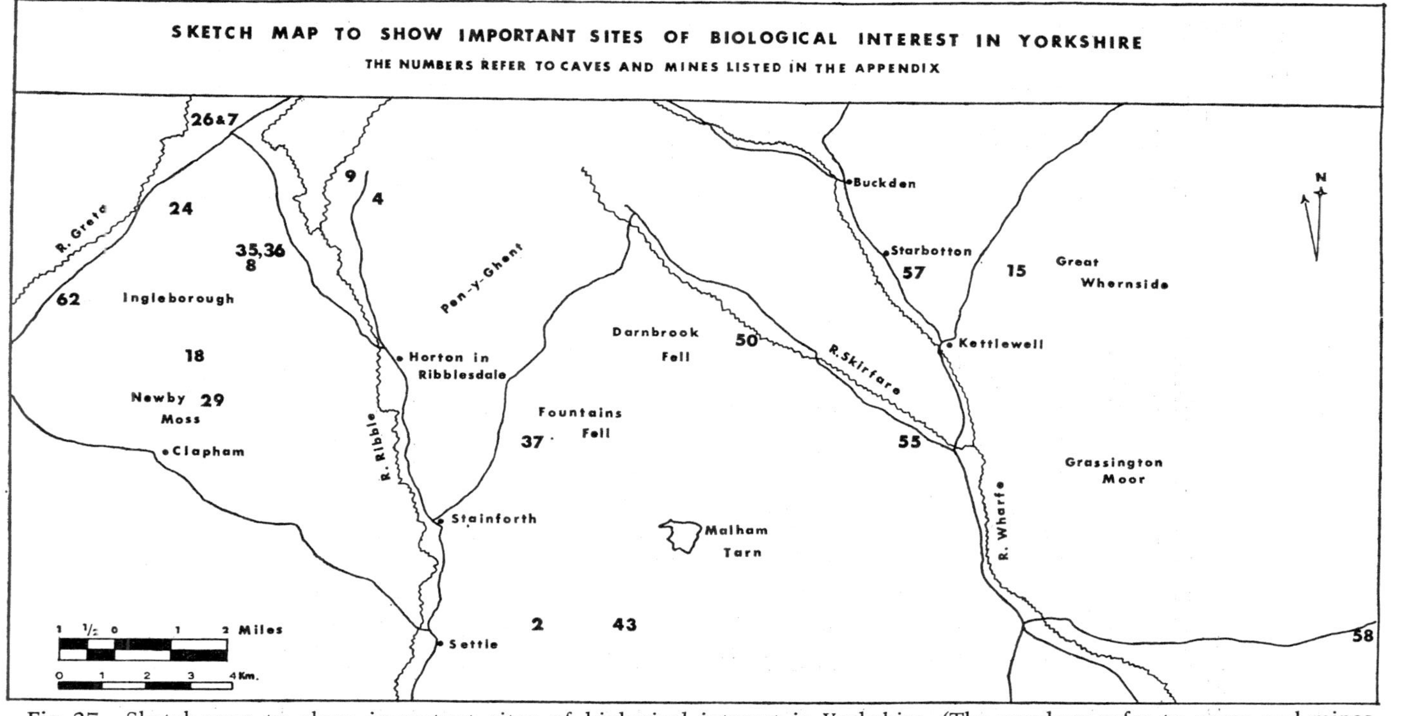

Fig 27 Sketch-map to show important sites of biological interest in Yorkshire. (The numbers refer to caves and mines listed in Appendix 1.)

Yorkshire		NGR	Entrance altitude (m)
24.	Great Douk Cave, Chapel-le-Dale	SD 747.770	340
25.	Greenwood Pot, Newby Moss	SD 728.737	360
26.	Gunnerfleet Cave—Upper, Whernside	SD 756.797	300
27.	Gunnerfleet Cave—Lower, Whernside	SD 756.797	300
28.	Holme Hill Cave, Ribblehead	SD 785.803	310
29.	Ingleborough Cave, Clapham	SD 754.710	260
30.	Keld Head Rising, Kingsdale	SD 695.765	250
31.	Kingsdale Head Cave, Kingsdale	SD 695.775	300
32.	Kirkdale Caves, Ryedale	SE 677.860	60
33.	Lost Johns System, Leck Fell	SD 670.786	360
34.	Lower Ling Gill Cave, Ribblesdale	SD 800.785	330
35.	Long Churn—Lower, Ribblesdale	SD 771.755	350
36.	Long Churn—Upper, Ribblesdale	SD 771.755	350
37.	Magnetometer Pot, Fountains Fell	SD 849.697	400
38.	Marble Steps Pot, Gragareth	SD 680.770	390
39.	Meregill Hole, Chapel-le-Dale	SD 740.757	390
40.	Mongo Gill Hole, Greenhow Hill	SE 096.632	370
41.	Nippikin Hole, Leck Fell	SD 667.800	330
42.	Old Ing Cave, Ribblesdale	SD 806.768	370
43.	Pikedaw Calamine Mine, Settle	SD 875.640	500
44.	Priscilla Level, Gunnerside	SD 936.014	420
45.	Providence Pot, Kettlewell	SD 993.729	400
46.	Rift Pot, Ingleborough	SD 760.728	410
47.	Rowten Cave, Kingsdale	SD 697.780	360
48.	Runscar Cave, Ribblehead	SD 765.797	310
49.	Scald Bank Cave, Ribblesdale	SD 796.789	340
50.	Scoska Cave, Littondale	SD 915.724	300
51.	Sell Gill Holes, Ribblesdale	SD 811.743	350
52.	Silverdale Gill Pot, Silverdale	SD 838.704	380
53.	Simpsons Pot, Kingsdale	SD 696.778	380
54.	Sir Francis Level, Swaledale	SD 939.002	350
55.	Sleets Gill Cave, Littondale	SD 959.693	260
56.	Springs Wood Level, Starbotton	SD 958.743	300
57.	Storrs Common Cave, Ingleton	SD 703.732	200
58.	Stump Cross Cavern, Greenhow Hill	SE 098.636	350

Yorkshire		NGR	Entrance altitude (m)
59.	Trollers Gill New Cave, Appletreewick	SE 071.624	260
60.	Turn Dub Rising, Ribblesdale	SD 797.748	240
61.	Twin Bottom Scar Cave, Malham	SD 877.642	480
62.	White Scar Cave, Ingleton	SD 712.745	260
63.	Wilsons Cave, Ribblesdale	SD 772.757	350
64.	Yordas Cave, Kingsdale	SD 705.791	310

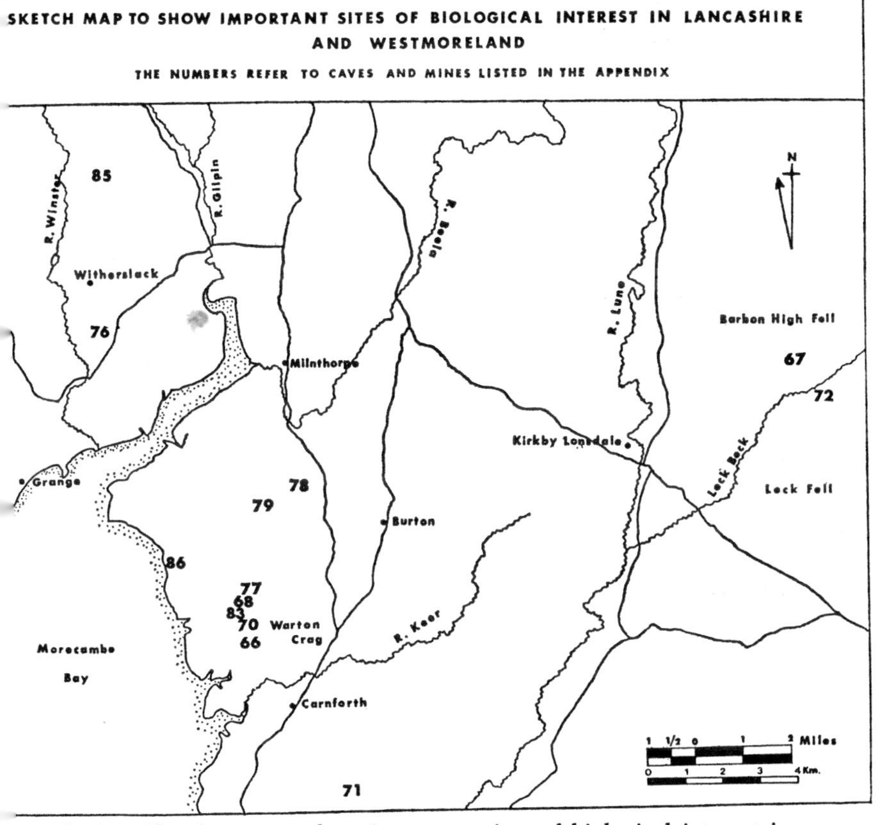

Fig 28 Sketch-map to show important sites of biological interest in Lancashire and Westmorland. (The numbers refer to caves and mines listed in Appendix 1.)

Lancashire and Westmorland		NGR	Entrance altitude (m)
65.	Aragonite Band Mine, Warton	SD 482.738	17
66.	Barrow Scout Mine, Warton	SD 485.724	15
67.	Bull Pot of the Witches, Casterton Fell	SD 662.813	300
68.	Crag Foot Mine, Warton	SD 483.739	65
69.	Daylight Hole, Lindal Moor	SD 253.763	80
70.	Dog Holes, Warton	SD 483.730	55
71.	Dunald Mill Cave, Nether Kellet	SD 516.684	80
72.	Ease Gill Caverns, Casterton Fell (County Pot)	SD 675.805	310
73.	(Pool Sink)	SD 676.806	330
74.	(Top Sink)	SD 681.811	350
75.	(Far Series)	SD 676.810	330
76.	Fairy Cave, Witherslack	SD 434.827	12
77.	Grizedale Wood Drainage Level, Warton	SD 482.741	8
78.	Hale Moss Cave, Beetham	SD 500.775	30
79.	Hazel Grove Caves, Beetham	SD 499.770	25
80.	Henning Valley Cave, Lindal	SD 247.763	90
81.	Jack Scout Mine, Silverdale	SD 459.738	5
82.	Lancaster Hole, Casterton Fell	SD 664.807	100
83.	Moss House Mine, Warton	SD 480.738	10
84.	Pool Bank Cave, Witherslack	SD 434.877	55
85.	Pylon Pot, Lindal	SD 244.762	100
86.	Red Rake Mine, Silverdale	SD 456.753	2
87.	Sea Wood Copper Mine, Furness	SD 295.736	22
88.	Windy Scout Mine, Warton	SD 480.735	35

Appendix 2 Fauna recorded from Hypogean and related zones in the North of England

(Numbers after the species refer to the caves and mines listed in Appendix 1.)

PLATYHELMINTHES

TRICLADIDA

Dendrocoelidae

Dendrocoelum lacteum (Mueller O. F.), 33, 56

Planariidae

Phagocata vitta (Duges), 9, 18, 61, 65, 68, 78, 83

Crenobia alpina (Dana), 37, 41, 46, 56, 72, 77

MOLLUSCA

GASTEROPODA

Basommatophora

Limnaeidae

Limnea truncula ta (Müller), 71

Stylommatophora

Arionidae

Arion ater L., 8

Arion hortensis Ferussac, 23, 84

Helicidae

Helix hortensis (Müller), 68

Limacidae

Agriolimax reticulatus (Müller), 4, 23

Zonitidae

Oxychilus helveticus (Blum), 2, 17

LAMELLIBRANCHIATA

Eulamellibranchia

Sphaeriidae

Pisidium personatum Malm, 79

Pisidium sp, 50

ANNELIDA

OLIGOCHAETA

Archioligochaeta

Enchytraeidae

Enchytraeus albidus Henle, 56, 77

Tubificidae

Tubifex tubifex Mueller, 62

Ilyodrilus sp, 30

Nai elinguis Müller, 62

Neoligochaeta

Lumbriculidae

Stylodrilus heringianus Claparede, 30

Lumbricidae

Allolobophora chlorotica (Savigny), 29, 51, 55

Bimastus eiseni Eisen, 62

Dendrobaena veneta Michaelson, 18

Eiseniella tetraedra f *typica* Savigny, 58

Lumbricid sp, 7, 12, 17, 27, 56

ARTHROPODA

CRUSTACEA

Ostracoda

Cypridae

Candona sp, 60

Copepoda

Harpacticoida

Canthocamptidae

Canthocamptus praegeri Scoufield, 24

Canthocamptus sp, 56

Cyclopoida

Cyclopidae

Acanthocyclops bisetosus (Rehberg), 67

Acanthocyclops viridis Jur, 29, 62, 69, 71

Eucyclops agilis Koch-Sars, 60, 69

Macrocyclops fuscus Jur, 29
Paracyclops fimbriatus S. Fischer, 56, 60
Cyclops sp, 56
Malacostraca
Syncarida
Bathynellidae
Bathynella natans Vejdovski, 24, 62
Isopoda
Trichoniscidae
Trichoniscus pygmaeus Sars, 68, 77
Trichoniscus pusillus Brandt, 79
Trichoniscoides saeroeensis Lohmander, 68, 77, 83
Andronicus dentiger Verhf, 69, 77, 78, 80, 83, 88
Oniscidae
Oniscus asellus L, 1, 17, 36, 41, 70
Porcellionidae
Porcellio scaber (Latr.), 70
Amphipoda
Gammaridae
Rivulogammarus pulex (L.), 4, 5, 6, 9, 10, 15, 16, 18, 20, 23, 26, 29, 31, 34, 35, 36, 37, 40, 46, 48, 49, 50, 55, 56, 58, 62, 69, 72, 76, 77, 79, 82, 84, 85
Gammarus duebeni Lillj, 86
MYRIAPODA
Symphyla
Scoliopendrellidae
Symphylella isabellae (Grassi), 83, 86
Diplopoda
Polydesmidae
Brachydesmus superus Latzel, 68, 72
Polydesmus augustus Latzel, 8, 77, 78
Glomeridae
Glomeris (*Eurypleuroglomeris*) *marginata* (Villers), 63
Craspidosomidae
Polymicrodon polydesmoides (Leach), 23, 29, 68, 78, 80, 84
Blaniulidae
Archiboreoiulus pallidus (S. G. Brade-Birks), 84
Blaniulus guttulatus (Bosc), 77, 78, 84

Iulidae

Tachypodoiulus niger (Leach), 50, 58

Cylindroiulus (Aneulobiulus) parisiorum (Brol. et Verh.), 8

Cylindroiulus (Aneulobiulus) latestriatus (Curtis), 69

Chilopoda

Lithobiidae

Lithobius forficatus (L.), 80

INSECTA

Collembola

Hypogastruridae

Hypogastrura purpurescens (Lubbock), 58, 62, 69, 83

Hypogastrura schaefferi emucronata gp. 62, 67, 68, 76, 79, 83, 86

Hypogastrura denticulata (Bagnall), 83

Anurida granaria (Nicolet), 62, 71, 83

Neanura muscorum (Templeton), 69, 83

Onychiuridae

Onychiurus ambulans (L.), 58, 71

Onychiurus fimetarius gp, 12, 27, 29, 58, 62, 68, 83, 87

Onychiurus schoetti (Lie Pettersen), 14, 25, 29, 33, 37, 43, 50, 58, 62, 68

Onychiurus arans Gisin, 69, 80

Onychiurus sp cf *dunarius* Gisin, 68

Isotomidae

Folsomia candida gp, 28, 29, 62

Folsomia quadriculata (Tullberg), 29

Folsomia fimitaria gp, 54

Folsomia cf *agrelli* (Gisin), 3

Folsomia sp, 18

Isotoma notabilis Schäffer, 18

Isotoma olivacea–violacea gp, 27, 35, 51, 62

Isotoma violacea Tullberg, 47

Isotoma (Vertagopus) arborea (L.), 29

Agrenia bidenticulata (Tullberg), 18, 35

Isotomorus palustris (Müller), 56

Entomobryidae

Lepidocyrtus curvivollis Bourlet, 83

Lepidocyrtus sp, 18

Pseudosinella alba (Packard), 18, 64
Tomoceridae
Tomocerus minor Lubbock, 8, 12, 19, 54, 56, 58, 68, 69, 76, 77, 85
Neelidae
Neelus sp, 56
Megalothorax minimus (Willem), 18, 62
Sminthuridae
Arrhopalites caecus (Tullberg), 62
Arrhopalites pygmaeus (Wankel), 18, 25
Dicyrtoma fusca (Lucas), 64
Ephemeroptera
Leptophlebiidae
Leptophlebia marginata (L.), 49
Paraleptophlebia submarginata (Stephens), 9
Habrophebia fusca (Curtis), 21
Baetiidae
Baetis rhodani (Pictet), 23
Baetis sp, 4, 5, 24, 28
Siphonuridae
Siphonurus lacustris Eaton, 16
Ecdyonuridae
Rithrogenia semicolorata (Curtis), 39
Heptagenia lateralis (Curtis), 4, 20, 23, 36, 47
Ecdyonurus venosus (Fabr.), 29
Ecdyonurus sp, 4, 47
Plecoptera
Nemouridae
Nemoura cambrica Stephens, 24
Leuctridae
Leuctra fusca (L.), 72
Leuctra hippopus Kempny, 24
Leuctra inermis Kempny, 18
Perlodidae
Perlodes microcephala (Pictet), 73
Diura bicaudata (L.), 15

Perlidae
Dinocras cephalotes (Curtis), 9, 15, 18, 20, 28, 39, 49, 54
Chloroperlidae
Chloroperla torrentium (Pictet), 9, 39
Hemiptera
Heteroptera
Miridae
Stenoderma holsatum (Fabr.), 50
Veliidae
Velia caprai Tamanini, 9, 20, 23, 49, 50, 71, 77, 83
Homoptera
Cercopidae
Philaenus spumaris (L.), 50
Cixiidae
Paracixius distinguendus Kirschbaum, 50
Mecoptera
Panorpidae
Panorpa germanica L, 50
Trichoptera
Philopotamidae
Philopotamus montanus (Donovan), 28, 47
Wormaldia sp, 75
Polycentropidae
Plectrocnemia conspersa (Curtis), 8, 15, 16, 29, 56, 77
Plectrocnemia geniculata McLachlan, 4, 46, 49, 63
Plectrocnemia sp, 13, 28, 35, 51
Polycentropus sp, 36, 37, 62
Limnephilidae
Stenophlyax permistus McLachan, 44, 59, 77, 83
Potamophylax sp, 75
Lepidoptera
Plusiidae
Scoliopteryx libatrix L, 1, 50, 56, 64, 73
Hydriomenidae
Operophtera brumata L, 56
Geometridae
Triphosa dubitata L, 2, 50, 56, 81, 86

Diptera
Tipulidae
Limonia nubeculosa Meigen, 23, 55, 63, 76, 84
Limonia sp, 56
Ormosia (Ormosia) pseudosimilis (Lundst.), 58
Limnophila (Elaeophilia) trimaculata (Zett.), 20, 21
Limnophila (Elaeophilia) submarmorata (Verall), 47
Lipsothrix remota (Walker), 55, 56
Pedicia (Crunobia) staminea (Meigen), 27
Trichoceridae
Trichocera maculipennis Meigen, 15, 17, 18, 40, 43, 45
Trichocera sp, 85
Psychodidae
Clytocera ocellaris (Meigen), 63
Culicidae
Culex pipiens L, 1, 6, 19, 50, 51, 56, 71, 77, 83, 84, 86
Chironomidae
Pentaneura longimana (Staeger), 36
Pentaneura (Pentaneura) binota, 47
Anatopynia (Macropelopia) notata (Meigen), 27
Anatopynia (Macropelopia) nebulosa (Meigen), 30
Prodiamesa olivacea (Meigen), 30
Hydrobaenus sp, 47
Tanytarsus sp, 30
Nanoclaudius (Eukiefferiella) brevicalcar Kieff, 26
Thienemannella clavivornis Kieff, 35
Microspecta sp, 8
Diamesia tonsa Meigen, 51
Chironomid sp, 62, 71
Ceratopogonidae
Culicoides delta Edwards, 11
Culicoides sp, 64
Simuliidae
Simulium (Simulium) monticola Fries, 47
Simulium (Simulium) sp, 18
Mycetophilidae
Speolepta leptogaster Winn, 6, 8, 17, 18, 19, 41, 42, 44, 47, 56, 62, 66, 68, 76, 77, 84, 86

Exechia contaminata Winn, 84
Exechia sp, 6, 86
Exechiopsis subulata Winn, 76, 84
Exechiopsis sp, 77
Rhymosia fasciata Meigen, 84
Rhymosia sp, 77
Bradysia sp, 7, 12, 22, 29, 38, 43, 58, 69, 72, 76, 77, 80
Tarnia fenestralis Meigen, 84

Cecidomyiidae
Miastor sp, 12

Phoridae
Triphleba antricola Schmitz, 20
Megaselia sp, 84

Empididae
Clinocera (Weidemannia) rhynchops Nowicki, 39

Helomyzidae
Scoliocentra caesia (Meigen), 56
Scoliocentra scutellaria Zett, 78
Scoliocentra villosa (Meigen), 76, 83, 84
Helomyza serrata (L.), 2, 6, 15, 17, 40, 50, 55, 56, 68, 71, 77, 79, 83, 84, 86

Sphaeroceridae
Crumomyia nigra (Meigen), 2, 17, 50, 56
Stratioborborus nitidus (Meigen), 80
Limosina sylvatica (Meigen), 19, 80
Limosina sp, 78
Copromyza roserii Rhodani, 84
Copromyza suilliorum Halliday, 6
Copromyza nigra Meigen, 6
Trichiapsis equina Fallen, 61, 78
Leptocera sylvatica Meigen, 77
Leptocera flaviceps Zett, 77, 84

Calliphoridae
Calliphora sp, 8

Muscidae
Orthellia caesarion Meigen, 6
Limnophora exuta Kow, 47
Hydrothea irritans (Fallen), 50

Cordiluridae
Scatophaga stercoraria L, 50

Coleoptera
Lathridiidae
Enicmus testaceous Steph, 43

Carabidae
Nebria brevicollis (F.), 33, 43
Nebria gyllenhali (Schaen), 6
Bembidion guttula (F.), 71
Trechus secalis Paykull, 50
Trechus fulvus Dej, 77
Trechus sp, 68
Loricera pilicornis F, 50
Pterostichus madida (F.), 17
Agonium muelleri (Herst), 9

Dytiscidae
Hydroporus aquaticus L, 52
Hydroporus ferrugineous Steph, 18
Hydroporus obsoletus Aube, 66
Agabus guttatus (Paykull), 4, 13, 18, 27, 48, 56, 58, 62, 63, 67, 77

Hydophilidae
Hydraena sp, 47

Silphidae
Choleva augusta F, 1
Choleva glauca Britten, 84

Staphylinidae
Lesteva pubescens Mann, 18, 19, 23, 37, 62
Lesteva luctuosa Fauv, 26
Geodromicus plagiatus var. *nigrita* (F.), 37, 72
Ancyphorus aureus (Fauv.), 12, 18, 29, 45, 55, 58
Lathrobium geminum Kraatz, 71
Dianous coerulescens (Gyll.), 18
Quedius mesomelinus Marsham, 18, 58, 72, 78
Quedius sp, 68
Atheta graminicola Gravenhurst, 71
Olophrum piceum (Gyll.), 24
Omaliinae larva, 12

Cantharidae
Cantharis sp, 83
Elateridae
Hypolithus riparius F, 9
Helodiidae
Helodes minuta or *marginata*, 47

ARACHNIDA

Araneae
Nesticidae
Nesticus cellulanus (Clerck), 8, 65, 70, 77, 79, 84
Nesticus sp, 44, 56, 76
Argiopidae
Meta menardi (Latr.), 1, 51, 52, 56, 59, 70, 71, 77
Meta merianae (Scopoli), 6, 8, 9, 21, 32, 35, 50, 56, 76, 77, 84
Meta sp, 13, 64, 74
Linyphiidae
Porrhomma convexum (O. P. Cambr.), 8, 9, 12, 47, 51, 69
Trachynella nudipalpis (Westring), 6
Tmeticus affinis (Blackwell), 9
Linyphia triangularis (Clerck), 50
Bathyphantes gracilis Bl, 52
Labulla thoracia Wider, 84

Opilionida
Phalagiidae
Mitopus morio Fabr, 50

Acari
Veigaiaidae
Veigaia herculaneus (Berl.), 8
Veigaia transisalae (Ouds.), 12
Gamasolaelaptidae
Euryparasitus emarginatus (Koch), 57
Parasitidae
Eugamasus magnus (Kram.), 29
Eugamasus loricatus (Wankel), 18
Eugamasus tragardi (Ouds.), 18
Eugamasus sp, 54, 58
Pergamasus minor Berl, 54
Pergamasus robustus (Ouds.), 25

Rhagiidae
Rhagidia gigas (Canestrini), 83
Rhagidia intermedia Koch, 18, 68
Rhagidia spelea (Wankel), 18, 22, 58
Rhagidia terricola (Koch), 18, 29
Rhagidia longipes Tragardh, 69
Rhagidia diversicolor (Koch), 83
Coccorhagidia sp, 25
Tydeidae
Tydeus sp, 62
Eupopidae
Eupodes variegatus Koch, 68
Linopodes motatorius (L.), 12, 68
Halacaridae
Soldanellonyx chappuisi Walter, 62
Calyptostomidae
Calypstoma lynceum Berl, 84
Acaridae
Coelognathus dimidiatus (Herm.), 62
Trombidiidae
Microtrombidium sucidum (Koch), 83

Appendix 3 Flora recorded from Hypogean and related zones in the North of England

BORRINS MOOR CAVE, Ribblesdale (NGR SD 771.754 OD 380m)

Bacteria

Bacillus cereus Frankland & Frankland
Bacillus mycoides Fluger
Bacillus megaterium de Bary
Bacillus subtilis Cohn
Bacillus sp
Sarcina lutea Schroeter
Flavobacterium spp
Achromobacter sp
Micrococcus spp (white)
Micrococcus spp (yellow)
Coliforms

Fungi

Penicillium sp
Mucor sp
Yeast sp

Bryophyta

Aphozia riparia (Tayl.) Dunn
Eurynchium confertum (Dicks) Milde
Isopterygium elegans (Hook) Lindb
Preissia quadrata (Scop.) Nees

DAYLIGHT HOLE, Furness (NGR SD 254.763 OD 80m)

Bacteria

Bacterium (short rods)

Actinomycetes

Streptomyces sp

FIRESTONE MINE, Middleton-in-Teesdale (NGR NY 947.270 OD 300m)

Fungi

Basidiomycetes

Polyporus sulphureus (Bull.)

Fomes annosus Fr

Fomes ?laccatus

Clitocybe sp

LONG CHURN CAVE, Ribblesdale (NGR SD 771.755 OD 350m)

Bryophyta

Lophocolea cuspidata (Nees) Limpr

Mnium punctatum Hedw

Pteridophyta

Asplenium adiantum nigrum L

SCOSKA CAVE, Littondale (NGR SD 915.724 OD 300m)

Fungi

Imperfecti

Hirsutella dipterigena Petch

SPRINGS WOOD LEVEL, Starbotton (NGR SD 958.743 OD 300m)

Fungi

Phycomycete—unidentified species with non-sporing, aseptate hyphae

Basidiomycetes—unidentified species with non-sporing, aseptate hyphae

Pseuhohiatula sp

Imperfecti

Hirsutella sp

Beauveria globulifera (Speg.) Pic

Beauveria bassiana (Bals.) Vuill

Aspergillus sp

Paeciliomyces sp

A Dermatophyte—may be *Trichophyton* sp

10

A Review of Archaeological Work in the Caves of North-West England

A. King

The limestone scars, Malham Cove and Gordale, are as popular today as they have ever been—improved road communications have facilitated the transport of the curious. This has been the situation in Craven at least since 1753 when the Keighley–Kendal Turnpike Act was passed and many portions of the road surfaced. Further west the Lancaster–Kendal turnpike had been completed the previous year.

The families who backed these schemes had already played their part in changing the agricultural scene and they joined forces later in the many railway ventures of the north-west during the mid-nineteenth century. The association of landowner and civil engineer was important in speleology and has to be recognised. Many of the first descents of the vertical shafts were made by their joint efforts such as that at Alum Pot in 1848. Unfortunately the confidence, the muscle and the piece-work payments of the railway companies were employed in the early cave excavations, and all interpretations of their reports must bear this in mind. When the British Association work at Victoria Cave began on 21 March 1870, three men 'excavated this day 3′ 6″ right and 4′ 6″ left of the datum and 8ft. forward to a depth of 4 feet', a total of 256ft^3 (7·2m^3); the following day it snowed, and little work was completed (Jackson, 1870).

Victoria Cave was not the first cave in the study area to attract the antiquarian, for some of the north Lancashire caves had been mentioned in this context by West in 1774, but it was the news of the finds from this cave in 1839 that seems to have triggered off the widespread cave hunting of the 1840s and 1850s. Techniques were developed rapidly by the teams of excavators, and funds of information on glaciation, animal remains and artifacts were built up, much of it coming from France and Germany.

Whilst we appreciate the significance of names like Darwin on the cave committees, and men like McKenny Hughes, Arthur Lyell, Adam Sedgwick and R. H. Tiddeman joining together to excavate a rock shelter at Cave Ha on Buckhaw Brow (Giggleswick Scar), their insight into the problems of Pleistocene geology was far greater than into those concerning archaeology. When Tiddeman (1875), in the Third Report of the Victoria Cave Exploration Committee to the British Association for the Advancement of Science, questioned the accepted view that between the two glacial periods there was 'an era of depression beneath the sea' (Dawkins, 1874), he was querying the traditional great deluge.

Innumerable caves were completely cleared of their archaeological levels and some of the details of the stratigraphy and finds were recorded. Collections were built up by many societies, field clubs and individuals, but regrettably many have been lost or destroyed and at the present time numerous cave finds have to be exhibited without provenance. Equally disconcerting to the present writer is the fact that the archaeological opinions and conclusions of the Victorians have hardly ever been questioned.

PALAEOLITHIC PERIOD

The sands and gravels of southern Lancashire and Cheshire were deposited after 30,000BP and it seems likely that the downwasting of the ice freed virtually all of northern England, other than the Lake District and possibly some adjacent Pennine high land, by 10,000BC. Barnes *et al* (1971) have shown that epipalaeolithic man was hunting elk (*Alces alces*) in the Fylde about a millennium later. Some areas in the north were free of ice before the beginning of the Allerod interstadial, zone II of the forest history zones. On the limestone plateau, at least in the Malham Tarn neighbourhood, grasses, sedges, pine, willow and juniper had been joined by the tree birches (Pigott & Pigott, 1959), and the area was inhabited by brown bear (*Ursus arctos*), horse (*Equus caballus*), red deer (*Cervus elaphus*) and hare (*Lepus timidus*). From Dog Holes Cave, Warton, Jackson (1910) obtained evidence for both Arctic (*Dicrostomyx torquatus*) and Norwegian lemming (*Lemmus lemmus*), clearly a late Pleistocene relic.

Recently workers at Kirkhead Cave have focused attention on to the cave sediments and have suggested that the lenses of gravel, cryoturbation and frost wedges at the top of the laminated clay sequence indicate the cold period of the Late Devensian, Cirque, glaciation, pollen zone III. From the base of the succeeding cave earth they obtained a number of worked flints, at least one of which seems to have Palaeolithic characteristics.

One difficulty at Palaeolithic level at the present time is a trans-Pennine one; a considerable amount of work has been done in the Magnesian Limestone caves and rock shelters of the Cresswell Crags, in Nottinghamshire, the type site for the late Upper Palaeolithic Cresswell point flint industry. C^{14} dates for this horizon range from 6850 $\pm$ 300BC, at Mother Grundy's Parlour, Cresswell, to 7990 $\mp$ 115BC at Anston Stone Wood, in South Yorkshire (Lavell, 1970), but Cresswell is situated well outside the limit of maximum ice of the Devensian, and should have been inhabited earlier than the north-western caves which are within the limit. The interrelationships of these geographically separate deposits remains to be worked out in detail.

In addition to the Kirkhead flints, late Upper Palaeolithic evidence has been found in Victoria and Kinsey Caves and Calf Hole, Skyrethorns. There the Rev E. Jones reported that a deep bed of fine yellow sand covered the cave floor and in this were found teeth of mammoth, rhinoceros and an implement of antler

Page 185

(*right*) *Ancyrophorus aureus* Fauv. Bar scale = 1mm (*below*) *Porrhoma convexum* (Westring). Bar scale = 1mm

Page 186

(*above*) The entrance to Victoria Cave in 1870. The original opening is behind the wall above and left of the archaeologists (*below*) The entrance to Victoria Cave in 1970. The notch at the extreme top left is all that was open in 1870

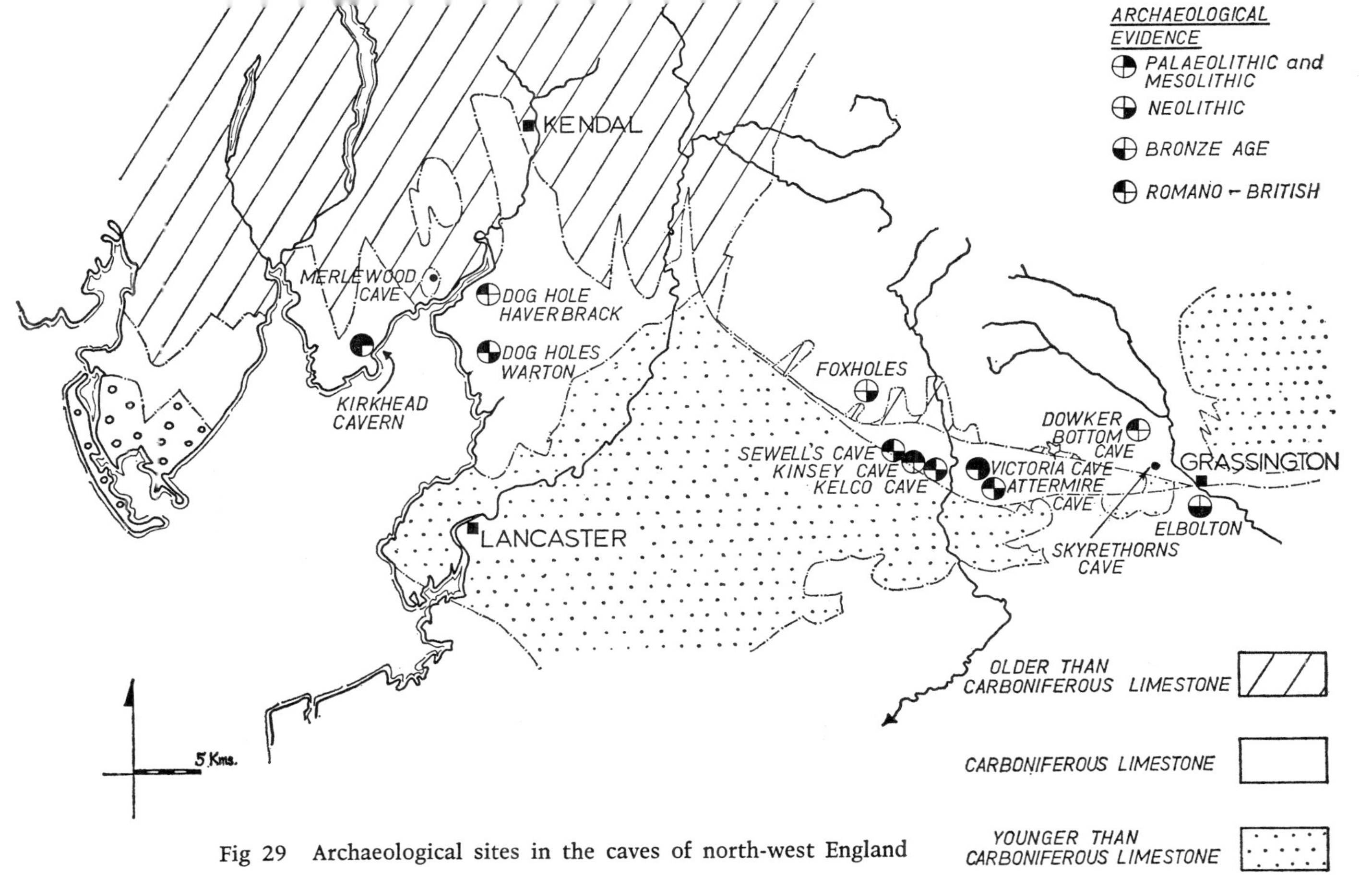

Fig 29 Archaeological sites in the caves of north-west England

with a canine of wild boar set in it (Kendall & Wroot, 1924).

The Abbé Breuil considered the cylindrical reindeer antler rods from Victoria Cave to be at least as old as Magdalenian 5 (Breuil, 1921), and to these must be added the broken point from Kinsey Cave; this has a tapering squared section, the corners being rounded, and longitudinally it has been scored with paired lines (Jackson & Mattinson, 1932). A reindeer antler find with a much better marked longitudinal groove, found in 1931 at Victoria Cave, is either a curved lance point or fish spear prong (Fig 30 *d*) (Jackson, 1945). The facet on the broader portion, on the inside of the curve, must be the basal facet needed for hafting. Being curved it cannot be a simple harpoon and it is envisaged as one of perhaps three prongs on a lance or leister. This point is virtually identical with some of the examples from Petersfels (Garrod, 1938), the only apparent difference being the fact that the German examples are straight and are described as javelin heads.

Some flints of Cresswellian character have been found in both Victoria and Kinsey Caves. Their numbers are small, but this may be attributed to the early cave excavation techniques (Fig 30).

Microlithic flint and chert flakes of the Mesolithic period are common at numerous sites on the Pennines, where they wash out from beneath the peat, but they are usually non-geometric forms. Walker (1956) was able to show that those found at the Stump Cross site (not in the cave) came from a position above but close to the zone VI/VII *a* boundary. By this time hazel had spread over the limestone plateau and the larger forest trees, elm and oak, had become well established.

Microliths are rare in cave collections, though a few have been found in Jubilee Cave, 300m north of Victoria Cave; more noteworthy are two finds from Victoria Cave itself, the Azilian harpoon (found at a depth of 4m outside the cave in 1870) (Fig 30 *c*) and a pointed rod with perforated expanded base—another harpoon head (Fig 30 *b*).

NEOLITHIC PERIOD

The Neolithic cultures of the third millennium BC had progressed beyond the subsistence level of their hunting, fishing and collecting predecessors. In the East and North Ridings the stone and earth long barrow mounds are clearly community efforts, and they usually cover communal multi-cremated burials (Manby, 1970). Inhumations, of extended or crouched bodies, are less frequently found and grave goods are notably absent; Manby has suggested that occupation refuse serves the same ritual role. On the limestone of north-west England, six long cairns have been identified on the western side of the Vale of Eden and one at Skelmore Heads in Furness. Elsewhere, the present writer considers the limestone caves obvious alternative entrances to the other world and though they were not erected by a large labour force, the ceremonies associated with burial would no doubt have been as elaborate and of equal significance. The limestone corridor was used as a trade route for Langdale stone, graphite and flint, linking Cumbria and the west coast with the east of the country, throughout the Neolithic and Bronze Age periods; whilst no settlements of this period have been found locally it is considered that the better drained soils would have attracted the immigrant farming people.

Table 10 lists the caves from which most of the Neolithic evidence has been gleaned, and it is clear that the amount of cultural material found is very small. The issue is to try to decide if the evidence is sufficient to show occupation by the living or the dead.

No complete or restorable vessels have been found. The ceramic evidence is normally a few scattered pieces of the mid-brown or dark grey, gritty Peterborough ware, decorated with impressed thick cord 'maggots' or jabbed triangles on the shoulder and rim. Sherds from Elbolton and Lesser Kelco suggest all-over decoration of triangular imprints. Rim fragments of a large diameter bowl from Sewell's Cave are decorated on the inside with impressions

Table 10 *Archaeological material from the Neolithic period*

Cave	Pottery	Lithic evidence	Skeletal material
Attermire	Peterborough and Beaker ware	One broken axe	—
Dowkerbottom	—	Flint	Scattered human remains at low levels, child in stalagmite grave
Elbolton	Peterborough	—	Four complete in sitting position, and eight other portions
Foxholes, Clapdale	'Neolithic pottery'	Two leaf-shaped and two tanged and barbed arrowheads	Disarticulated burnt bones and two skulls
Lesser Kelco	Peterborough	One broken axe	Disarticulated bones
Victoria	Poss. Grooved ware	One adze, one broken axe	Several skeletons found by Farrer, pre-1864
Dog Holes, Warton	Beaker ware	Flint	More than twelve
Sewell's Cave	Peterborough and Beaker ware	Broken leaf-shaped arrowhead	Disarticulated bones
Jubilee	Peterborough	—	—

made with the articular end of a bird bone, whilst the outside has diagonal grooves; deeper grooves are seen on pottery from Victoria Cave. The lithic evidence consists of broken polished axes, none of which have been examined petrologically, together with 'an adze of greenstone' from Victoria, a part of a leaf-shaped arrowhead from Sewell's Cave, and two leaf-shaped points from

Foxholes, in Clapdale, associated with two later tanged and barbed examples.

For the purpose of the present review it is proposed to ignore all skeletal evidence with the exception of the human bones, even though the material would give some insight into the number of species to be found locally. It is felt that the subject has been fully covered recently by Jackson (1962).

Artifacts of bone are rare. Two 'skewer' pins found at Foxholes (Broderick, 1924) and the six from Elbolton, two with slight projections on their sides, should be noted as possible type fossils. It was in the latter cave that four skeletons were found in an upright crouched position together with the remains of eight others. All bodies were enclosed by a 'semi-circular wall of rude masonry'. The Foxholes rock shelter was faced with a stone wall 1·2m high running 11m, 1·5m inside the overhang drip line, with transverse walls running to the cliff. The northernmost enclosure contained over a hundred pot sherds together with burnt human bone; in other parts of the site portions of two skulls were found (Broderick, 1924). In the western gallery of Dowkerbottom Hole, Farrer (1865) found a small grave, 30cm long by 20cm wide cut into the hard stalagmite in which were the remains of an infant. About 75cm of soft undisturbed stalagmite covered the interment.

Clearly the caves had a sepulchral use similar to some of those in other Carboniferous Limestone areas in North and South Wales, Derbyshire, and possibly the Mendips (Piggott, 1953 *a*). The practice of burying articulated and disarticulated skeletons without cremation is possibly a reflection of geology and ecology. On the Mesozoic hills, mortuary houses were constructed of timber, and burnt beneath the barrows; in the caves, stone slabs were utilised for building uncapped cists, but fires seem still to have played a part in the ritual. There is evidence of peat burning in Elbolton Cave.

Generally below the level of the charcoal in Dog Holes, Warton Crag, Jackson found portions of eleven lower jaws, several perfect skulls, flint flakes and a red deer antler pick. The earliest ceramic evidence obtained was late Neolithic Beaker ware, decorated with a trellis pattern of incised squares probably applied with a comb. Cord-impressed Beaker ware has been obtained from Attermire and Sewell's Caves, and whilst there are no C^{14} dates for Secondary Neolithic or Beaker burials from the North West, one has been established from a cave site, Antofts Windypit, Helmsley, in the North Riding, of 1800BC $\mp$ 150 (Radiocarbon 1960, 2; 28). There, a cord-decorated beaker was associated with a flint knife and scraper, the date being obtained from hazel charcoal.

BRONZE AGE

During the Bronze Age in the north of England the climatic optimum encouraged the expanding population to settle on previously ignored land, in particular on the Millstone Grit. Possibly the limestone soils were of less value in the drier conditions: whatever the reasons, there is a paucity of Bronze Age material in the caves. Material was found in Kirkhead Cave but translating Victorian terminology is difficult; the 'fine looped chisel (too commonly called a celt)' is a socketed axe, and it seems likely that a spearhead and possibly a flat bronze axe were found with 'immense quantities of human bones' (Morris & Smith, 1865). A small bag-shaped socketed axe was found in Victoria Cave, but it was associated with bronze Romano-British fibulae and may have been scrap metal. Fairy Holes Cave, at Whitewell in the Bowland Forest, a short passage in a thin limestone, has yielded five sherds of middle Bronze Age pottery. It is also clear that this cave had been closed by building two stone walls across its entrance.

THE ROMANO-BRITISH EVIDENCE

Without the stratigraphical details of early excavations, it is impossible to date many of the finds. Bone, antler, bronze and iron were used for the manufacture of everyday articles like combs, rings and knives, whose forms were virtually unaltered throughout the Roman period. For this reason, study has been concentrated on the more rapidly evolving objects of fashion and trade, on the brooches, pottery and coins.

Pre-Roman Iron Age finds are very rare in the North West. In a recent study of Romano-British metalwork of the Settle neighbourhood, which included surface and cave finds, two items, a mirror handle and a brooch, were dated to the first half of the first century AD (King, 1970). Extending the study area to the west coast and Furness, the total number of finds could be doubled, but no more would be from caves. The two Romano-British brooches, one from Kelco Cave and one from Attermire Cave, probably pre-date the Roman conquest of the North West, but the majority of brooches, forty-six out of fifty-four, in the work mentioned above, dated to the second quarter of the second century. The dragonesque and trumpet brooches, finely cast and sometimes with red, blue or green enamel, were manufactured in the North and belong to this period, as do the disc brooches embossed with the late La Tène tricelis scroll work.

Material, similar to that obtained from the Settle caves, was obtained from Dowkerbottom Hole, but unfortunately it was grouped with some of the finds from Langcliffe Scars when it was sold to the British Museum by Jackson (British Museum Register, Nos 48, 5–11, 1–84). Amongst this collection is a 6cm diameter lens-shaped ingot of lead; it is very close to the average size of Roman and Romano-British metallurgical crucibles from the north of England (Tylecote, 1962). Three Roman lead 'pigs' have been found between Grassington and Pateley Bridge, the earliest dated to AD 81 (Raistrick, 1934). Accepting AD 74 as the date of the defeat of the Brigantes at Stanwick, possibly a year or two more would have been necessary to overrun the Dales area and to consolidate victory with forts and roads; but it is too much to expect them to have prospected the minerals of the area and have mines functioning within five years. It seems more realistic to suggest that ores, and many of them near the surface would have been carbonates, were known and were being mined prior to the Roman invasion of the area.

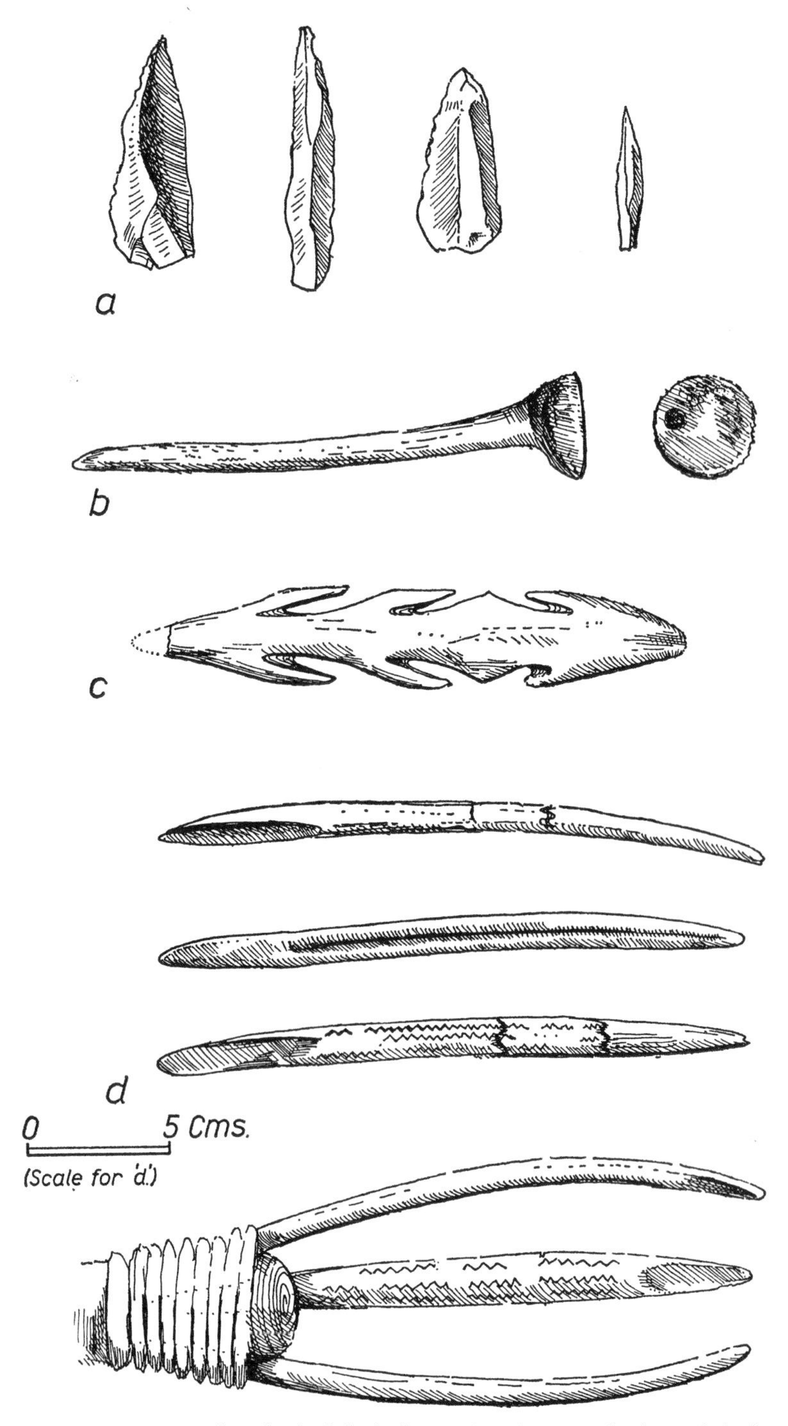

Fig 30 Some archaeological finds from Victoria Cave (*a, b, c, d, f, g*),

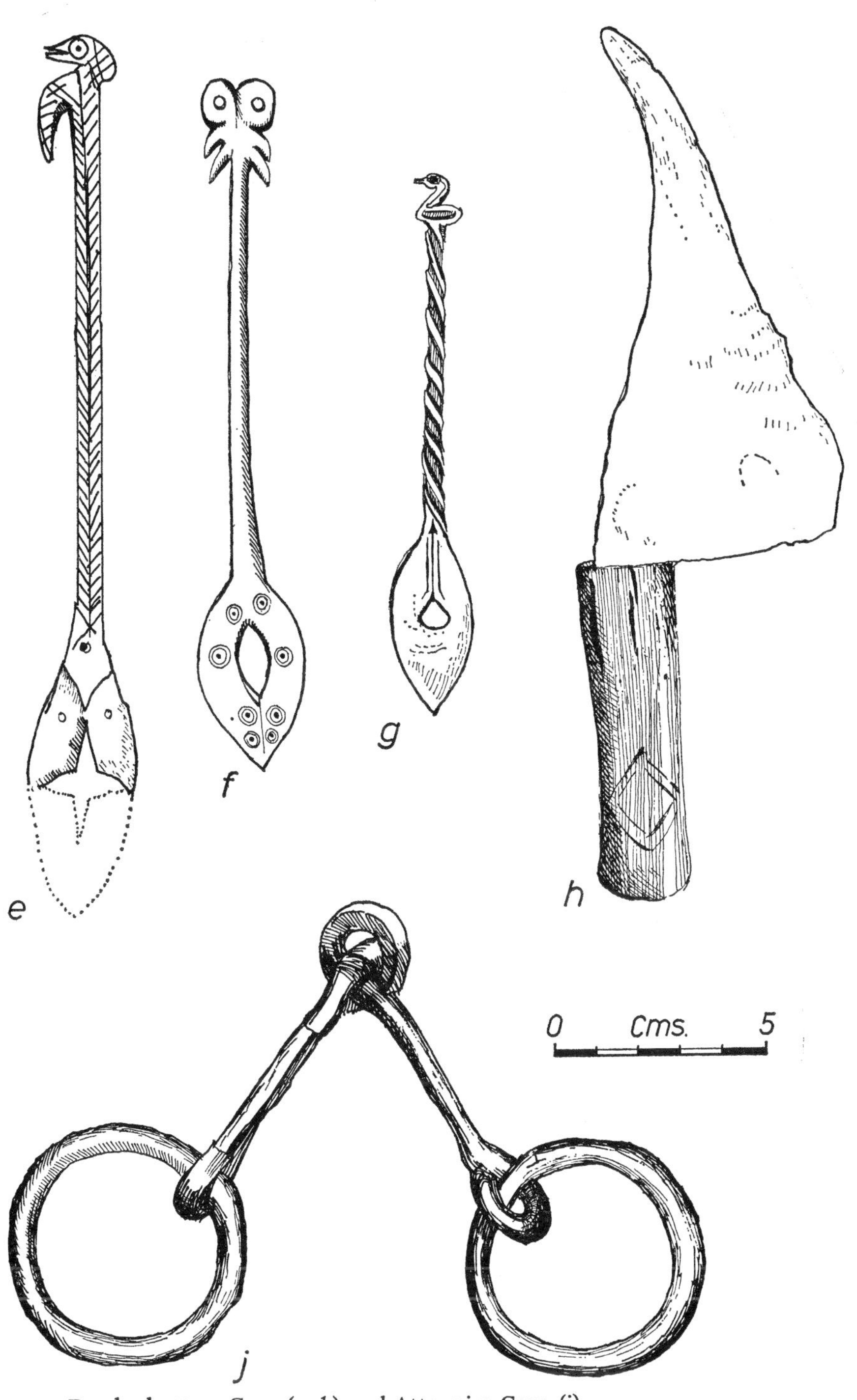

Dowkerbottom Cave (*e*, *h*) and Attermire Cave (*j*)

Enamelled bronzework, a 'pruning hook', bone and iron objects were found together with a balance, which had a bronze beam and two scale pans, in Dog Holes Cave, Warton Crags. Samian and other Roman pottery was also found there and in the neighbouring Fairy Hole; Romano-British material, which includes iron studs, an iron axehead, and a bronze ring, was obtained from Haverbrack Cave. In 1863, Morris obtained an iron axehead, trefoil-shaped fibula, an enamelled pin and a coin of Domitian (AD 81–96) from Kirkhead Cave (Smith, 1865). But generally the north Lancashire and Westmorland cave associations of this period are rather poor; no silver or gilded bronzes have been found, nor the suite of worked bone objects represented in the Dales.

The most common of the worked bone finds were described by Boyd Dawkins as spoon-shaped fibulae (Dawkins, 1874). They are finely carved and decorated pieces having cultural links with the Somerset and Scottish Iron Ages (Fig 30). The same decorative motifs are used on some of the long-handled weaving combs but other items of the domestic textile industry like spindles and shuttles are plain. Spindle whorls are made from a variety of waste materials, usually sherds of pottery or knobs of bone, but some were cast in lead specifically for the job. The intriguing 'spoons' are all perforated with a circular, square, triangular, lens-shaped or stellate hole, which appears from the high polish on the perforation to have been functional. Wool was spun and, from the evidence of the loom weights, woven too. Possibly the spoons were used for drawing and twisting a stronger yarn than could normally be spun from the distaff. A strong twist warp would be needed for the vertical looms.

As far as the metalwork sequence is concerned, and this includes coins, there appears to be a gap in the record taking in the second half of the second century and the beginning of the third. But it is not yet known whether this is echoed by the pottery finds as no specialist study has been made. First- and second-century Samian ware has been found in virtually every excavated cave, and also third- and fourth-century fine and coarse wares.

The cultural assemblages show clearly that the native population was not as poverty stricken as has been suggested. Periods of upheaval and change did occur but the writer doubts that the caves were 'more spacious, more durable (or) more comfortable than the average native hut' (Collingwood, 1969, p 186).

A scientific study must differentiate between caves, rock shelters and shafts and to try where possible to consider the caves' pre-excavation form. The two photographs of Victoria Cave illustrate the obvious fact that a cave becomes more spacious as it is excavated (page 186). The photograph taken in 1870 shows the cave being opened up for the British Association excavations; it was through the walled-up opening at higher level that the earliest collectors gained access to 'three large ill defined chambers filled with debris nearly to the roof' (Dawkins, 1874, p 81). Roach Smith records Jackson's account of the cave's discovery in 1838. It is sufficient here to mention that some of the cave was floored with a stalagmitic crust for 'it was on this crust I found the principal part of the coins, the other articles being mostly embedded in the clay' (Smith, 1848, p 70). The cave had been sealed, otherwise accumulation would have continued; Jackson himself says that the entrance to the inner cave had been walled up.

Other sites in the Settle neighbourhood, Kelco, Kinsey and Sewell's Caves, are southerly-facing rock shelters with incomplete excavation reports. The most interesting was the latter site which yielded one complete, and one incomplete, short Roman sword together with a piece of possible plate mail; six skulls were found and also the skull of a young pig, but because of disturbance it was not possible to say if the skeletal material was Romano-British or earlier. Sir Arthur Keith, in his report, did suggest the cave had been used as a burial place with new burials disturbing the older ones (Raistrick, 1936).

The floor of the spacious Kirkhead Cave was described as a compound of 'bones, earth and charcoal' (Morris & Smith, 1865). Yet again the stratification was confused with Bronge Age material, but bearing in mind evidence from Ogof yr Esgryn, Dan yr Ogof (Mason, 1968), and elsewhere, it seems as though some caves and rock shelters continued to be used for burial. (The writer has drawn no conclusions where only skulls appear in reports—they could be the result of selective digging or, being the most resistant portions of the skeleton, they could be the penultimate stage of solution. On the other hand, they may have a more sinister or cult role.)

There appears to be enough positive evidence to suggest that some sites do have ritual significance with or without burial. Attermire Cave is likely to have been one of the 'ten other caves within a mile of' Victoria Cave examined in 1838 by Jackson and subsequently worked by Farrer (Smith, 1848). An enlarged cleft 6m up a sheer fault face, it has yielded a wealth of material including the metalwork of a native chariot, ie the tyres, nave hoops, nails and lynch pins, and a wrought iron Roman hanging lamp-stand. It is probable that the chariot was dismantled before being placed in the cave, unlike the many chariot graves where the complete vehicle is placed in the grave with the burial. An offering of at least ten chariot-wheel tyres and bridle bits were found in the magnificent votive deposit from Llyn Cerrig Bach, Anglesey (Fox, 1946). To the north of our area chariot parts and a Roman hanging lamp-stand have been found in some of the southern Scottish votive deposits (Piggott, 1953 *b*). In addition to making votive offerings into pools and lakes the Celts put similar objects into wells and springs and into pits and shafts (Ross, 1967).

The mire at Attermire was drained at the turn of the century; up to that time the scree below the cave would have run down to the water's edge. It was from this scree that an iron bridle bit (see Fig 30 *j*) was found in 1970 together with the punches and small 5cm anvil of a bronze smith; the bit matches examples from Llyn Cerrig Bach.

Dog Hole, Haverbrack, Dog Holes on Warton Crag, and Dowkerbottom Hole are all shafts into virtually horizontal limestone pavements. (Dowkerbottom Cave, ie the opening into the east passage, was not unblocked until Poulton's work in 1881.) The shaft entrance to Dowkerbottom was 10m by 4·5m and the top of the charcoal level was 21m below ground level, though a mound of rubble and stones reached to within 6m of the surface. Two galleries extend east and west about 200m and though there are chambers up to 12m high it cannot be considered a habitation.

Dog Holes, Warton Crag, has a roughly rectangular mouth 2m by 2·5m and about 7m deep; galleries extend a total of about 50m north and west. Human remains from the shaft were solely skull fragments, some limb bones being found in the north passage and in the 'Bone Cave', the number of individuals represented being about twenty. Jackson (1910) thought the skeletal material could have been Bronze Age but the associated finds are predominantly Romano-British; he rejected the 'theory of refugees' and thought that cave life formed a feature of Romano-British civilisation.

The most recent shaft excavation, that at Dog Hole, Haverbrack, was a continuation of work by Jackson in 1912. The cave entrance is only 2m by 1·5m and it is at least 10m deep. No stratification or occupation layers were found but Romano-British bronzes and iron work were found with animal and human skeletal material (Benson & Bland, 1962). A minimum of twenty-three individuals were represented, eleven of them being considered between 12 and 25 years old. Many animal bones, especially of ox and sheep, exhibited cut marks which have been interpreted as possible butchery evidence on food species. The authors concluded their report by suggesting the shaft had filled naturally with material washing in from shallow graves and settlement sites on the plateau surface, though earlier they had considered the possibility of the cave being a ritual pit. The plateau measures about 150m square and is now bare pavement.

From the evidence of any one cave or shaft, it is difficult to reach a satisfactory interpretation but the writer feels there is sufficient information available, from the three shafts, Attermire and Victoria Caves, to indicate their use as votive sites in the Romano-British period.

The inhabitants of the limestone country used the caves as sepulchres in the later part of the Neolithic. The ceremonies were associated with death, and the caves developed to the stage when careful burial was no longer essential. The sites became ritual sites, and the bodies (or heads) and the extremely rich deposits, which contrast so markedly with the finds from excavations of the adjacent settlements, can no longer be accepted as occupation debris. It is inconceivable that gilded and silver fibulae or coins would be left on any cave floor by its inhabitants. The finds must be associated with the other highland zone votive deposits. They represent the finest material available, the intensely Brigantian fibulae, the late La Tène decorated bonework and the imported Samian ware. Generally military equipment is absent and where early Roman pieces have been found they may be considered battle trophies, but the Romans too were conscious of the local gods, cults and superstitions.

The use of cave shrines continued locally at least until the ninth century, for Northumbrian sceats of Eanred, Aethelred II and Vigmund have been found at Merlewood Cave in north Lancashire, while sceats of Eadberht, Eanred and Aethelred II, together with a number of ninth-century strap ends, have been obtained from Attermire Cave. A recent British Museum acquisition, the best piece of Anglo-Saxon metalwork from the caves, is another ninth-century bronze strap end inlaid with niello in the Trewhiddle style, found in Smearside Cave, Stainforth, about 1910.

None of the caves appears to have become a Christian shrine but in terms of future work it is something that could be kept in mind. In fact the more one reads excavation reports the more obvious it becomes that re-appraisals of British cave material are due. It is clear these will be of greatest value when associated with excavation.

11

The Caves and Karst of the Morecambe Bay area

P. Ashmead

The caves in the limestones around Morecambe Bay are mostly dry and fossilised since the karst has long since lost its impervious capping, except in the extreme south-east near Kellet. The major caves lie close to the valleys and depressions which separate the high isolated blocks of limestone; they are thinly scattered across a wide region extending from Furness to Lunesdale (see Fig 31). There are not the deep potholes of the limestone further east, but the caves are none the less important, for they provide evidence of a type of phreatic development almost unique in the north of England.

Many of the caves are partially filled with sediment, and the details of the surface karst have also in many places been masked or modified by subsequent phases of erosion. Consequently there have been many difficulties with the geomorphological research which has only recently been started in the area. Many problems remain, though the evidence so far accumulated has defined some unusual and distinctive patterns of cave development.

GEOLOGY

The area (see Fig 31) forms an almost continuous belt of Carboniferous rocks fringing the Ordovician and Silurian core of the Lake District extending from Furness in the west to Kirkby

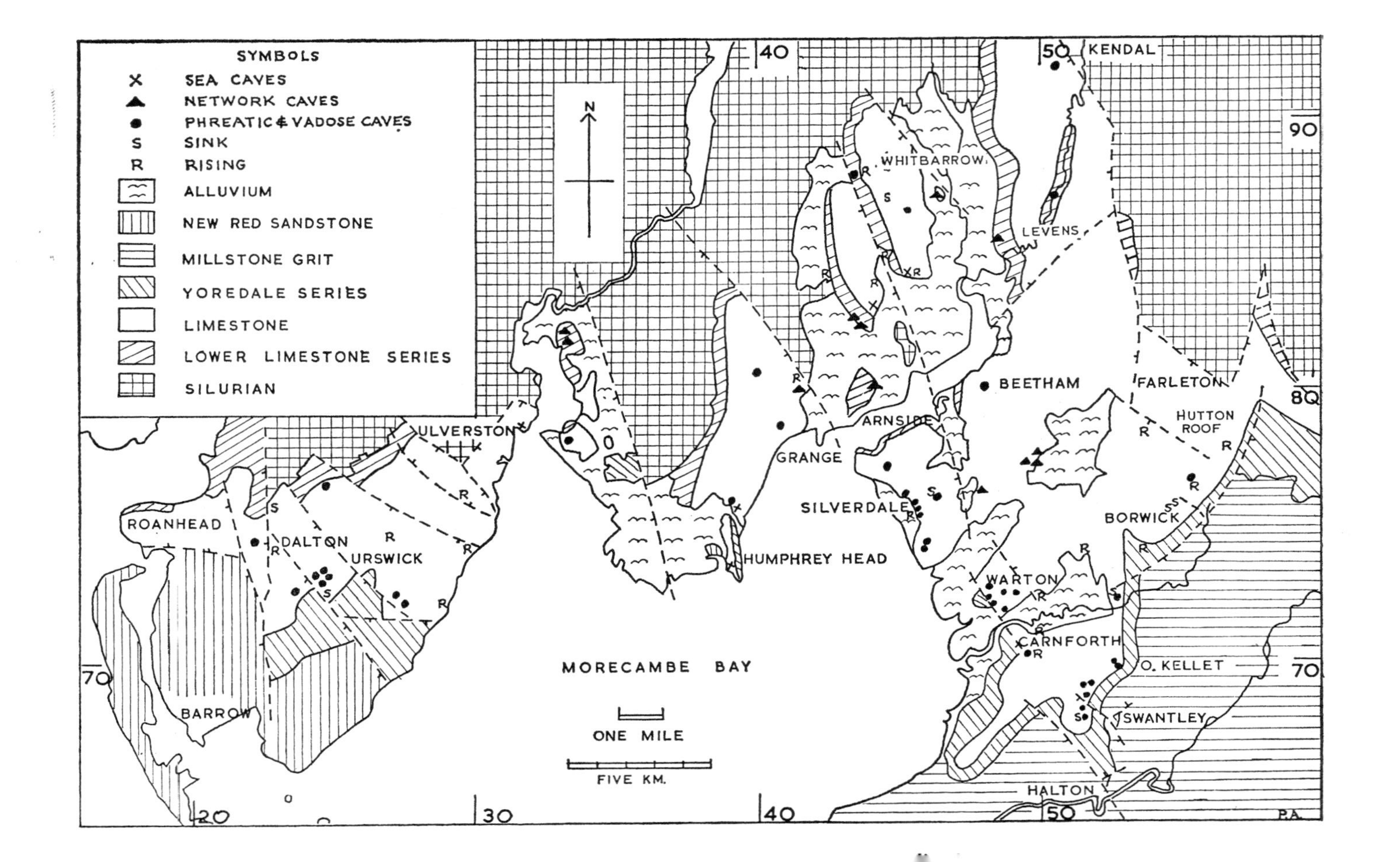
SYMBOLS
SEA CAVES
NETWORK CAVES
PHREATIC & VADOSE CAVES
SINK
RISING
ALLUVIUM
NEW RED SANDSTONE
MILLSTONE GRIT
YOREDALE SERIES
LIMESTONE
LOWER LIMESTONE SERIES
SILURIAN
KENDAL
WHITBARROW
LEVENS
BEETHAM
FARLETON
HUTTON ROOF
ARNSIDE
GRANGE
SILVERDALE
BORWICK
HUMPHREY HEAD
WARTON
CARNFORTH
O. KELLET
SWANTLEY
HALTON
ULVERSTON
ROANHEAD
DALTON
URSWICK
BARROW
MORECAMBE BAY
ONE MILE
FIVE KM.
P.A.

Page 203

(*above*) Horizontal networks of phreatic tubes at Hazel Grove, Morecambe Bay district
(*below*) Glacially scoured grooves in the limestone of Hale Fell, Morecambe Bay district

Page 204

(*left*) The main stream in Ibbeth Peril Cave (*below*) One of the Minaret Passages in Lancaster Hole

Lonsdale in the east where it is separated from north-west Yorkshire by 3km of Lower Palaeozoic rocks. The general radiating dip away from the Lake District gives rise to a continuous upward succession of Carboniferous rocks in a southerly direction, from the basement beds to the Millstone Grit moorland districts of Lancaster.

The general lithological character of the lower zones of the Carboniferous Limestone are constant over large areas. Dolomites and porcellanous micrites are characteristic of the C_1 zone, and oolitic limestones form a major part of the C_2 zone. The standard, massive limestone is limited to the S_2–D_1 zones and is similar to the Great Scar Limestone of north-west Yorkshire. Shales are more frequent in the C_2 and S_1 zones and occur with sandstones in the D_2 zone. The upper beds of the D_1 show a marked difference in lithology; north of an east–west line through Bolton-le-Sands they are of massif facies, whilst south of that line they are of basin facies. Along the line there is a marginal reef development which has been pointed out (Hudson, 1936 *a*) to be a continuation of the Craven Reef Belt. A similar variation occurs in Furness.

As a result of the resurvey of South Cumberland and Furness (Dunham & Rose, 1941), the limestone was divided into seven lithological units (see Table 11).

On the northern massif the limestone rests on highly folded and eroded Silurian and Ordovician rocks. The top of the D_1 is taken to be the top of the massive, white and grey, limestone and is followed by a development which is usually termed the Yoredale Series. However, this shows considerable lateral variation and is diachronous.

On the Reef Belt there are very marked unconformities and non-sequences. In the Kellet area, the shales and limestones are cut out, and the Pendle Top Grit of E_1 age rests directly on D_1 limestone (Hudson, 1936 *a*). Marginal reefs are developed (eg at Dunald Mill Quarry) and the limestone was eroded prior to the deposition of the D_2 shales. Similar structures are recorded in

TABLE 11 *The Carboniferous Limestone succession in Furness*

D_2	Gleaston Series	Black shales; thin sandstones; dark limestones
D_1	Urswick Limestone	Cream well-bedded limestone with oolite horizons; several shale bands
S_2	Park Limestone	White and massive; no shale bands
S_1	Dalton beds	Thinly bedded with shale bands
C_2	Red Hill Oolite	Light grey and massive
C_1	Martin Limestone	Fine grained with porcellanous beds
Z	Basement Beds	Sandstones, shales and conglomerates

Furness (Smith, 1920) and are analogous to those occurring near Settle (Hudson, 1936 *b*).

In the southern sedimentation basin no beds below the D_1 are exposed. In D_2 times the Bowland Shales were limited by a shoreline along the line of the Reef Belt, and are not found further north (Moseley, 1954). The shales are followed by the E_1 grit facies which overlaps higher and higher horizons northwards.

The structure has been only partially elucidated by previous research (Marr, 1916; Turner, 1935, 1949; Hudson, 1936 *a*). As can be seen from Fig 31 the area consists of several limestone blocks separated one from another by fault-guided valleys floored with alluvium. In Furness the limestone is considerably faulted (only the more important faults are shown on the map). The effect of faulting is to downthrow each block to the west whilst the prevailing dip is to the east or south-east. Folding is slight in the north but becomes more complex southwards giving rise to domes and basins which in turn give way to north-east–south-west pitching folds on the outer flank of the Ribblesdale Fold Belt.

Along the Reef Front (ie Gleaston and Bolton-le-Sands) Carboniferous folding and faulting accentuated the separation of massif and basin (Hudson, 1936 *a*). At the close of Carboniferous times the region was elevated and block and basin structures developed. Turner (1935) has shown that the north–south fault belts are Late Variscan, and along these the rocks are locally vertical or even reversed (Garwood, 1912).

In the Morecambe Bay area considerable erosion took place at the end of Carboniferous times and New Red Sandstone overlaps lower horizons in a northerly direction. It overlies the Carboniferous Limestone at Clitheroe, and near the Lake District the New Red Sandstone rests directly on rocks of Ordovician age.

Pre-Pleistocene cavern development

During the long time interval between the Carboniferous and the Pleistocene periods the area formed land undergoing erosion from time to time.

Channelling and potholing has been observed between S_1 and S_2 limestones in Cumberland (Eastwood, 1953) and again at Warton Crag. On the Reef Front, erosion occurred in D_1–D_2 times at Swantley Reef Knoll, where pipes are filled by sandstone and Millstone Grit boulders.

The principal feature of Furness at the end of Carboniferous times was a plateau of limestone, undergoing erosion, extending around the High Haume anticline with a north–south fault scarp between Park and Dalton. In south Westmorland a further plateau extended from the north–south Cark to Haverthwaite Fault, on the west, to the Silverdale monocline, on the east, where it dipped under a shale cover.

The longest period of erosion was during the Permian, at the time of the Hercynian uplift. By this time the area existed as a dissected plateau with fault scarps not unlike the present physical features of north-west Yorkshire. During this period, caves could have developed with swallow holes on the plateau surfaces extending to cave systems beneath.

Some of these cave systems were infilled with haematite iron ore and now form the bodies known as sops. The origin of these

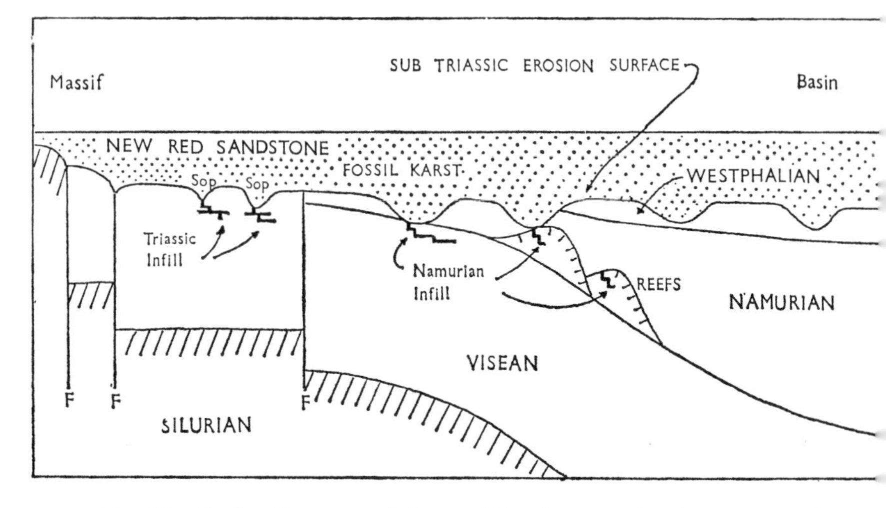

Fig 32 Carboniferous and Permo-Triassic cave development in the Morecambe Bay district

is obscure, but where the limestone was overlain directly by New Red Sandstone the permeability of the latter would allow meteoric waters to percolate through the joint and cave systems beneath. Infilling of voids and metasomatic replacement of the walls has resulted in the iron-ore bodies, now largely worked out. Textural studies of the iron ores have not readily distinguished metasomatism from natural cave infilling which was later subjected to metasomatism. Primary metasomatism is generally found as vein-like ore bodies along faults and joints which may vary in width from a few centimetres to many metres. There are horizontal extensions from the veins known as flats and sometimes whole beds are entirely replaced. Certain beds have been found to be susceptible to replacement. The most important criteria involved is that metasomatic replacements do not contain sediments since there have been no voids—for example, Daylight Hole at Lindal-in-Furness. In more recent geological times, glaciation has eroded some ore-bodies and solution cavities occur above ore-bodies with glacial infill.

Large irregular pockets of ore, sops, were found at Roanhead and Dalton, and consisted of steep-sided hollows lined with haematite containing cores of brecciated New Red Sandstone. Smith (1920) pointed out that these were originally swallow holes, and some of them were found to connect with each other by typical water-worn passages partially filled with ferruginous sand, such as Pickshaft Cave (NGR 245763) just west of Lindal-in-Furness. Occasionally unfilled caverns were encountered by miners on Lindal Moor. The walls of the Henning Valley mines show solutional features, covered with a skin of clay, and parts are filled with sand and powdery ore. More conclusive evidence of natural caves infilled during this period comes from the Arnside–Carnforth area (Moseley, 1969). Here haematite is found in a ferruginous sandstone within pre-existing cavities. The suggested ore genesis of Moseley favours hydrothermal deposition of the iron, but the iron may have been derived from the overlying sandstones. Much of the brecciated sandstone includes particles of haematite. Nevertheless, the bedded layers of sandstone suggest that they were water laid. Later faulting has caused some recrystallisation and dolomitisation (Moseley, 1969). Laminated beds of ferruginous sandstone occur in Crag Foot Mine (NGR 484735). They appear to fill phreatic passages which show rock pendants and low arched ceilings. Both the infillings and solution caves have been affected by faulting which is not later than Miocene. The sandstone in the caves has been shown to be identical to the Namurian sandstone of E_1 age of the Ings Point outlier. In the Pleistocene further solution has occurred, as testified by the upper layer of glacial material in places overlain by till or flowstone.

To summarise, the period of erosion at the end of Carboniferous times involved karst processes and cavern development. The present topography, although modified by Miocene faulting and glacial sculpturing, represents an exhumed Permo-Triassic topography. Caves attributable to this period of erosion are extensive and can be recognised by their characteristic infills, which, though ferruginous, can in many cases be differentiated from metasomatic ore-bodies.

The subconical limestone hills have been described as relics of a tropical karst developed during Tertiary times (Corbel, 1957). However, the form of most of the limestone blocks is closely related to geological structure, and Parry (1960) has indicated that the limestone was not exposed to subaerial erosion until the Pleistocene. At present, there is no conclusive evidence for either conical karst or caves which have developed during the Tertiary periods.

Pleistocene marine erosion levels

The highest erosion surface to affect the limestone in the Morecambe Bay district is found at the 275m (900ft) level. Consequently the surfaces above this are not considered here, as their development did not involve solutional erosion.

Several marine erosion levels, the highest at 210m (690ft), have been described by Parry (1960) as of Pleistocene age. Certainly they show a marked agreement with Pleistocene eustatic sea levels recorded elsewhere (Zeuner, 1946), and may possibly be correlated. They are, moreover, of a fragmentary nature. But superimposed upon these must be recognised the Late Glacial high sea levels, caused by isostatic depression of glaciated regions, the terraces of which are much fresher in appearance, being little modified by subsequent erosion.

An erosion surface at 245m (800ft) in the Lune Valley at Kirkby Lonsdale can be traced almost to Morecambe Bay, at Hutton Roof Fells. It is considered to be pre-glacial by Parry (1960), and by the writer in a discussion on the ages of erosion surfaces at Casterton Fell (see chapter 14). A lower surface at 170m (550ft) can be similarly traced; this was thought to be pre-Devensian by McConnell (1940), and Early Pleistocene by Parry (1960).

Inner marginal breaks of slope ascribed to the Plio-Pleistocene marine transgression occur at an altitude of 210m (680–700ft) at

Dalton Crag (Parry, 1960). The summits of several other limestone blocks show a marked accordance between 200m (650ft) and 215m (710ft). Minor still-stands are marked by notches at a series of altitudes between 175m and 130m (570ft and 430ft), which may possibly be grouped together as related features of the 180–170m (600–550ft) subaerial surface inland. Remnants of a stage at 110–115m (350–380ft) are fairly common, being well-developed notches and valley side benches. A stage at 100m (330ft) is recorded, and remnants can be seen cutting across tilted limestone and sandstone in the Kellet area. A further stage is recorded at 90m (290ft), but is so extensively covered by deposits of the piedmont ice sheet of the lowland that it becomes difficult to interpret. The surface is no doubt compound, forming a group together with a lower feature of 55–60m (180–200ft) (Parry, 1960). The latter is very extensively developed.

Very little is known of the older glaciations of the district, the latest glaciation having obliterated, or reworked, their deposits. The Last Glaciation (Devensian) is most completely known.

During the Last Glaciation the area was overwhelmed by ice to a considerable thickness. This retreated in several well-marked stages which can be traced northwards from the Cheshire Plain, and overflow valleys formed at each intermediate stage.

Marine erosion levels of 45, 30 and 15m (150, 100 and 50ft) are recorded by Parry (1960). The latter two are cut into the drift of the Last Glaciation and are therefore interstadial, whilst the 45m surface is overlain by drift of the Last Glaciation, and Parry postulates an interglacial origin. There is evidence to support pro-glacial lakes, in the form of strand lines and alluvial fans at various levels between 55 and 7m (180 and 25ft) directly associated with moraines and overflow valleys of the Last Glaciation. It is possible that the lakes once formed a continuous unit, Lake Morecambe, at a succession of levels, each one held back by an ice-choked Irish Sea.

PATTERNS OF PLEISTOCENE CAVE DEVELOPMENT

In the Morecambe Bay area there are characteristically three types of caves—phreatic networks; abandoned vadose caves; and abandoned sea caves.

The area is dominated by large tilted blocks of limestone reaching altitudes of up to 260m (800ft), separated one from another by broad fault-guided valleys, floored by alluvium or estuarine sediments. The limestone is rarely horizontal but forms long dip slopes generally facing east or south-east, with scarps facing west or north-west. The extensive escarpments have been modified by glacial erosion; though some scarps are on faults of only Tertiary age, others may be exhumed Permian features, as at Humphrey Head.

The steeper dip slopes are characterised by rundkarren and some rinnenkarren, most spectacularly at Hutton Roof, while the more gentle slopes consist of bare pavements with areas of outstanding knolls, separated by lower irregular grassy bowers littered with transported blocks and piles of stones.

The enclosed hollows are in many ways the more interesting karstic features. They occur mostly in the valleys but are also found at higher altitudes. The rarity of shakeholes, which only occur in abundance at Kellet, is due to the very limited extent of the glacial drift on the limestone. Most depressions other than the swallets may be described as dolines and uvalas. There are a great number of deep dolines with precipitous edges, some of which contain small tarns. Joints and bedding planes on their perimeters show solutional features, and breakdown has also occurred. A particularly large single doline occurs at Deepdale near Yealand, and the coalition of dolines has occurred at Haweswater.

The largest basins appear to have developed from several closed hollows and may possibly be true karst poljes. Hale Moss is a good example, being bounded on the west by a discontinuous crenulated cliffline. Most of these poljes occur in chains, and

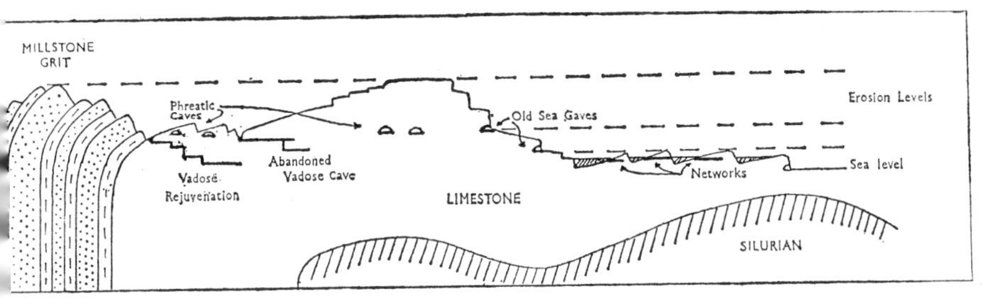

Fig 33 Schematic diagram of Pleistocene cave development in the Morecambe Bay district

whilst a few might be fault guided, many are not. They appear to have interglacial origins, but have been modified by glacial erosion. They are floored by lacustrine sediments and pollen analytical work shows that they were receiving glacial outwashes in zone 1 (Oldfield, 1960). Some of the lowerlying basins were infilled during the Flandrian marine transgression (Gresswell, 1958), although most of the inland peat mosses were high enough to avoid this.

THE PHREATIC NETWORK CAVES

Four main series of phreatic network caves are known and it would seem that there is every prospect of finding others in the many similar topographic situations. They all have broadly similar morphology and are not confined to any particular lithological type. Roudsea Wood caves are in oolitic limestones, while Hazel Grove caves lie in massive standard limestones. Essentially they are found in cliff edges and residual knolls of either the very wide estuarine mosses or the enclosed hollows which were lakes in Late Glacial times. They are all restricted within a narrow vertical range of 2–3m, and are horizontal even though the limestone bedding may be tilted by as much as 25°.

It is suggested that the basins were solutionally eroded in lacustrine or estuarine environments during interglacials, whilst further erosion occurred under late- and post-glacial conditions.

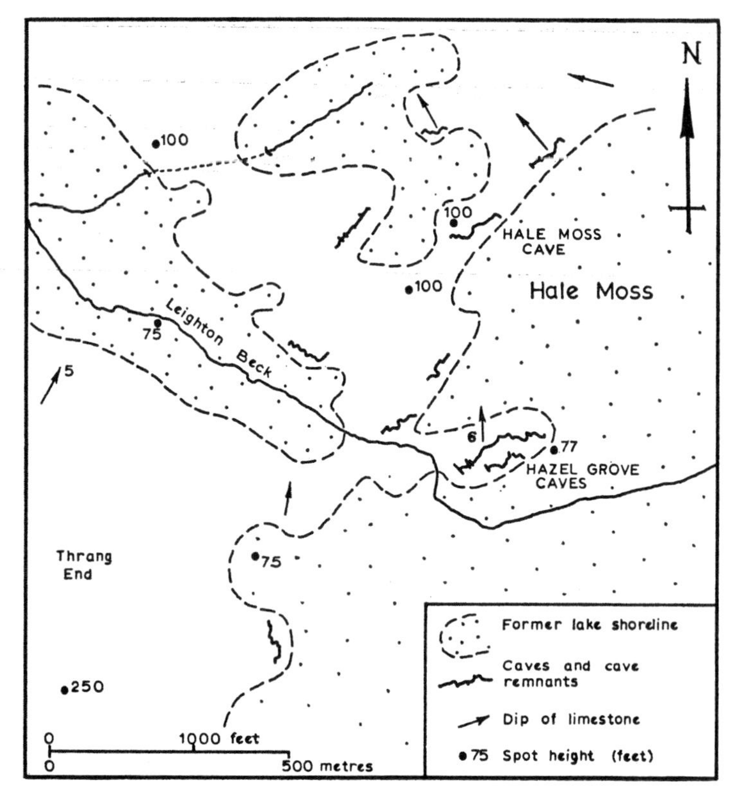

Fig 34 The poljes and caves of the Hale Moss area

The caves originated and evolved by processes integral with solutional widening of the depressions.

Many solutional forms are found within the caves showing that vadose activity has been minimal. Some of the features such as anastomoses, spongework, rock pendants, rock bridges and razor-edged dividing walls indicate an almost complete lack of mechanical corrasion.

The Hale Moss Caves

Hale Moss is situated 8km to the north of Carnforth and occupies a low-lying area of approximately 4km². On the west the flat ground is bounded by a series of small scarps in massively bedded Urswick Limestone. The limestones have a gentle dip of 6° to the north-west and are intensely jointed, as many as five sets being conspicuous.

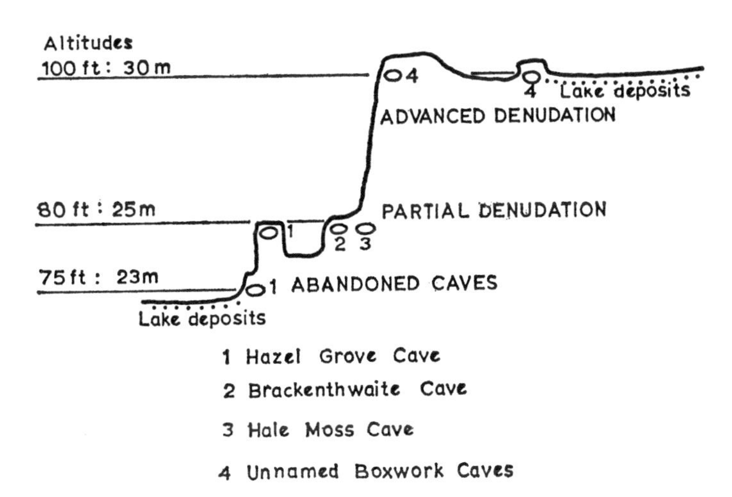

Fig 35 Diagrammatic cross-section showing levels of development and subsequent erosion of the caves around Hale Moss

The caves lie at altitudes between 23m and 25m OD. There are two fairly large systems and several smaller ones. The longer of the two large caves is Hazel Grove Cave, containing almost 300m of passage running approximately parallel to the Moss edge with numerous entrances adjoining the main trunk passage. Hale Moss Cave has been explored over a length of 200m, with its main passage low for nearly all its length and terminating in peat infill (see Fig 36).

Above and to the west of Hale Moss, islands of limestone stand up above a further level area similar to that of Hale Moss. The remains of former caves, resembling those of Hale Moss, can be seen along the edges of these limestone ridges. They are, however, almost denuded, and remains survive only of roofless cave

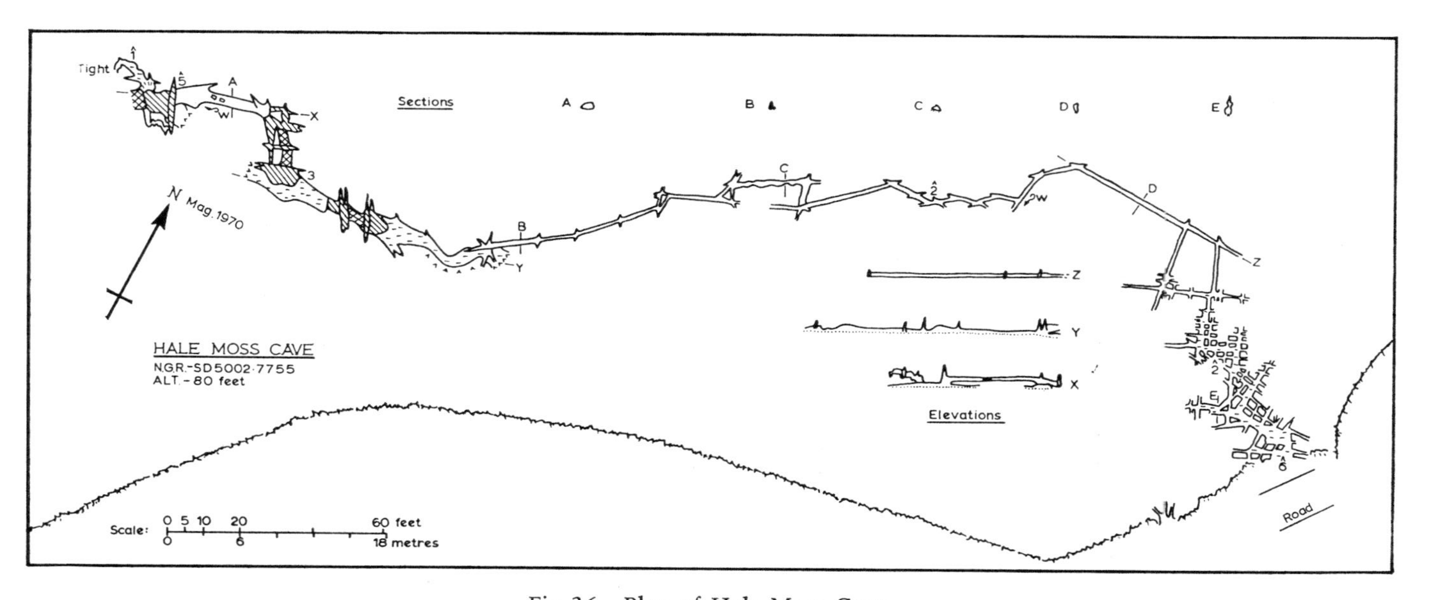

Fig 36 Plan of Hale Moss Cave

networks. Small network caves are also found at the edge of the adjoining Brackenthwaite Moss (Ashmead, 1969).

The cave development can be separated into three distinct stages. The initial stage appears to be the dissolving of limestone along exposed bedding planes, at first as anastomosing channels. This has been later enhanced by solution at intersections of joints, and in selected areas probably of greater porosity or joint density. The second stage is the rationalisation of the flow pattern with development of a trunk passage. The third stage is represented by peaty mud infilling the passages, and the collapse of roof and wall blocks which, if continued, eventually destroys the cave.

In Hale Moss Cave, joint-determined passages link areas of development on bedding planes. Hazel Grove Caves contain two distinct levels of development, relating to two past water levels. This water-level control is particularly clear in the Hazel Grove Caves because the limestone is locally dipping at about 6°. Though each passage is formed on a bedding plane or joint, the overall pattern of development is horizontal, cutting across the dip of the limestone, and clearly related to past water surfaces in the adjacent, now dry, lakes. Ice expansion may possibly have assisted the breakdown which has now destroyed parts of the caves. The infilling of peaty mud is probably the result of inwash accentuated during a past more pluvial climate.

Roudsea Wood Caves

Roudsea Wood Caves, on the edge of the Leven Estuary, consist of two main systems of passages totalling 600m in length. They are formed in low hills of thinly bedded oolitic limestone dipping between 12° and 25° eastwards under Holker Moss, a raised peat bog overlying marine clays.

The main passage takes the form of a continuous tunnel of oval or arched section, with subsidiary joint-determined passages forming small networks which decrease in density inwards from the moss edge escarpment (see Fig 37). Although the limestone

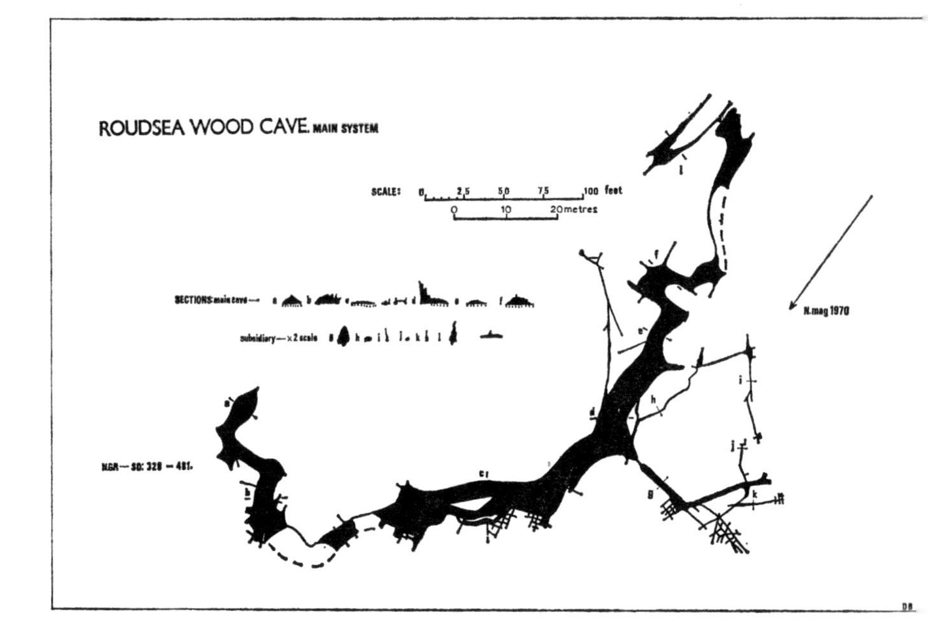

Fig 37 Plan of Roudsea Wood Cave

dips at over 15° the cave passages are horizontal—due to the water-level control of their development.

During the retreat of the glaciers the area around Roudsea Wood consisted of small islands standing above a pro-glacial lake or estuary. These islands of limestone were thus open to solutional attack along bedding planes and joints. The joint network diminishes in intensity away from the escarpment; corrosion was greatest in a broad band roughly parallel to the limestone cliffs. Solutional widening has been more intense at joint intersections, and solution along the same joint from opposite directions occurs with constricted areas in between. Re-solution of collapsed blocks and infilling with loamy peat represent the post-glacial, final, stage of development.

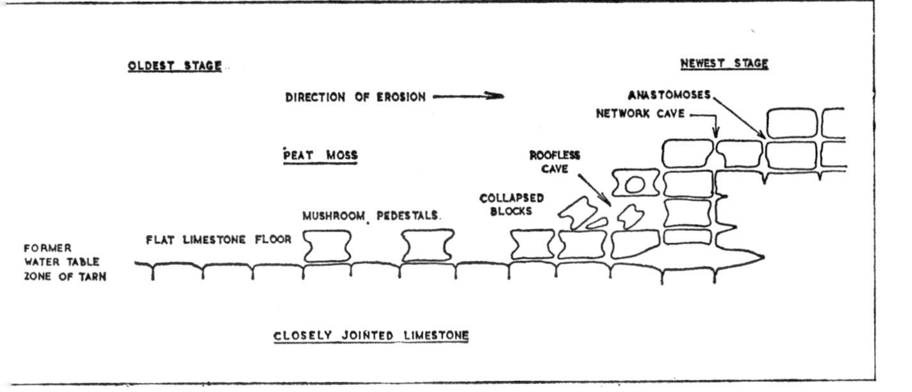

Fig 38 Diagrammatic section showing erosion processes on polje margins

The abandoned vadose caves

A number of caves in the Morecambe Bay area have phreatic origins, but have been subjected to considerable vadose modification.

These caves lie in glacial overflow or melt-water channels, and their abandonment relates to the temporary nature of these channels. The vadose-modified caves are all close to the boundaries of impermeable rocks adjacent to the limestone, while small

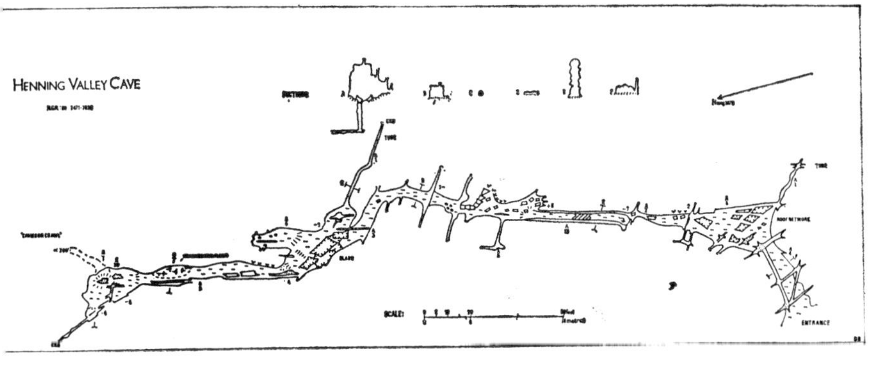

Fig 39 Plan of Henning Valley Cave

purely phreatic caves are found some distance away from the lithological boundaries.

All the caves have basically a similar phreatic origin consisting of small networks of tubes and joint rifts, or cavities aligned along single joints. These initial phreatic phases of the caves are similar to those of the phreatic networks described above; they originated at a time when the base level was higher than at present. Rejuvenation has resulted in either complete abandonment, or vadose modification in situations where surface water has been available, such as in the melt-water channels. Further rejuvenation has left dry abandoned sections showing distinctive vadose features.

The caves of Furness

Henning Valley is an old melt-water channel situated west of Lindal, and the cave of the same name is located at the edge of a small abandoned quarry.

The highest parts of the cave form an intricate network and in places the major abandoned stream passage lies hidden by collapsed blocks and sandy infill. The terminal chamber is quite spacious while beyond is only a constricted muddy stream passage at a lower horizon.

Mud lapies are very common, and include sets of individual grooves, parallel and regularly spaced, and also forms with a dendritic pattern. They occur on the lower sloping ledges of wall pockets. On the upper overhanging parts of the pockets the grooves extend from spongy wrinkled areas of mud and there is a definite line of demarcation. The glaciofluvial infill consists of gravel overlain by laminated sands.

Henning Valley Cave initially transmitted a stream which resurged somewhere near the present entrance, having arrived in the presently known cave through a phreatic lift in the floor of the terminal chamber.

The entrance to Stainton Cavern is unfortunately buried beneath a huge pile of quarry overburden. It was an important cave, possessing not only solutionally widened joints and loops but also a long 'gothic arch' phreatic trunk passage suggestive of a former higher base level; several other caves nearby consist of fragments of phreatic loops, probably part of the Stainton Cavern system.

The caves of the Grange and Arnside areas

The large outcrop of limestone each side of the River Kent estuary have very few significant caves, other than the phreatic networks of Hale Moss, described above.

Pool Bank Cave in Whitbarrow Scar is an active resurgent cave located at the base of the limestone. The active part is extremely small and normally flooded, and there are also two higher fossilised levels. In the latter, calcite formations blanket a glacial breccia which is being eroded by the present stream.

Fairy Hole, at Warton, consists of small joints leading from a larger chamber, and appears to be a former resurgent cave at the 80m (270ft) level. There is a pronounced rock terrace at the same altitude along this side of Warton Crag. Within the cave, stream-laid calcareous sands are covered by a layer of boulder clay.

Dog Holes at Warton, Heathwaite Cave and Haverbrack Cave all show phreatic origins but there is also some abandoned vadose development at the 60m (200ft) level. They have all been extensively filled with Pleistocene deposits.

The caves of the Kellet area

The earliest phase of Dunald Mill Hole, near Nether Kellet, was a primary phreatic network of tubes in solutionally widened joints. Below the passages of this stage are scalloped, partially walled, oxbows at the 80m level. Further rejuvenation has resulted in the high aven in the Third Chamber, and the vadose incision of the present stream passage.

Just south of Over Kellet, the High Roads Quarry has revealed two small caves. High Roads Cave I consists of two levels, an

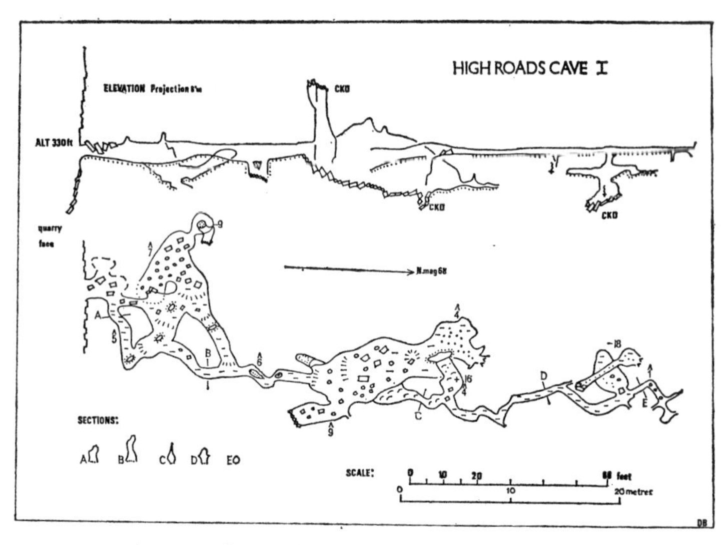

Fig 40 Plan and section of High Roads Cave I

upper phreatic network, and a lower abandoned vadose passage connected, at several points, to the upper passage by solutionally widened joints. High Roads Cave II is of similar morphology, and contains several rock bridges.

The caves of the Kellet area exhibit the threefold development of phreatic initiation—vadose modification; infill; and collapse. The phreatic tubes and other features of solution are fully developed, and indicate a past base level in this area of about 80m (270ft), phreatic features in most of the caves occurring up to this height. All of them are situated in a series of valleys which have been utilised and modified as glacial overflow channels with floors at 85m (275ft). Originally they were probably fed by small streams from the Millstone Grit outcrops to the east.

There are deposits of at least one full glacial stage in all of the caves, and the last glaciation has certainly fragmented those of them now situated near the tops of hills and completely out of phase with the present geography.

In Dunald Mill Hole late glacial melt-waters, assisted by the re-invasion by the post-glacial underfit stream, have maintained its development. The other caves of the area are completely abandoned.

ABANDONED SEA CAVES

Old sea caves associated with marine-eroded notches may be seen on the seaward side of all the major limestone blocks. All occur at the marine erosion levels referred to by Parry (1960).

Near the south-west corner of Whitbarrow Scar is Whitbarrow Cave, at an altitude of 115m (380ft). Developed in a small fault, it has an extensive notch on either side. Remnants of a glacial breccia are visible in its eastern wall, and its south-western aspect favoured its preservation from later glacial erosion. On Warton Crag is another small cave, Harry Hest Hole, a marine-eroded fissure cave developed at the same 115m level. It too has a well-developed notch nearby.

No further sea caves have yet been found at altitudes higher than 30m (100ft), largely due to the subsequent glaciations which would have destroyed any caves once developed on these older marine erosion levels.

On Whitbarrow the lower sea notches are present, but poorly developed due to their sheltered position at the extreme north-eastern end of Morecambe Bay. Nevertheless the area is of great interest as the succession of sea levels and marine benches can be followed upwards to the highest surface. Quarrying of frost scree breccia has revealed a well-developed notch at 30m (100ft).

Kirkhead Hill forms a conspicuous headland west of Grange-over-Sands and Kirkhead Cavern at 30m (100ft) is a well-known archaeological cave. Marine-eroded notches occur at 15 and 30m (50 and 100ft) below a platform at 40–45m (130–150ft). The headland overlooks a plain 5m (18ft) above sea level which was filled in during the post-glacial transgression (Gresswell, 1958). Recent excavations at Kirkhead Cavern suggest that the cave was enlarged by marine erosion during the Devensian (Würm 2a/b Interstadial) and a sequence of deposits are found dating from Middle Devensian to the present.

About 4km south-west of Kirkhead is the Humphrey Head peninsula. The 15m (50ft) level is represented by several small old sea caves and marine notches, and above is the cave known variously as the Grand Arch or Edgars Great Chapel. A natural bridge spans the cave and the blowhole extends for 12m to a wave-cut platform at 45m (150ft).

At Silverdale the post-glacial transgression is recorded both by wave-cut notches in solid rock and glacial drift, and several small caves occur. The latter are phreatic in origin but have been modified by marine erosion of the post-glacial transgression. Glacial fill has been overlain by calcite, and later eroded by wave action.

On Warton Crag several old sea caves are found along the fault-guided south-western scarp edge. Most are at the 15 and 30m (50 and 100ft) levels. An extensively developed notch also occurs at the higher level.

The headland of Heysham is situated to the west of Lancaster and consists of Millstone Grit folded in a series of elongated domes. It possesses morphological features similar to the headlands in limestone, for marine erosion benches occur, cut into solid rock and overlain by drift at the 15 and 20m (50 and 65ft) levels. There are also several small sea caves aligned on small faults at the 5m (18ft) level.

The evidence so far accumulated shows that the marine features described above are widespread and not only of local significance. The sea caves are consistent with the marine erosion facets recorded by Parry (1960). They give evidence of former sea levels, and hence base levels, which in turn have controlled the phreatic cave development further inland and its subsequent vadose rejuvenation.

THE MODERN KARST HYDROLOGY

The major rivers of the Morecambe Bay area almost reach sea

level before they flow into the limestone regions, and thus the estuaries penetrate well inland, with the limestone hills forming the interfluves.

Rainwater enters and percolates through the limestone in a diffuse pattern due to the absence of an impermeable capping. Several perched aquifuges form spring lines, many with tufa deposition. Such resurgent streams have short surface courses before sinking again, and are so youthful that caves have rarely formed. Due to the prevalent dip of the limestones the majority of springs and risings occur inland, on the west of the escarpments.

In Furness the Silurian slates, of the High Furness Fells, support small streams which sink at the edge of the limestone. One is Poaka Beck which flows through an immature cave to reappear at Yarl Well, in Dalton-in-Furness. In the Kellet area, where shales and sandstones occur adjacent to the limestone, small immature caves are forming at present such as Capernwray Cave, and the lower, now inaccessible, reaches of Dunald Mill Hole.

The River Kent is at present excavating a gorge across its limestone outcrop. Large undercuts, and underground water courses below the river bed, have not yet sufficiently developed to take all the water. Otter Holes (NGR 506863, near Levens) has a length of 250m, and is only explorable in drought conditions.

The inland springs are very small and it is doubtful whether they have large caves behind them. Many of the springs are on the coast line, and their waters rise from small tubes or joints at the foot of the limestone cliffs. Some appear from cave-like openings of small dimensions where present-day wave action has enlarged the joint, echoing the similar manner in which the older sea caves developed.

The natural rifts in Red Rake Mine, at Silverdale, are at sea level and the water level within them varies with the tides. Some unique mud formations which exist here may be due to this constant fluctuation of water level. The formations occur as a series of horizontal semi-circular discs of mud attached to the walls by their edges and arranged in layers along horizontal mud furrows. The perimeters of the mud discs have a white dentiform fringe which may be crystallised sodium chloride or other salts derived from sea water. This hydrologic connection between the rifts and the open sea appears to be restricted, for in wet weather the rifts flood and drain via a temporary spring in front of the mine entrance.

Also, at the present time, small phreatic networks are actively developing around the edges of some of the inland tarns, notably at Haweswater and Meathop.

Dye-testing in the Morecambe Bay limestones has not yet been carried out. Consequently there is insufficient evidence to indicate the degree of karst drainage integration, though morphological features do infer that this is very low.

12

The Caves of Dentdale

M. K. Lyon

The karst of Dentdale represents a transition zone between the classic Craven area to the south, with its massive outcrops of Great Scar Limestone, and the rhythmic succession of limestones and shales of the Yoredale country to the north. The slight northerly dip of the strata, seen so clearly on the southern slopes of Gragareth and Whernside, continues through to Dentdale. Although the valley is relatively low, 135m (440ft) OD by Dent village, only the top of the Great Scar Limestone outcrops along the entire floor of Dentdale, and its major tributary, Deepdale. Thus the main cavernous rocks lie near to, and below, the valley floor level, only thin Yoredale limestones seaming the slopes above (Fig 41).

This simple pattern is broken where the Dent Fault crosses the valley, immediately west of Gawthrop village. This is a major, and complex, fault, running northward along Barbondale and across Dentdale. It has caused considerable disturbance in the Carboniferous rocks on the down-throw, upvalley side of the fault. The Great Scar Limestone has been turned upwards parallel to the line of the fault, so that it is locally vertical, but the folding lessens in intensity as the disturbance line is left behind. This area also exhibits subsidiary faulting oriented north-west–south-east. With Ordovician shales and mudstones on the downvalley side of the Dent Fault, the point where the River Dee crosses it marks the lowest possible resurgence level. The major spring near Barth

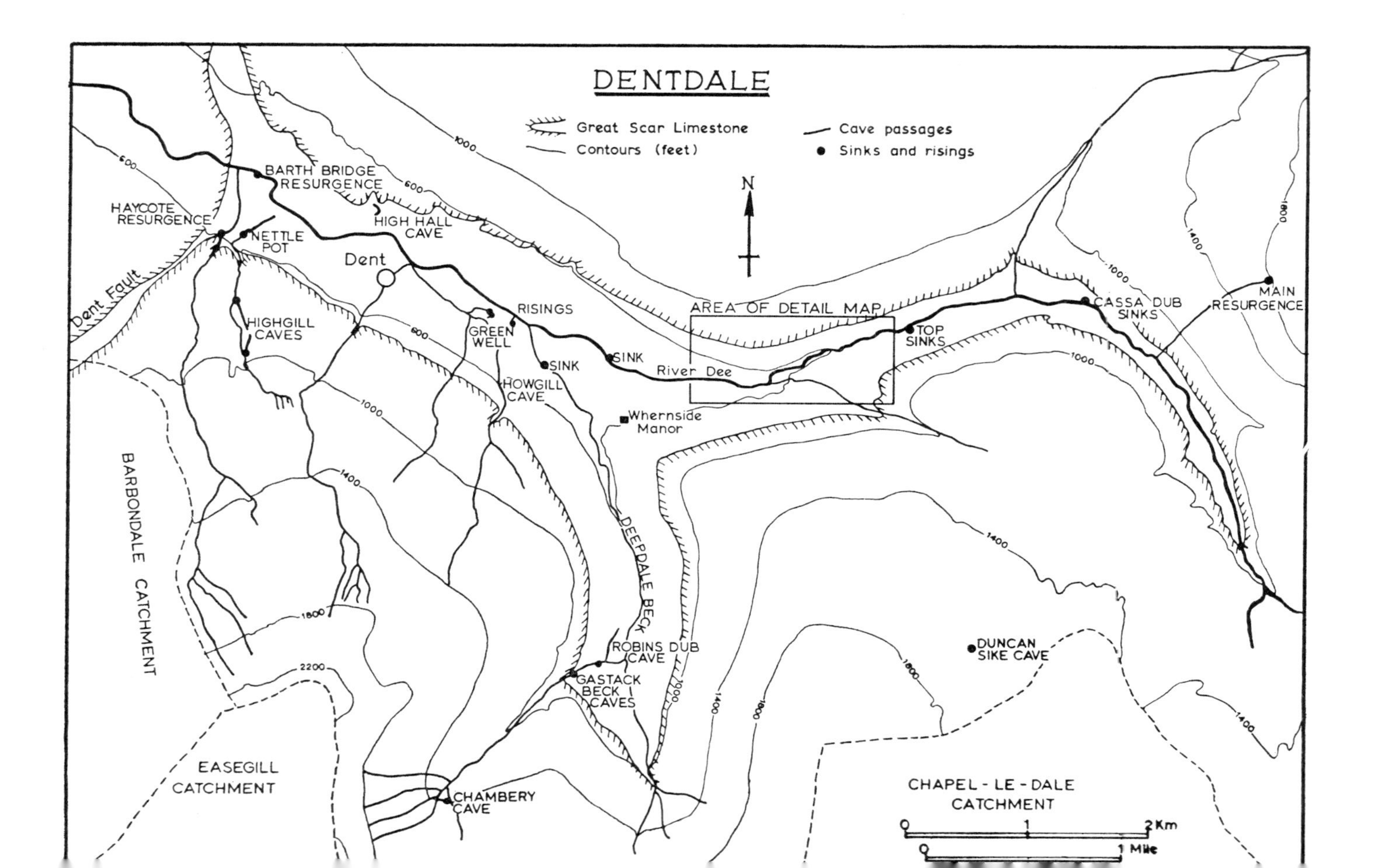
DENTDALE
Great Scar Limestone
Contours (feet)
Cave passages
Sinks and risings
N
BARTH BRIDGE RESURGENCE
HAYCOTE RESURGENCE
NETTLE POT
HIGH HALL CAVE
Dent
Dent Fault
HIGHGILL CAVES
GREEN WELL
RISINGS
SINK
SINK
HOWGILL CAVE
River Dee
AREA OF DETAIL MAP
TOP SINKS
CASSA DUB SINKS
MAIN RESURGENCE
Whernside Manor
DEEPDALE BECK
ROBINS DUB CAVE
GASTACK BECK CAVES
CHAMBERY CAVE
DUNCAN SIKE CAVE
BARBONDALE CATCHMENT
EASEGILL CATCHMENT
CHAPEL - LE - DALE CATCHMENT
600
800
1000
1400
1800
2200
0
1
2 Km
1 Mile

Bridge, with a mean flow about 140l/sec, is within a few metres of this level (125m, 410ft).

No caves have yet been entered in the fault-line limestones, but they undoubtedly exist. The resurgence at Haycote, in Gawthrop village, may be backed by a cave system over 1km long and deeper than 100m. At present the resurgence is used as a water supply; exploration is hence prohibited, and water tracing inhibited. This could be a site of major importance in the chronology of cave development, for the horizontal tunnels of the Barbondale caves and Bull Pot of the Witches could be continued here at lower levels.

Dentdale contains the whole range of Yoredale limestones, although in practical terms the Hawes and Gayle beds are best considered part of the Great Scar Facies. Above this level there is a 30m thick shale band, followed by the Hardraw Limestone. This, and all the succeeding limestone beds, contain significant caves. Best known are the High Gill Caves. The upper entrance, in Simonstone Limestone, is fed by Brown Beck, which resurges from the Main Limestone by the old 'occupation' road. The stream, entering through open joint passages, reaches the base of the limestone almost immediately, and flows down the 8° dip along the top of the underlying shale, through breakdown chambers and a stream passage. After 200m the passage is choked, where it meets one of the minor faults. This allows the stream to drop through to the Hardraw Limestone, and it is from this bed that the water reappears. It continues for 300m, where it intersects another fault, and falls 7m to regain the same bed. It then resurges high in the side of a surface ravine. Upstream of the waterfall, only the flat roof is of limestone, the passage having developed almost entirely by corrasion in the underlying shale. In section it is a wide trench, typically 2m wide, and nearly as deep.

The Middle Limestone has a number of unentered risings, but one cave indicates the type of passage that probably exists elsewhere. Glacial erosion at the head of Gastack Beck, a hanging-valley tributary to Deepdale, has bisected an old passage within it. The longer section has been invaded by a recent stream which enters through a number of small avens that give the passage its name—Chambery Cave. These are connected to the main passage, triangular in cross-section, with its base on the shale, and completely joint-oriented, in textbook Yoredale fashion.

The Undersett Limestone forms a wide bench on the northern flank of Whernside, engulfing all the streams that reach it. Only one has been followed underground, in Duncan Sike Cave. The Main Limestone has yielded no caves on the Dentdale slopes of Whernside, although it contains the very extensive Greensett Caves 1km further east. 5km north, across the Dale on Great Knoutberry Fell, an extensive exposure of the Main Limestone contains numerous shafts, but little horizontal development. The main resurgence for these is below Pikes Edge, where, in times of flood, a flow of 1cumec may be ejected through massive blockfall.

The caves of the Great Scar Limestone group into two major hydrological systems, excepting some outlying caves that resurge before, or on reaching, the valley floor.

Most important is the Upper Dentdale System (see Fig 42). Only a fraction of its passages has yet been discovered (Long, 1971), largely due to the regular flooding that fills virtually every passage and blocks all the entrances with boulders and rubbish; the entire system is below river level. The River Dee sinks in its bed at many points from Cassa Dub downstream, resurging, in dry weather, at The Popples, a line of bedding plane risings over 3km from, and 75m below, the first sinks. In fairly high-water conditions a through flow time of twenty hours has been recorded.

The first sinks occur 1km after the river meets the limestone. A number of corroded joint openings drop 1–3m into an extensive bedding development, running under the north bank. Water entering here is not seen before the resurgence. The next sinks, 2km downstream, follow a similar pattern, but their water is seen underground, issuing from the sump in the upstream passage of Ibbeth Peril No 1 Cave. A bedding below Slit Sink, where the

Dee undercuts the road, is indeterminately wide, and very low, with marked joint development aligned at 34°. The joints are infilled with bands of insoluble cherty material that hang as grilles from the roof, forming incomplete barriers in the passage. Downstream again the small passage of Scythe Cave passes underneath the river-bed. It is oriented down the northerly dip, so that, like every other known inlet to the system, it leads into the northern side of the Dale.

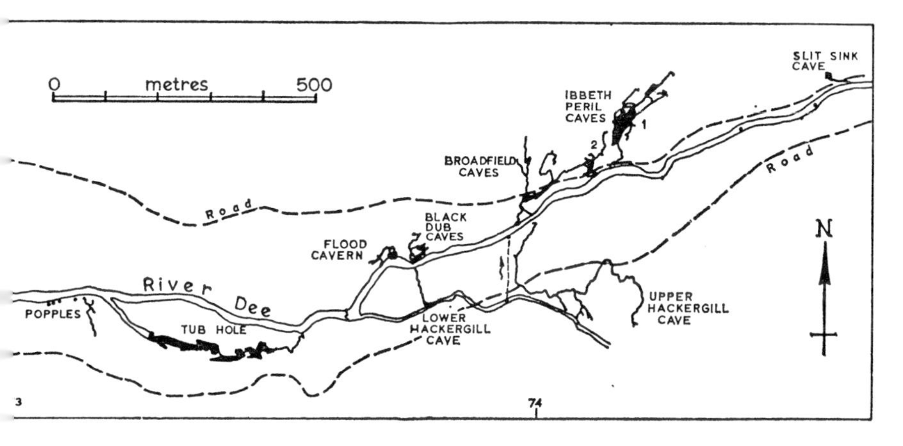

Fig 42 The caves of the Upper Dentdale system

At Ibbeth Peril, a plunge pool below a 5m step, a down-dip passage opens immediately behind the dry fall. Initially a bedding crawl, it gradually enlarges into a vadose trench, 2m high, leading to a very large collapse chamber. The roof spans 40 by 80m, and has broken away in enormous slabs which lie over the floor, 5–12m below. At the far side water enters at roof level from an up-dip, essentially phreatic passage, the top half of which continues across the roof of the chamber to a calcite choke. The stream falls on to the blocks, and down into lower levels, where the normal bedding is disrupted by local dips of up to 28°, shale beds up to 2m thick, and zones of highly shattered limestone. The stream runs through to a roomier passage, trending back towards the river-bed, but, as with every other inlet, it ends in a sump before the main underground flow is reached.

No passable connection is known between the two Ibbeth Peril Caves, even though Number 2, another north bank inlet, has its entrance only a few metres from Ibbeth 1. Ibbeth 2 is however connected to two inlet passages with entrances further down the river-bed, and this whole complex comprises the Broadfield Caves. The passage connecting the inlets, the Old Link Way, is a low, indeterminately wide bedding development running parallel to, and 6m lower than, the surface river-bed. It is a very old, phreatic passage, that was probably the channel of main underground flow before the valley was cut down to give the present resurgence level. The present inlet, Jordan's Passage, is developed in the bed above the Old Link Way, and then it drops into, and continues at, that horizon. In contrast, the main Broadfield Inlet is trenched down below the level of the link passage. Both passages trend down-dip (the link passage approximately follows the strike), and reach the line of massive shales and steeply dipping beds first seen in Ibbeth 1. Both end only 1m or so above resurgence level, in static sumps.

Upper Hackergill Cave opens from the bed of Hackergill, on the south side of the Dale. Two entrance passages quickly trench down to meet a 5m high canyon passage, the main streamway. Downstream it splits vertically, the upper level becoming calcited and blocked. The lower level continues for 250m, to a level 5m below the Dee river-bed where it is choked by collapse and resultant silting. The water continues, to reappear 10m away in the Hackergill Inlet to Broadfield Cave, a continuation of the same passage. It appears that the Hackergill–Broadfield–Ibbeth Peril complex of inlets were developed before the river-bed cut down to its present level, and that water sinking in the river-bed is now modifying pre-existing passages.

Downstream of the Broadfield Caves the surface river-bed drops rapidly through a rocky gorge. Resurgences deliver water from the south bank, and inlet caves (Lower Black Dub, Flood Cavern,

Hell's Cauldron Cave) channel this water under the north bank again. In the Hell's Cauldron area, a change in direction of dip slopes the beds gently into the south bank, and it is probable that the underground flow passes from the north to south side of the Dale below here.

The low flow resurgences—The Popples—eject water from the south bank. A tributary dry-stream bed enters from the south. It originates at Tub Hole, the main flood resurgence for the Upper Dentdale System. In dry weather a boulder slope leads down 3m to a static water level that continues for much of the length of the cave, in a wide passage of over $10m^2$ cross-sectional area. The cave curves round, and ends not far from the river-bed, below Hell's Cauldron, the main passage ending in a large sump. In dry weather Tub Hole is completely by-passed by the drainage, but in flood conditions the normal resurgences back up and overflow into Tub Hole, which fills completely and ejects a flow of 3cumec or more with great force.

The Middle Dentdale System has sinks in the bed of the River Dee below High Chapel (at 723864), and also below Cage in Deepdale Beck (at 718863) in the same bed of limestone (see Fig 41). These sinks are only 3m above the risings, which are both in the Dee river-bed below the confluence with Deepdale Beck, and just to the south at Dent Keld. The latter drain via a separate stream to join the Dee below Dent. Only higher sinks in the How Gill and Yellow Gill may lead to penetrable cave; most of the system is at or below saturation level, and probably approximates to conditions in the Craven area when valley floors were 100–150m higher than present levels.

Outlying areas of significance include the lower reaches of Gastack Beck, where it falls steeply to join Deepdale. A number of caves are known including Robin's Dub Cave. Water sinks through the river-bed in a faulted area, and then flows down-dip in a low-bedding cave. Downstream the water has cut a trench in the floor, and the passage continues for over 300m before becoming impassable. The passage is very finely decorated, and runs for all its length at a higher level than the surface stream-bed; the water eventually resurges in the bank of Deepdale Beck.

The stream issuing from High Gill Cave, above Gawthrop, sinks again on reaching the Great Scar Limestone. It falls 5m through avens into a network of vadose trenches (Nettle Pot Cave) developed entirely at one bedding horizon. The direction of the trenches is controlled by the dip, and also by chert banding in joints, a feature of many Dent caves. The streamway changes from a trench to a low-bedding development at the lowest level, and resurges immediately below the Dent–Sedbergh road.

Future work in Dentdale will reveal more about its speleogenesis. Nearly all the caves have been only recently discovered and the Upper Dentdale System in particular now merits a major study.

13

The Caves of Barbondale and the Dent Fault zone

J. R. Sutcliffe

The short but unique valley of Barbondale lies mainly in Westmorland, whilst Yorkshire claims its upper reaches. In 6km, Barbon Beck descends from 300m (950ft) to 180m (300ft) above sea level, before joining the River Lune near Kirkby Lonsdale. The hills on either side rise to 600m (2,000ft) and the Dale owes its existence to the Dent Fault—a part of the main fault system which defines the western edge of the Pennines.

The line of the Dent Fault is coincident with upper Barbondale but then swings south across Casterton Fell to meet the Craven Faults beneath Leck Fell. Whereas the latter fractures have uplifted the Carboniferous rocks of the Pennine fells, leaving younger sediments on the low ground to the west, the Dent Fault has thrown the opposite way. There has also been considerable compression along this line, so that the Carboniferous strata have been contorted and turned up at their margin. The resulting faulted monocline is complicated by minor folds and fractures and, in Barbondale and on Casterton Fell, the influence of the fault belt on the Carboniferous beds appears to extend for up to 250m from the main fracture.

On the west side of Barbondale, Coniston Grit of Silurian age forms the high ground of Middleton Fell which rises in a very steep and uniform hillside with considerable scree. Making a sharp contrast, the slopes of Barbon High Fell and Crag Hill on the east are gentle except near the bottom, where the upturned beds of the fault belt consist of Great Scar Limestone. The less steep upper slopes are composed of alternating sandstones, shales and limestones of the Yoredale Series, masked to a large extent by superficial deposits of boulder clay and peat, except where these have been removed by the streams. Several picturesque gills have been cut through the lower slopes and the limestone can be seen dipping into the hill at an increasing angle as the fault is approached. Much of the rock is veined with white calcite which has infilled innumerable tension cracks; traces of other minerals, mainly baryte, occur close to the fault and the remains can be seen of small trial workings, presumably in unfruitful searches for lead.

Most of the known cave development of Barbondale is associated with the deep-cut gills. Tectonic action, prior to the ingress of water, has opened the joints and caused minor thrust movements along inclined shale beds. As a result, numerous passages have been readily opened up, many of them very narrow and many to be soon choked. In addition, several long trunk passages of phreatic origin have been found which appear to follow the strike along the minor synclines bordering the fault zone. Away from the fault belt, the general dip of the strata is gently north-eastwards, and this has influenced the drainage from more recent sinks higher up in the Yoredale limestones.

A high proportion of the catchment area of Barbon Beck (called Barkin Beck in its upper reaches) lies on the Carboniferous rocks, rising to the grit-capped Crag Hill. On a map of the area, an unusual feature of the drainage stands out. All the streams are seen to deviate suddenly from their courses straight down the fall line to flow for some distance south-westwards before reverting to their original direction. These SW–NE sections are strikingly parallel and probably reflect the direction of flow of ice during the main glaciation. Morainic deposits aligned with the flow would constrain the post-glacial streams until they were breached by recessive development of the gills on the lower slopes.

Prior to the Ice Age, the early Pleistocene valley followed the weakness of the Dent Fault through the Bull Pot col, and a narrow strip of upturned limestone was exposed along its floor. The highest level in Bull Pot of the Witches may well represent the only known relic cave of this period in the vicinity, pre-dating the whole of the Lancaster Hole–Ease Gill system. During an early glaciation, the breaching of the Silurian ridge of Barbon Low Fell took place, followed by substantial deepening of Barbondale. The valley floor was lowered to about 230m (775ft) above sea level, and cave development appears to have been roughly graded to this horizon during the subsequent inter-glacial period. It has been pointed out that similar levels have been observed on Casterton Fell (Ashmead, 1967) but this may be coincidental.

THE AYGILL CAVES

Bull Pot of the Witches is an important cave system controlled by, and in close proximity to, the Dent Fault. A considerable length of passage is contained within a belt only 40m wide, aligned along the strike, with three main phreatic horizons. The lowest passage transmits the present-day Aygill stream southwards at about 50m depth, whilst the content of fluvio-glacial fill provides evidence to the antiquity of the higher levels.

Several chambers of considerable size have developed in the upper passages where inlets enter from a series of choked sinks. These now carry only superficial, wet-weather drainage but once acted as successive points of engulfment of a major stream. The wide pothole entrance is one such sink which has been enlarged by collapse into a passage 20m below the surface. The small stream, which today occupies the shallow valley leading down from the col, sinks beneath the north wall of the open pot to reappear as a waterfall in a complex chamber below. This is situated at the northern end of Burnett's Passage, an abandoned phreatic tube some 30m below the entrance. On the west side of the chamber, where the bedding of the limestone starts to turn up steeply towards the fault, a narrow descending rift extends northwards along the plane of an inclined shale bed. Much of the rift is heavily calcited and, in a wider section above, the sloping floor has facilitated the growth of some large gours of a depth unusual for caves of this region. Below, the rift opens out into the main active streamway and wall scalloping suggests that the rift once carried an upward flow into Burnett's Passage and thence, still under pressure, to a resurgence initiated during an earlier phase of the cavern's development. Burnett's Passage terminates 'downstream' in a massive choke near to a surface depression, rock-walled on its up-valley side. A large cave mouth may lie hidden here beneath the drift deposits which fill the valley down towards Leck Beck Head, where the water now reappears.

It appears that deep phreatic conditions prevailed for a considerable period during ponding behind restricted outlets to the Bull Pot system. Later, vadose development in the neighbouring Lancaster Hole system facilitated an easterly flow along a bedding plane, resulting in drainage away from the fault zone. 90m of drowned passage separate the open caves and, although no quantitative comparisons have been made, it appears probable that part of the Bull Pot water has found another path. Further phreatic development along the fault belt is almost certainly proceeding and will eventually short-circuit completely the drainage route to Leck Beck Head.

Although considerations of modern hydrology and topography link Bull Pot of the Witches with the caverns of Casterton Fell, geological and morphological controls have provided an affinity with Barbondale. This statement is even more valid in the case of Aygill Cavern which lies on the north side of the surface watershed but constitutes a direct upstream continuation of the Bull Pot system. A distance of 240m lies between the sumps in the two caves with negligible drop in altitude. The Aygill Cavern streamway is a wide, phreatic passage with an arched roof, whilst above it lies an interesting fossil series which extends downstream of the sump before becoming choked. In Bull Pot, however, no comparable upper level has been recognised north of Burnett's Passage. The Fossil series has phreatic origins but later vadose invasion at a number of points has modified and complicated its layout. The passage stems from the upper part of the main active inlet which joins the streamway 140m upstream from the sump. The lower part of this inlet consists of a magnificent swirling canyon 15m high, in which the stream has cut down rapidly in a series of cascades. Several abandoned inlets enter the roof of the canyon and one of these leads back to the excavated sink which constitutes the entrance. The upper levels of the Entrance Series are largely vadose in character and their rapid down-cutting might have resulted because the low-level main drain to Bull Pot was already in existence, fed from sources no longer apparent.

The north-easterly trend of the cavern does, in fact, continue for some distance under Barbon High Fell but there is a change in its character. Beyond a low, silt-floored phreatic tube, in which water backs up in times of flood, lie a series of cavities enlarged along beds inclined at about 45°. Two shattered chambers have developed by blockfall resulting from the undercutting of a shale bed in the synclinal zone where it turns steeply up towards the fault. This shale bed has controlled development in a long inlet which continues the NNE trend of the cave. New Year Passage is a large phreatic rift passage with very slight gradient, which dwindles to a constricted crawl where it approaches the top of the shale. Water has obviously backed up behind the constriction but the meagre, present-day stream carries only a fraction of the flow which once utilised this passage. The stream first trickles from a massive boulder choke which is situated well beneath the overlying Yoredale rocks. On the surface, Hazel Sike sinks into Yoredale Limestone not far away and, although the gentle northerly dip would carry it away into Barbondale, the possibility that a local fault has allowed water to penetrate the sandstones and shales into New Year Passage cannot be ruled out. Beyond the choke, the cave may well continue for some distance along the flanks of Barbon High Fell, guided by the slight syncline adjacent to the fault belt, to originate in inlets from pre-glacial sinks, now hidden beneath the drift. Aygill Cavern has, in the past, drained a much greater part of Barbon High Fell than it does now, despite the fact that the Aygill stream itself probably once flowed to Cow Pot and directly into the Lancaster Hole system. The remainder of the pre-glacial catchment area has since been captured by the gills further up Barbondale.

The main fracture of the Dent Fault lies virtually along the line of Barkin Beck for more than 1km up the dale. The Beck has cut 3m or so into the drift which lines the base of the U-shaped valley, but bedrock is rarely seen in the stream itself. Above the road on the east side, the steepening slope rises to a small scar near the 275m (900ft) contour, and eases above, where a line of shakeholes marks the upper limit of the Great Scar Limestone outcrop. Wet weather run-off from moor-drains soaks away in some of the holes to enter an old cavern system which trends with the strike, approximately 30m beneath.

THE BARKIN GILLS CAVES

The entrance to Barbon Pot has formed by collapse into an aven close to the shale-limestone boundary. The 20m shaft leads to an initial series of chambers with breakdown ceilings and, mainly false, boulder floors. Beneath these, the main passage continues southwards with alternating open sections and collapse areas. It is essentially an abandoned phreatic drain of some importance, and the small modern stream is mainly entrenched in a narrow floor-slot. Remnants of false floor, banks of gravel and the solutional marks of old static water levels indicate a complex history since its initial rejuvenation. Nearly 200m from the entrance, the passage enlarges to a chamber decorated with fine straw stalactites and a well-preserved section of roof spongework, above the floor of blockfall in boulders. A short branch passage, oriented towards the valley, has a large phreatic tube rising from its wall, before it is completely choked. This was possibly part

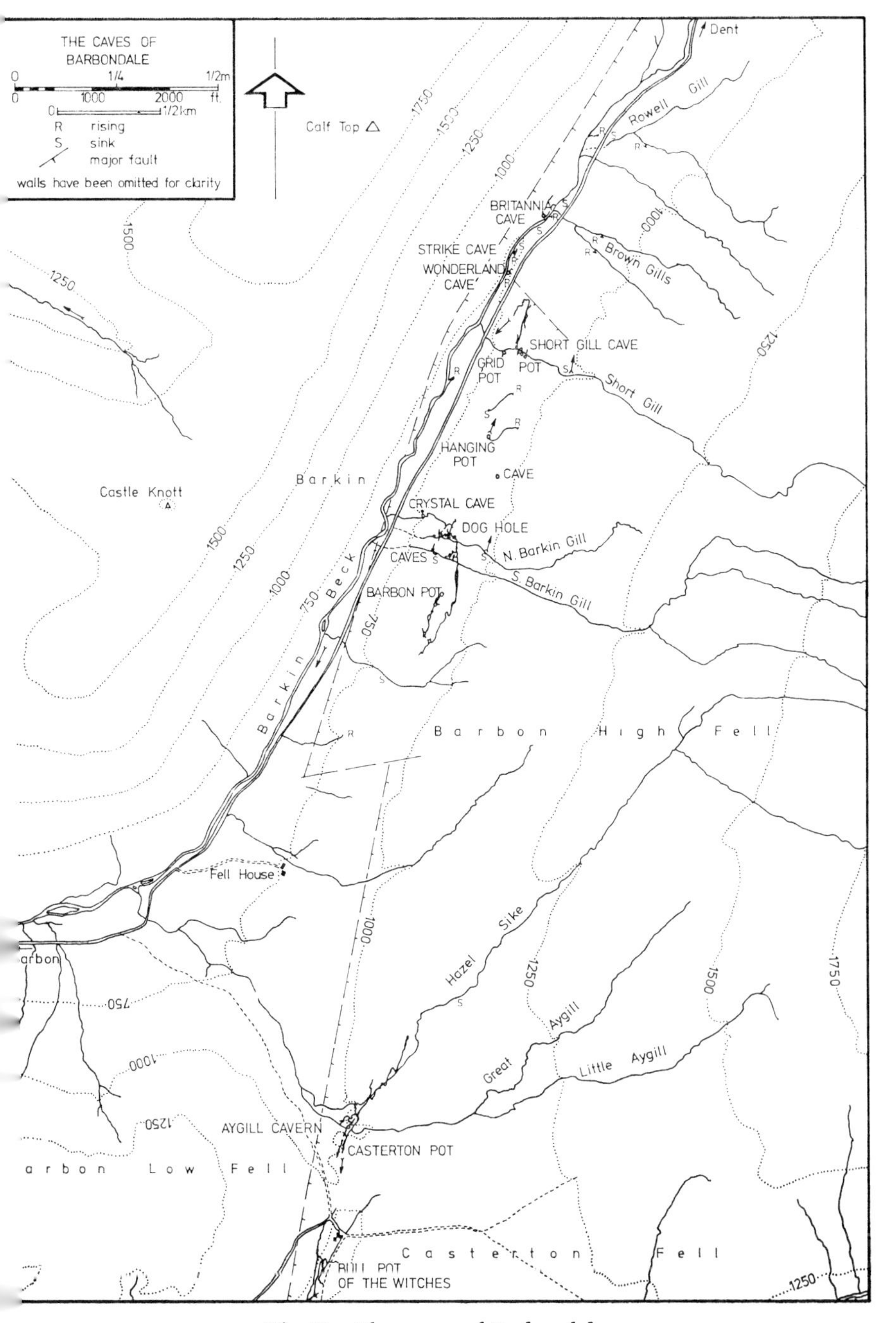

Fig 43 The caves of Barbondale

of the main phreatic outlet for both Barbon Pot and South Passage, in the adjacent Dog Hole system.

Barbon Pot is undoubtedly one of the oldest known caves in Barbondale. Its source and destination are both uncertain and the Main Passage has clearly passed through several phases of development since its phreatic tunnel became established. In addition to the blockfall and minor vadose development which have modified its gently graded profile, there have been periods of total submergence with deposition of silt in a slow-moving flow, followed by removal of this fill by a large, free-running stream. This phase is presumed to have resulted from a lowering of the valley level by glaciation. The inwashed gravel deposits of this period have in places been overlain by further layers of silt and the cave is thought to have been drowned once more when its outlet became blocked by glacial material.

A kilometre north of Fell House, two narrow wooded gills cut steeply through the lower limestone slopes of Barbon High Fell. In their lower reaches, the Barkin Gills are less than 40m apart but both are associated with appreciable streams. Sinks and caves occur and a large resurgence—Crystal Cave—lies at the foot of the hillside not far from the North Barkin Gill. A considerable proportion of the total flow can now be absorbed by sinks which have developed higher up in the Yoredale Series, and this is carried down-dip away from the Barkin Caves catchment to resurge further up Barbondale. The remainder of the flow is absorbed in the middle, incised, parts of both gills by various sinks which fluctuate in their effectiveness. Only on rare occasions do surface streams reach Barkin Beck in the valley bottom.

In the South Gill, Holly Tree Sink has a low, phreatic passage at a depth of 8m which collects water from several vertical rifts beneath the stream bed. The stream drains northwards along the strike and is seen in Barkin Cave below the other gill, and then in Crystal Cave. The upper levels of Barkin Cave consist of quite roomy phreatic cavities following the strike which connected with Crystal Cave in an earlier stage of its development. The entrance is an open hole, 5m above, in the stream bed directly beneath a waterfall.

The obvious entrance to Crystal Cave is situated 100m from the road at the foot of a partially quarried limestone face. 35m of low, phreatic passage runs mainly along the strike of beds dipping at 50° into the hill. The roof lifts abruptly beyond, and upstream is a roomy vadose streamway which descends steadily, against the dip, from the level of a distinct roof passage some 7m higher than the entrance. This upper 'Gallery' has been enlarged by collapse, and extensive deposition of flowstone has followed, amongst which some fine crystal pools are still active. Above the First Waterfall, 60m from the entrance, the stream passage maintains an irregular cross-section, its contorted and calcite-veined rock eroded into numerous hooks, pockets and pendants.

The character of the cave changes once more where the passage turns south-east along the strike and is sumped where water enters from the lower Barkin Gill sinks. This flow is joined by a larger stream which descends two 5m waterfalls from a flooded bedding plane extending towards the higher sinks. The upstream continuation of this main stream can be reached through a high-level connection from Dog Hole, entered from the North Barkin Gill. A dry, constricted inlet, which enters between the two waterfalls, comes from Dog Hole but represents just one relatively brief stage in the development of this complex system. A short section of dry, tubular, passage (North Passage) crosses the streamway above the falls and may be a relic of a much earlier phase—predating Crystal Cave and leading to a pre-glacial outlet at about 230m (775ft) above sea level. Boulders choke the passage not far from a headless dry gully on the surface.

The Dog Hole entrance leads into a rift chamber linking gently inclined bedding planes and washed-out shale beds on several levels. Breakdown has modified the lower parts and stalagmite has blocked the older high-level continuation. Eastwards, a maze of breakdown chambers extend to a larger active passage at

Monsoon Chamber. This shattered blockfall zone, the Jungle, appears to be fairly local in extent and lies between two shale beds dipping gently into the hill and levelling out in the vicinity of Monsoon Chamber. In wet weather, water enters in numerous places from partially choked sinks in the North Barkin Gill.

The main upstream passage continues towards the South Gill sinks where two joint-controlled inlets branch off. The first of these, Foreleg Passage, contains an interesting abandoned level situated close to the Southpaw Sink system which can be compared to the upper parts of Dog Hole in age and character. It consists of a series of abandoned chambers on several levels, the lower ones containing gravel fill and flowstone and the upper ones being shattered blockfall cavities similar to the Jungle. The two zones would in fact appear to be situated in the same beds and roughly on the same strike line. Streams from sinks still active have undercut the chambers and opened up the shale bed which carries the flow through to the present inlets in Foreleg Passage.

It seems probable that Foreleg Passage originally continued downstream in a southerly direction and that capture of the inlets by Crystal Cave followed much later. An underground watershed now occurs in the vicinity of the junction with the second inlet (Hindleg Passage) where South Passage leads forward—downstream. The presence of substantial remnants of a collapsed stalagmite floor, amongst the water-worn boulders in the main streamway between the two inlets, lends support to the theory that South Passage was the former main drain for the South Barkin Gill water. It continues for well over 400m to terminate beyond the downstream end of Barbon Pot. The passage is at a similar level to the latter but lies further into the hill, beneath the impervious cover of the overlying sandstone and shales. Until its further reaches, the gradient is very slight and the cave probably developed in shallow phreatic conditions along the syncline behind the fault belt. For 100m the passage runs in a straight line due south as a broad, flat-roofed gallery. A mud bank slopes from roof level to the stream trench by the left wall and breakdown enlargement has occurred in places. A stream-cut section through a bed of laminated silt has revealed a lenticular deposit of coarse gravel within it, suggesting that a fast-flowing stream occupied the passage for a short time during a period of predominantly slow-moving, deep-water conditions.

A further 100m of gravel-floored phreatic passage leads to a series of constricted joint rifts. Beyond these, the stream cuts into a narrow vadose trench, beneath a wider upper rift. The walls of the upper section are continuously etched into a series of small horizontal ledges and grooves, which represent old static water levels dating to periods of partial drainage of the flooded system. After another 100m, the stream trench continues south where the upper rift turns west and adopts a tubular cross-section before ending in a low choked chamber. A small inlet stream is thought to come from Barbon Pot, the end of which is very close. An apparently extensive collapse area now masks the original continuation of both caves but South Passage may well have turned westwards through the final chamber of Barbon Pot on its way to a pre-glacial resurgence. The present-day streams combine in a twisting vadose passage beneath the choke, ending at a sump after 60m.

THE SHORT GILL CAVES

North of the Barkin Caves, the next hydrological system is associated with the Short Gill Rising—a large, impenetrable resurgence 500m north of Crystal Cave. Lying close to Barkin Beck its flow is derived partly from sinks further up the valley stream but a substantial contribution comes from sinks in, and to the south of, Short Gill. Between there and Crystal Cave are two unusual little hanging valleys which have formed parallel to the Dent Fault along the monoclinal axis where the upturned rocks of the fault belt flatten out. The effect of the fold is greater here than in the Barkin Gills. Ribs of vertically bedded Great Scar Limestone and Yoredale Sandstone are exposed in the ridge bordering the hanging valleys, whilst, on the inner slopes, nearly horizontal Yoredale Limestone is seen. Springs occur here and the southern one is presumed to be fed from the higher sinks in the Barkin Gills, mentioned above. From it, an appreciable stream flows down to a shattered sink at the blind, lower end of the valley, some 50m above Barkin Beck (Hanging Pot). This water follows the near-vertical beds northwards along the strike to reappear at the head of the Main Stream Passage in Short Gill Cave.

Short Gill is similar to the Barkin Gills in that it cuts deeply into the steep limestone hillside, but the dip of the exposed limestone is here much steeper, ranging from 75 to 90°, because of the greater proximity to the fault. The lower beds of the Yoredale Series are also folded and, at the head of the main gill, a thin band of grit is seen as an almost vertical rib projecting from the banks. A considerable proportion of the stream sinks higher up in untilted Yoredale limestone to be taken away northwards to risings in Brown Gills.

Short Gill Pot has formed where a steeply dipping shale band has been washed out in the south bank. A 1m wide rift leads back beneath the stream and descends in steps to a depth of about 35m. Narrow chimney descents alternate with wider sections which extend horizontally to beyond the far bank of the gill. In fair weather, the sinks at Short Gill Pot can absorb all that remains of the flow, leaving the lower Gill substantially dry. The bed of the Gill drops rapidly below the Pot in a series of falls with incut rifts following the bedding into both walls.

The entrance to Short Gill Cave lies in a corner of the north wall where the sides of the Gill converge. Overhangs and projecting flakes make the Gill itself almost cave-like and one slender eroded rock-bridge does in fact make a complete span. Narrow rifts open out, at a depth of 10m, into a low, arched, phreatic passage floored with gravel and mud. Now normally dry, it originated when Short Gill was less deeply incised. A choked rift cave seen in the south bank of the Gill nearby has been just one of several earlier sinks. A flood stream has cut through the inwashed deposits and this channel enters a diverging passage which connects with further flood inlets but becomes choked downstream. In the main passage, which continues northwards, it is apparent that a period favouring the deposition of flowstone has followed the partial removal of the gravel fill. Sections with calcited false floor are interspersed by a series of impressive gours, some shallow and some almost 1m in depth, occupying the full width of the passage. Where a small inlet is met 110m from the entrance, a narrow vadose trench leads shortly to a chamber where the choked upper level comes in from the left. The rock floor slopes down into the roomy Main Stream Passage which carries a large, fast-flowing stream.

The water, entering from a constricted sump a few metres upstream, is presumed to be the combined flows from Hanging Pot and Short Gill Pot. A clean-washed floor descends rapidly past the junction and the downstream passage becomes tall and narrow, controlled by the high-angled bedding. A low oxbow in the roof represents the original continuation of the entrance passage and this fossil level continues for some distance down the cave, bypassing two sumped sections of the streamway. The Main Stream Passage beyond descends steadily and maintains good height through several zig-zags, cutting across the near-vertical beds. Before the Second Sump, the cave has a consistent northerly trend, following the strike in the upvalley direction, but beyond it turns south again. A NW–SE fault has been observed on the surface which would appear to run close to the Second Sump and, in this part of the cave, jointing in that direction can be seen. Beyond the zig-zags, the roof lowers and the fossil level is lost. It is probable that a roof passage turns off along the fault line towards an old, choked, resurgence at about 220m (750ft) altitude, or perhaps to Wonderland Cave which lies alongside Barkin Beck nearby.

For 75m the Short Gill streamway loses height but widens as it runs southwards with little gradient to the Third Sump.

The cave beyond is still unentered but further sumps are likely to occur with increasing frequency as the passage cuts across the beds towards the resurgence. It is possible that a minor fault has been partially responsible for the Third Sump and also the main upstream sump; marked calcite veining is seen crossing the entrance passages where they intersect a line drawn on the survey between the two sumps. This fracture zone is perpendicular to the Dent Fault line.

North of Short Gill, the geology of Barbondale becomes more complex, with subsidiary faulting bringing Yoredale rocks close to Barkin Beck and confining the Great Scar Limestone to a strip less than 100m wide. Cave development in the upper reaches of the dale is associated mainly with Barkin Beck itself. For a distance of nearly 1km, numerous sinks and risings occur and, in dry weather, much of this length of the stream bed is dry, as far down as the Short Gill Rising. Where the NW–SE fault meets the Beck, a spring occurs in the east bank and, just beyond it, in a projecting spur of limestone, is the open entrance to Wonderland Cave. The cave consists of a dry, mud-floored passage turning roughly parallel with the Beck, beneath which a low tube carries the stream. The main passage is choked after 30m, but running eastwards is a mud- and calcite-choked branch which could conceivably be part of an old outlet from Short Gill Cave.

In Wonderland Cave and upstream, the limestone is no longer striking parallel to the Dent Fault but dips at about 50° to NNE. It is possible that lateral displacement of the main Dent Fault line has occurred here because limestone appears for the first time on the west bank of Barkin Beck. Some way upstream is the only open cave system on that side of the stream—Britannia Cave. The stream flowing from it is derived from sinks further up the Beck, and its fissure passage is generally around 2m high for 80m with little rise in level.

Between Britannia and Wonderland Caves, a number of sinks take much of Barkin Beck underground beneath its left bank to resurge in various places downstream. A shallow gorge with small falls occurs on Barkin Beck itself, just above Britannia Cave, and marks a nick point at about 230m (775ft). Above there, where the ground starts to rise towards the Dentdale watershed, the stream has two main branches with no known caves. Small sinks near the col may well drain northwards into Dentdale.

The unique character of Barbondale, with its restricted outcrop of folded limestone along the Dent Fault, has made a study of its caves particularly interesting. The accessible caves can assist in mapping more accurately the geological structures which, in this fault zone, are still far from clear. In addition, the underground evidence is of undoubted importance in assessing the geomorphological history of such an area, and much work remains to be done in the Dale.

14

Development of the Caves of Casterton Fell

P. Ashmead

Casterton Fell is a triangular area adjacent to the Dent Fault, on the extreme western margin of the Askrigg Block. Its limestone is bordered on the north by the outcrops of the Yoredale Series, and the south-eastern side is marked by the valley of Ease Gill. Most of the water in the Gill sinks into a number of caves, and resurges at Leck Beck Head at the southern tip of the Fell. The caves comprise the Lancaster Hole–Ease Gill Caverns system, totalling 27km of interconnected passage, and therefore the longest cave system in England (Eyre & Ashmead, 1967).

SURFACE TOPOGRAPHY

In the north-west corner of the Fell lies the catchment area of Aygill and the cave of the same name. Further south, and adjacent to the Dent Fault, the entrance to Bull Pot stands as a wide open shaft in a shallow, almost dry, valley. Lower down the same valley some surface drainage is derived from the impermeable rocks to the west, and Leck Beck Head is situated almost at the confluence of the Bull Pot and Ease Gill valleys. This major resurgence is the rising for almost all the water sinking on Casterton and Leck Fells.

The main part of Casterton Fell is almost featureless due to the thick cover of boulder clay and peat overlying the limestone. A few shallow valleys carry drainage from the Yoredales into dolines on the limestone; the largest of these is the Cow Pot valley, and below the active sink of Cow Pot, the entrance shaft of Lancaster Hole lies in a small patch of pavement. Ease Gill flows from a peat-covered Yoredale tract, the resistant bands of Yoredale limestone forming small waterfalls. It reaches the Great Scar Limestone at Top Sink Pot. The west bank is featureless barren moorland, whilst to the east several tributaries flow underground, resurging as spouts into the increasing depths of the Ease Gill valley. In normal weather, the river is alive with many small channels of the tributary streams which eventually slip away underground. Along the floor and banks of the valley, the entrance to the tributaries of Ease Gill Caverns lie in close succession—Pool Sink, Borehole, Slit Sinks, Corner Sink and Swindon Hole. At the latter point, the river has cut down below the levels of the tributary caves thus bisecting them, leaving a number of fragmented outliers in the eastern cliff face, opposite County Pot, Oxford Pot and Rosy Sink. A small stream appears, only to descend the dolly tubs for a short distance, and sink in the deep pool of Cow Dubs, below the upper waterfall.

In wet weather Ease Gill is a raging torrent. Top Sink is submerged and the river flows down the valley, pouring into most of the normally dry cave entrances. Only County Pot lies above all the floods.

Below Cow Dubs the valley is wider and more open for over 1km. The entrances to Lower Ease Gill Pot and Kirk Pot both lie down narrow joints in the river-bed. Below the latter, the Gill develops into a narrow incised gorge. This contains a fine series of nick-point waterfalls and has intersected a number of old cave passage remnants. But these are all normally dry and Witches Cave is the highest of the frequently active flood resurgences; from there it is only a short distance down the rocky ravine to Leck Beck Head and the permanent flow of Leck Beck. At the onset of flooding a pulse appears at both Leck Beck Head and Witches Cave immediately after the flood wave reaches Kirk Pot. This does at least indicate the presence of an integrated phreas in this area during high stage conditions, though the hydrology at base flow suggests a number of discrete drainage routes.

THE CAVES

The Lancaster Hole–Ease Gill Caverns system is not only long but also very complex, containing numerous tributaries and different series relating to successive phases of development. The total depth of the system however is little over 100m. The high-level caverns which run almost the whole length of the system include many 15m diameter phreatic tunnels, deserted by the streams but containing large quantities of clastic fill and, locally, fine displays of stalactites and stalagmites. Their floors are commonly masses of breakdown blocks where they have been undercut by the later and lower passages, and, where the undercutting has been more thorough, some large boulder-strewn collapse chambers have developed. Beneath these dry tunnels the active main drain is essentially a deep vadose canyon, in many places over 20m high and mostly 3–5m wide.

One distinctive feature of the system is that the plan patterns of the high-level and low-level passages are almost exactly superimposed. Consequently the degree of undercutting and collapse is very high, and there are numerous connections between the levels, for example at Fall, Stake and Stop Pots. Both levels have dozens of inlet passages originating from the upper reaches of the Ease Gill valley. Many of the active, low-level, passages are open and provide the present entrances; in contrast the easterly ends of the high-level tunnels are mostly choked with clastic sediment. Further west the two main trunk passages extend to Stake Pot, together with various other levels of development (see Fig 45).

South of Stake Pot, the Earby '71 Series extends as far as the Ease Gill valley, but so far without any connection to the surface. These passages were only discovered after this account was written, so they are not considered below. They do however appear to continue the complex multi-phase pattern of cave development exhibited elsewhere in the system.

Almost below the Lancaster Hole entrance—a 30m shaft—the main drain sumps just after it is joined by the Bull Pot water from the north. It is next seen at Leck Beck Head. The high-level passages however continue further, both north towards Bull Pot and south towards the Graveyard (see Fig 46). Their main continuations are lost behind boulder chokes and mountains of inwashed sediment.

GEOLOGY

The caves of Casterton Fell are developed entirely in the Great Scar Limestone. The stratigraphy of the limestone is reasonably typical of the whole of the Askrigg Block, except that there appear to be fewer significant shale bands on Casterton Fell than elsewhere. Though several marker bands can be recognised, the Porcellanous Band has not been traced into the area. The *Cyrtina septosa* and *Chonetes comoides* Band outcrops at about 15m below the top of the D_1 zone, and forms a significant feature opposite County Pot, giving the beds a fissile structure.

The Carboniferous rocks are cut off on the west by the Dent Fault zone, which is a faulted monocline, and the rocks are tilted at a high angle along a narrow zone to the east of the fault. The fault cuts across the general strike of the beds and it affects different horizons. Parallel to the Dent Fault are small flexures and tear faults, which can be seen underground in Bull Pot and elsewhere. At Aygill the flexures affect the beds up to the Hardraw Scar Limestone.

The major joint direction is 310°, whilst a secondary series is aligned at 315°. Near the Dent Fault a third series is aligned at 350°. Each series have smaller bunched joints at right angles, whilst several tears and faults parallel to the Dent Fault are found in Bull Pot and Aygill Cavern.

GEOMORPHOLOGY OF THE SURFACE

The present work has entailed the examination of erosion levels

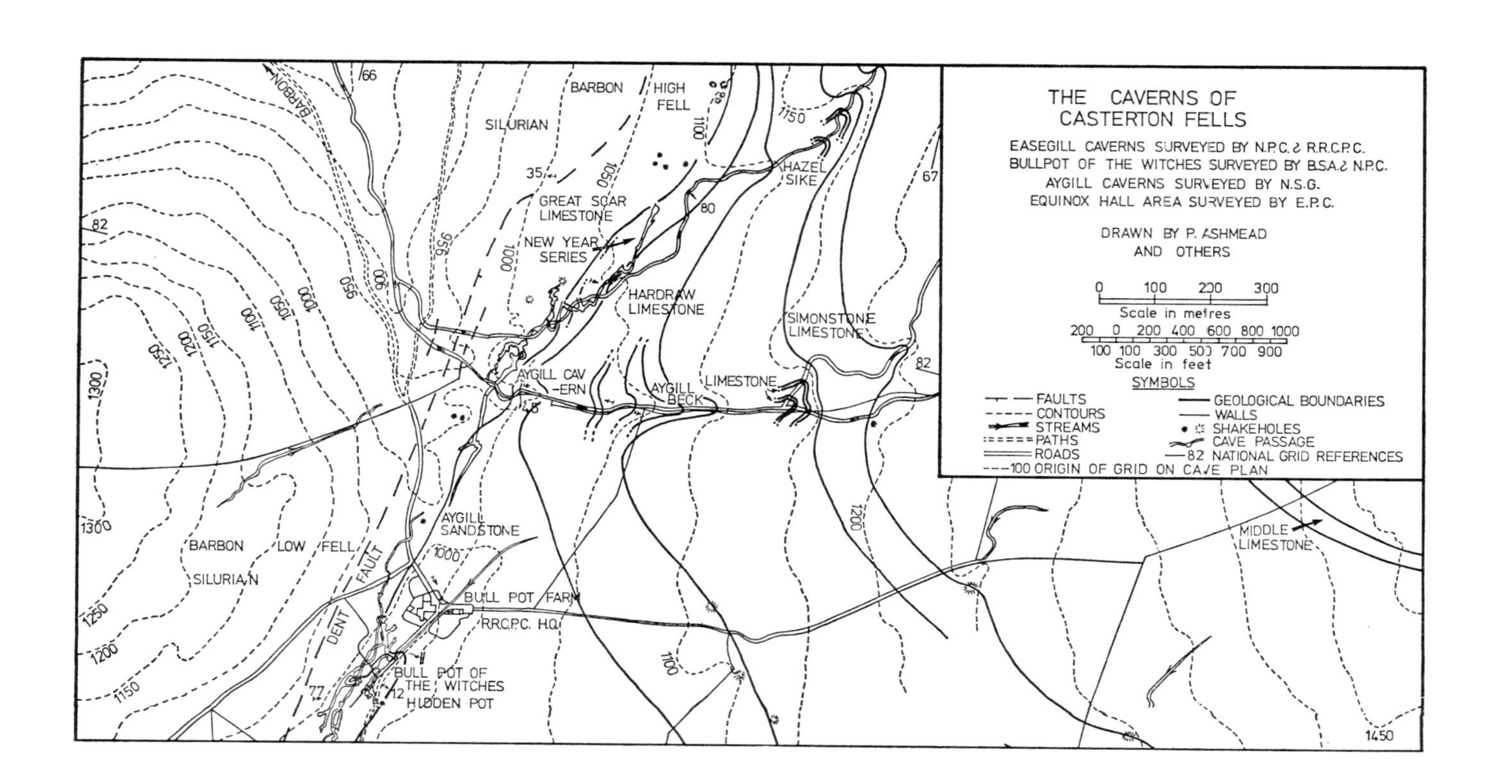
THE CAVERNS OF
CASTERTON FELLS
EASEGILL CAVERNS SURVEYED BY N.P.C. & R.R.C.P.C.
BULLPOT OF THE WITCHES SURVEYED BY B.S.A. & N.P.C.
AYGILL CAVERNS SURVEYED BY N.S.G.
EQUINOX HALL AREA SURVEYED BY E.P.C.
DRAWN BY P. ASHMEAD
AND OTHERS
0 100 200 300
Scale in metres
200 0 200 400 600 800 1000
100 100 300 500 700 900
Scale in feet
SYMBOLS
FAULTS
CONTOURS
STREAMS
PATHS
ROADS
100 ORIGIN OF GRID ON CAVE PLAN
GEOLOGICAL BOUNDARIES
WALLS
SHAKEHOLES
CAVE PASSAGE
82 NATIONAL GRID REFERENCES
BARBON
SILURIAN
BARBON
HIGH
FELL
GREAT SCAR
LIMESTONE
NEW YEAR
SERIES
HARDRAW
LIMESTONE
HAZEL
SIKE
SIMONSTONE
LIMESTONE
AYGILL CAV
-ERN
AYGILL
BECK
LIMESTONE
AYGILL
SANDSTONE
BARBON
LOW
FELL
SILURIAN
FAULT
DENT
BULL POT FARM
RRCPC HQ
BULL POT OF
THE WITCHES
HIDDEN POT
MIDDLE
LIMESTONE
66
67
82
82
80
35
77
12
1300
1250
1200
1150
1100
1050
1000
950
900
956
1750
1450

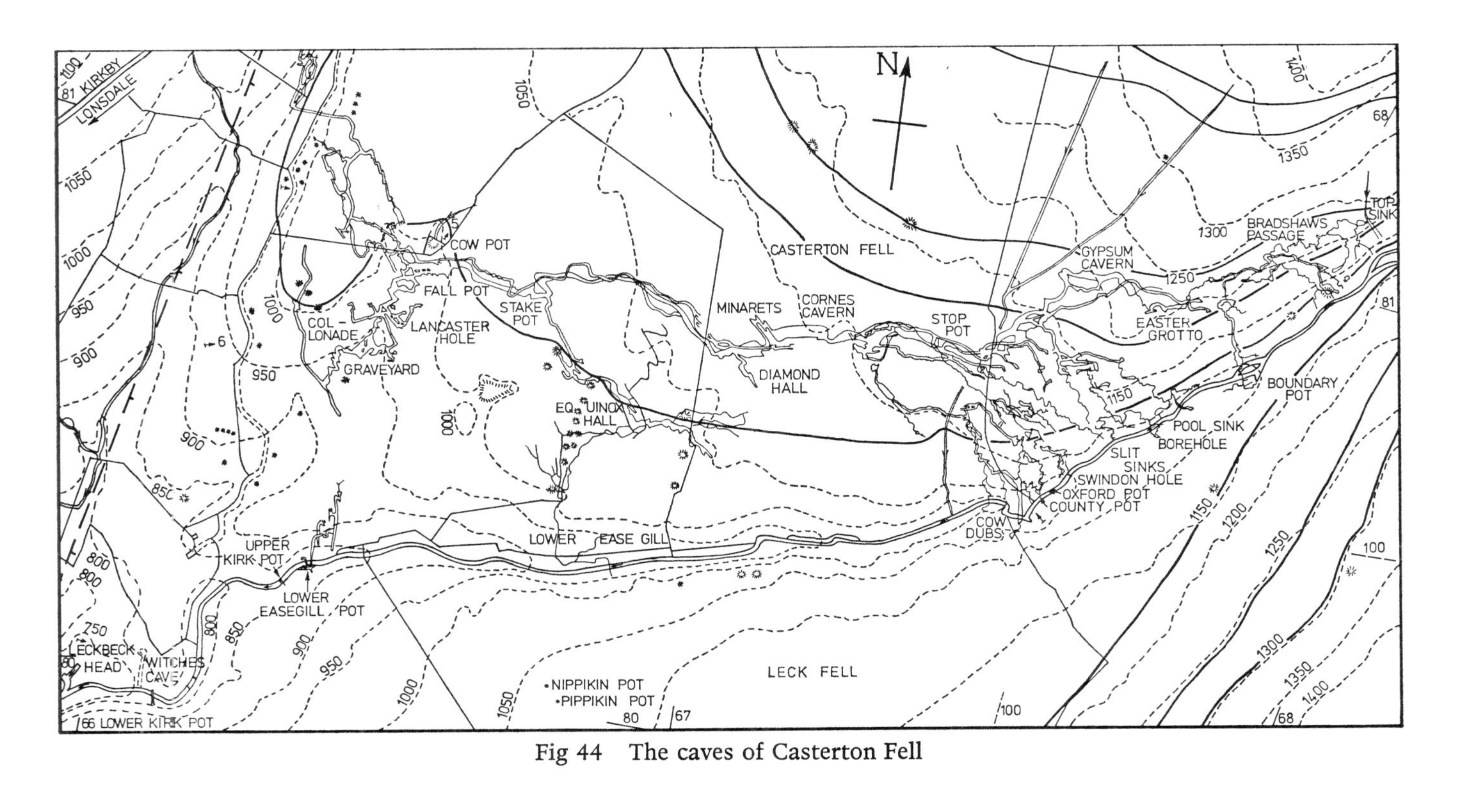

Fig 44 The caves of Casterton Fell

and geological mapping to eliminate structural coincidences. In an area composed of gently inclined strata some difficulty has been experienced in separating these accidental benches of structural origin from normal erosion facets. However, they are seen to cut across the steeply dipping rocks wherever they occur, and can be followed across the Dent Fault to the folded Silurian rocks. The erosion surfaces and related features coincide with those found in surrounding districts and with such a widespread occurrence it is fairly certain that they are the same. Comparison may be made with the Lake District (Parry, 1960; Hollingworth, 1937), Howgill Fells (McConnell, 1940), and north-west Yorkshire (Sweeting, 1950).

The highest erosion surfaces are well above the outcrops of the limestone and therefore unconnected to the karst development. A compound surface between 380 and 480m (1,250 and 1,600ft) occurs on Castle Knott, Barbon Low Fell, Brownthwaite Pike and on the north-west flank of Crag Hill. Several spurs of a fragmentary nature occur at intermediate heights, and it has not been possible to distinguish two main surfaces. In Barbondale, glacial action has removed any trace on the west, and on the east the surfaces are largely controlled by resistant Yoredale rocks. In the Lune Valley, spurs and steps at this height cut dipping strata. Rejuvenation stages are represented by a series of benches and waterfalls at 350–370m (1,150–1,200ft). In both Ease Gill and some tributaries to Barbondale, benches and nick points occur at 360m (1,175ft), though local structural geological controls diminish their significance. An extensive bench on Thorn Moor has its inner marginal break of slope cutting across Silurian rocks at the same altitude.

Many remnants of a widespread erosion surface are found at 320m (1,050ft). The cols at Bull Pot and at the top of Barbondale occur at this altitude, whilst shoulders between Garsdale and Dentdale, the Lune and Leck Beck, and Leck Beck and Kingsdale also have extensive steps at the same height. Parry (1960) describes it as an Early Pliocene partial peneplain.

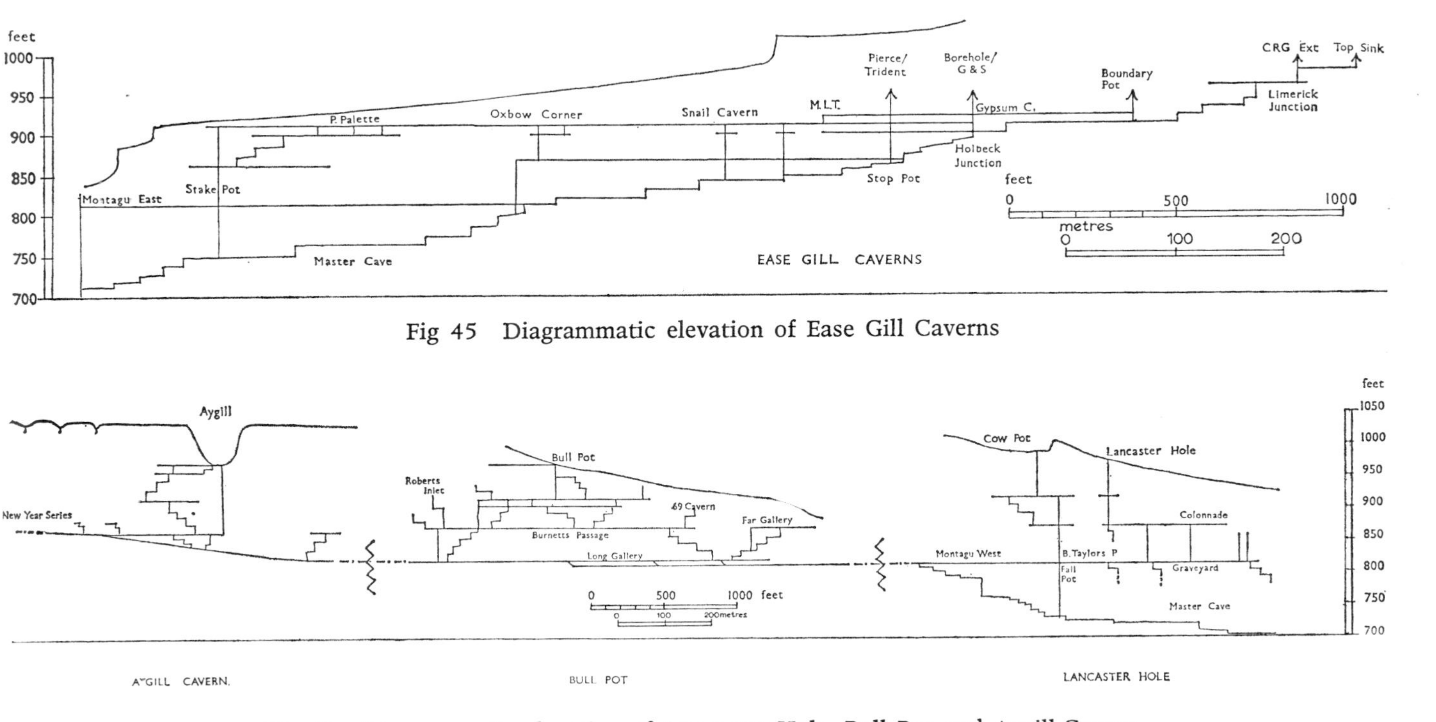

Fig 45 Diagrammatic elevation of Ease Gill Caverns

Fig 46 Diagrammatic elevation of Lancaster Hole, Bull Pot and Aygill Cavern

Several minor stages of rejuvenation are found in the form of benches and nick points, some of which are fossilised. The two most consistent and widespread are at 280 and 265m (910 and 870ft). Fine rejuvenation waterfalls are present in Barbondale, while a series of nick points occur in Aygill. A fossil nick point occurs in the Cow Pot Valley at 280m (910ft), and a buried nick point can just be discerned, in process of exhumation, in the bed of Ease Gill below Cow Dubs, at approximately 275m (900ft).

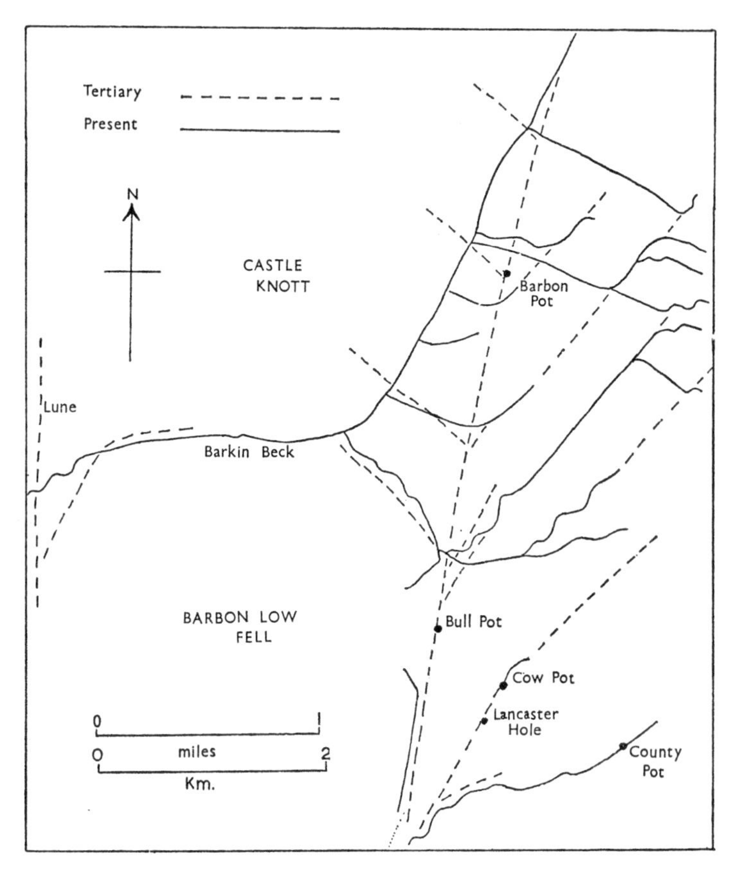

Fig 47 The Tertiary and modern drainage patterns of Barbondale and Casterton Fell

Extensive changes in the drainage pattern have taken place in the time interval between the development of the surfaces at 320 and 250m (1,050 and 825ft). River capture occurred on a large scale, Barbon Beck having been captured by the Lune and Aygill captured by Barbon Beck (see Fig 47). It is not known how many of these changes have been caused by glacial diversion, but the present patterns were certainly in existence prior to the latest glaciations.

The 250m (825ft) surface is an important one recognised in the Yorkshire Dales and elsewhere. It is found as an extensive bench in Barbondale and also as a bench at Leck Beck Head. It maintains a constant horizon in lower Leck Beck, above Springs Wood, over dipping Silurian rocks, and can be followed across the Lune Valley to Farleton Knott and Hutton Roof. A series of nick points, many marked by waterfalls, are found below the surface, extending down to the 180m (600ft) surface which can be followed down the Lune Valley.

No general agreement has yet been reached concerning the age of these various erosion surfaces, though those above the 250m level are considered to predate the main glaciation of the area.

There is a similar lack of absolute dating possible for the succession of glacial advances. Some glacial features may be ascribed to a relative succession, but in general the later glaciations have removed all traces, on the surface at least, of the preceding ice advances. Features due to the successive glacial advances within the Devensian include boulder clay up to altitudes of 440m (1,440ft), a distinct notch in Barbondale at 305m (1,000ft), and kettle moraine in the lower Leck Beck valley.

Fluvioglacial terraces can also be seen where the rivers have cut down through the extensive boulder clays on the retreat of glaciation in two, or possibly three, stages. Three terraces occur at Leck Beck Head, one of which rests on the 250m (825ft) surface. Each of the terraces can be followed to the Ease Gill Kirks.

On Casterton Fell, the two major erosion levels to affect the cave development form the 320 and 250m (1,050 and 825ft) surfaces. The higher of the two is found at the Bull Pot col where it cuts across steeply dipping *Lonsdalaea* Limestone and Aygill Sandstone. Deserted meanders and oxbows of the three Aygill streams are common, and are now cut down into gorges as a result of the rejuvenation to the 250m (825ft) surface. An important feature is the occurrence of an abandoned, captured, watercourse leading from Aygill towards Bull Pot.

On the fell, the surface is heavily drift covered, but in between the south-westerly trending trails of boulder clay, solid rock is exposed at several points where post-glacial streams have cut down, as at Lancaster Hole and Cow Pot. Elsewhere the drift is exceptionally thick (about 15m in places). It is significant that the present streams are in process of removing the boulder clay which plugs the pre-glacial valleys originating in eastern Barbondale. This gives rise to the possibility that pre-glacial sinks in these valleys may become exposed.

The 320m (1,050ft) surface may be seen at Howe Gill where the stream has removed the boulder clay. A waterfall marks the retreating nick point. The surface is visible directly above County Pot where it is overlain by more than 10m of boulder clay. It is evident that pre-glacial swallow holes are here also buried beneath a thick mantle of boulder clay. Oxford Pot contains the remains of an old fluted pothole directly below the rock surface. The 320m surface lies 30m below the top of the limestone here, and continues across to the benches of Leck Fell which contain the entrance to Pippikin Hole.

The 250m (825ft) surface is visible in the region of Hellot Scales Barn and Leck Beck Head, where it is dissected by deserted meanders of the Bull Pot valley and Ease Gill streams. Several blind headed valleys nearby suggest the possibility that the older fossilised resurgences of Ease Gill lie buried here. The surface is formed again above the Lower Kirk waterfall where Upper Ease Gill widens abruptly. The waterfalls are youthful rejuvenation features and have cut through fossil caves in several places.

The same erosional pauses and changes which led to the formation of these surface levels had a profound influence on the patterns of cave formation.

During the Pleistocene, the phases of cave development were contrasting and alternating. Melt-waters washed out earlier glacial infills and in some cases re-invaded the old cave passages; elsewhere complex changes in the drainage patterns developed during the interglacial periods. These successive stages may be distinguished by erosion of clastic cave-fills, infilling of channels in the fills and the alternation of stalagmite horizons with layers of sediment.

CAVE DEVELOPMENT IN CASTERTON FELL

Both Simpson (1935) and Myers (1948) reasoned that active explorable caverns contain predominantly vadose features. Myers thought that the water-filled phreatic features were developed at or just below a water table and not in a deep-seated environment. Atkinson (1936) postulated that Ease Gill caverns were predominantly of vadose origin, and that large caverns were the result of a mechanical breakdown as opposed to control by a surmised water table. Myers (1948) felt, however, that some high-altitude caves of phreatic morphology must have originated when the water table was very much higher, perhaps before glacial deepening. Sweeting (1950) extended these ideas and presented a theory of horizontal phreatic development at several successively lowering water tables. Waltham (1970 *b*) described several morphological features and separated these into a pre-main-glacial phreatic phase and a post-main-glacial vadose phase. The writer, whilst concurring with the latter, believes that in the Casterton Fell systems the sequence of development is capable of further subdivision, and has already postulated several pauses in the development and stages of infill, and attempted their correlation with stages of subaerial or glacial erosion (Ashmead, 1967). Evidence is given below to show that two major cycles of development have occurred in the Casterton Fell caverns. The first cycle developed from a series of sinks on the 320m (1,050ft) surface to a base level of

250m (825ft). The second cycle developed as a re-invasion and modification of the first cycle, down to its present base level.

In the Lancaster Hole and Ease Gill caverns system, where the bedding is nearly horizontal, the phreatic development is also predominantly horizontal. Along the line of the Dent Fault, where the bedding is locally steeply dipping, phreatic development is relatively sharply inclined, showing that water has been lifted to higher levels by way of series of now fossilised passages. Thus, in both cases there is geological control. But it is significant that in the latter case the water has been lifted to the same horizontal phreatic horizons as in the former case, demonstrating that geological factors alone are insufficient to explain these features.

Aquifuges such as shale bands may only be of temporary duration and Atkinson (1963) showed how shale bands impede vertical

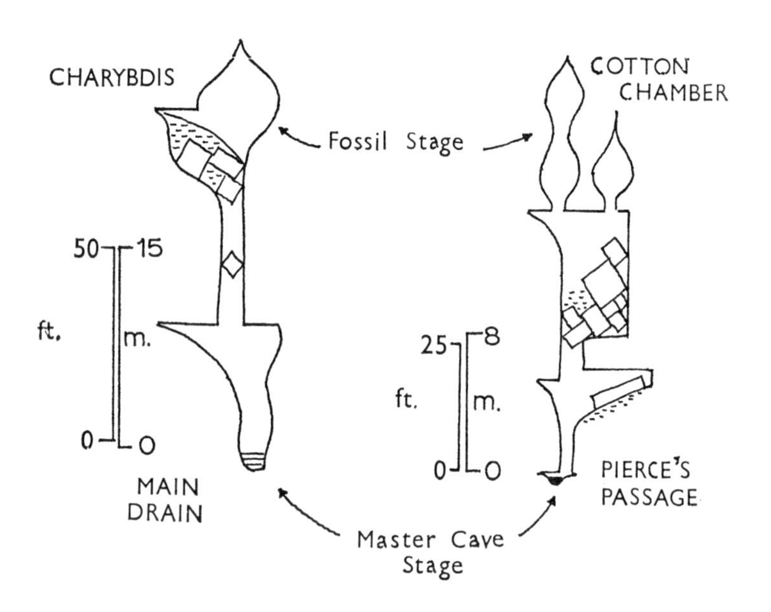

Fig 48 Cross sections through the multi-level passages at Charybdis, on the main high-level route, and Cotton Chamber, above Pierce's Passage

erosion. The blockfall is to a large degree due to undercutting on shale/limestone interfaces (Fig 48).

Vadose morphology

Below primary tubes and network phases, canyon forms cut through several beds, denoting continuous free downward movement. The Main Drain at Lancaster Hole is a fine example of vadose development with high- and low-level oxbows, cascades and dolly tubs (round rock basins connected by chutes of swift flowing water). The active Ease Gill inlets also present model vadose features notably out of phase with their high-level fossil counterparts, whilst the Bull Pot Main Drain shows much the same morphology at Waterfall Passage and Wilf Taylor's Passage. Fossilised shafts are often fluted showing former vadose erosion. The shafts mostly drop into high, rift-like, passages, and there may be a series of fossil pitches extending back upstream to the present active passage; this may descend by only a series of small steps

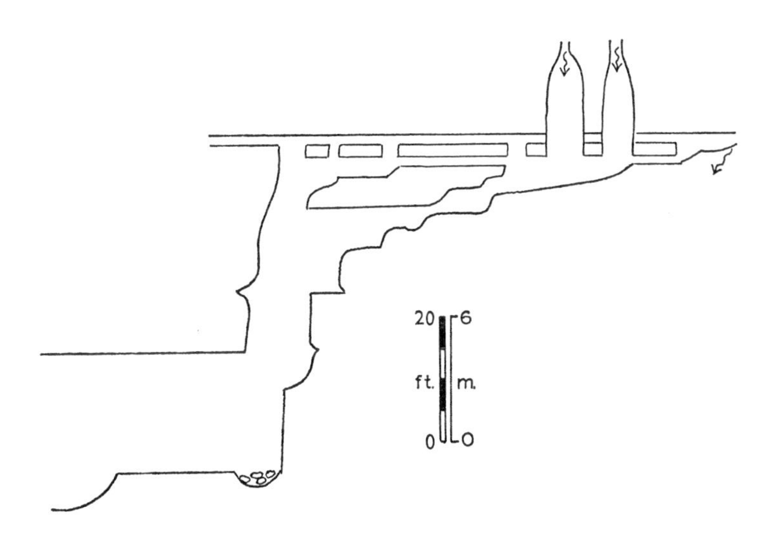

Fig 49 Retreat stages of vadose waterfalls in the Borehole

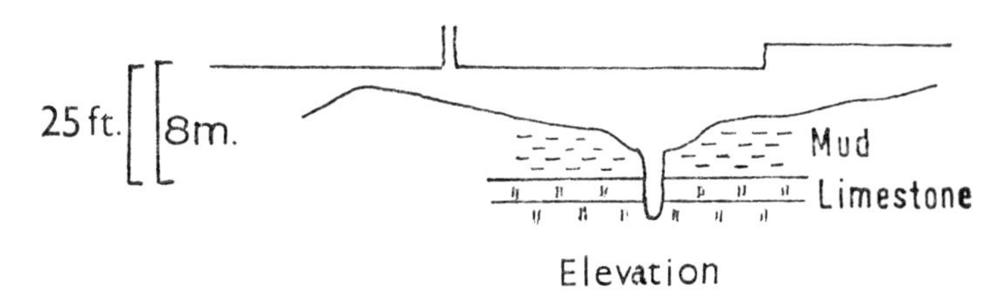

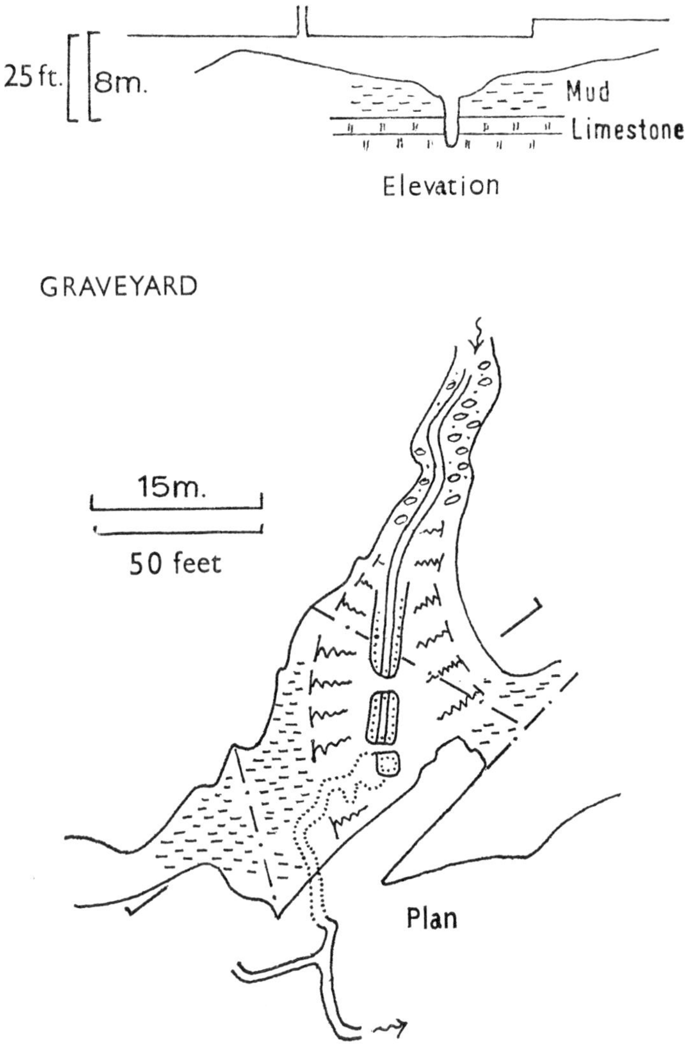

GRAVEYARD

15m.

50 feet

Plan

Fig 50 Plan and section of the multi-phase development in the Graveyard, Lancaster Hole

or cascades, whilst the upper abandoned level may continue for some considerable distance. This can be seen at Pierce's Passage

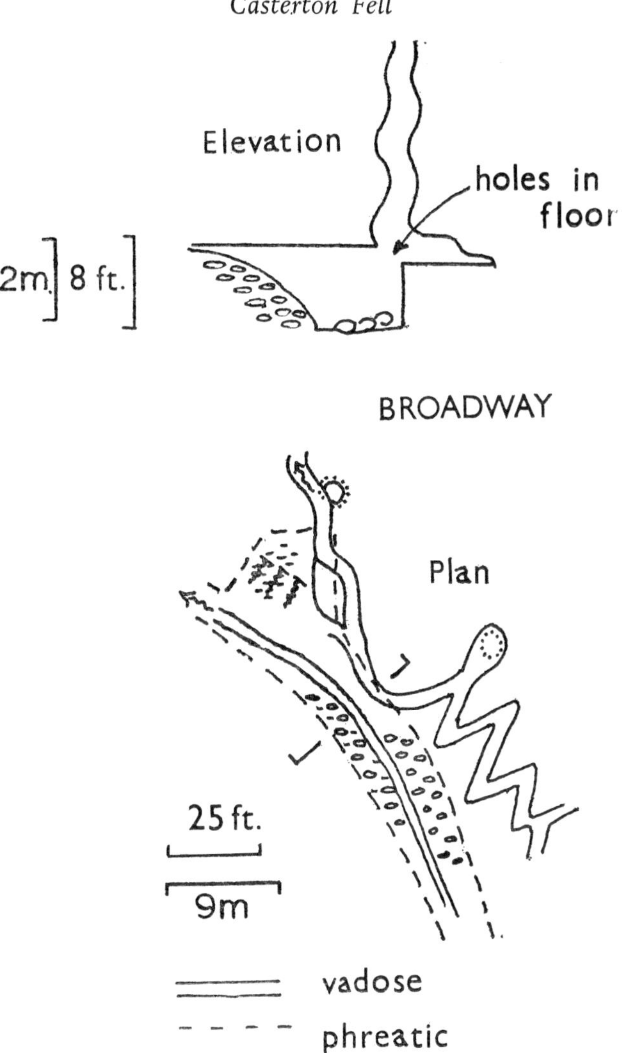

Fig 51 Plan and section of Broadway, in County Pot, where it is intersected by a younger vadose canyon

between Poetic Justice Pitch and Trident Passage where the upper level continues fossilised to Cotton Chamber Series and the Trident Roof Series (see Fig 48).

All the normal vadose features are to be found. The dominant forms are high-level meanders, as at Limerick Junction, incised meanders, and superimposed passages, ie bedding plane passages connected by narrow joint mid-sections. Many passages with vadose morphology show that they are actively re-invading existing passages, as in the Graveyard, where the new stream has cut down through the sediment infill and into the underlying limestone (see Fig 50). At County Pot the invading stream has successively opened joints through into Broadway, and abandoned its former course. Several examples of superimposed tubes occur in 'T' Piece Passage, where the upper primary phreatic tube has been cut down via a joint to a lower bedding plane; erosion has then proceeded laterally, possibly along a shale band. Commonly, the eventual outcome is breakdown of the wall blocks on either side of the mid-section. Both active and fossil vadose passages show these features. Large-scale breakdown occurs at Stop Pot, revealing complex superimposed levels; this is no doubt due to the vigorous undercutting of the Main Drain in more than one stage. Blockfall is a characteristic feature of the High Level Route where infilling has occurred after breakdown has taken place (see Fig 48).

Phreatic morphology

The caves contain many classic features of phreatic morphology, such as minarets, low-arched roof passages, tubes, solution rifts, rock pendants and anastomoses. These may occur in individual trunk routes or horizontal networks. Distinctive series of these form the downstream continuations of some fossilised vadose passages. The change from the vadose environment to the phreatic environment is seen where the lower ends of vadose canyons lead into wide low-arched phreatic passages with rock pendants in various stages of erosion, the most heavily eroded being upstream. Several of the phreatic trunk routes and networks are filled, to a greater or lesser degree, by extensive glacial sediments which would appear to have arrested cavern development.

Cycles of development

Small-scale morphological features (Bretz, 1942) have been used to separate the Casterton Fell caverns into distinct systems which show two major cycles of development.

Many of the abandoned upper levels show an initial primary tube and joint phase, subsequently entrenched by vadose waters. The former water-rest level, the former resurgence level, is marked by a change to phreatic morphology at 250m (825ft) in Lancaster Hole. The fossil resurgence at this altitude may exist near Hellot Scales Barn where there is a blind-headed valley plugged with glacial deposits.

In Bull Pot of the Witches, deep-seated phreatic waters appear to have ascended to this altitude from the Gour Chambers, but they have, in earlier times, ascended to higher altitudes in several stages. The highest, at 32 Cavern, may possibly have emerged at the head of the valley below Bull Pot Farm. Succeeding lower levels developed along Burnett's Passage and Long Gallery, the

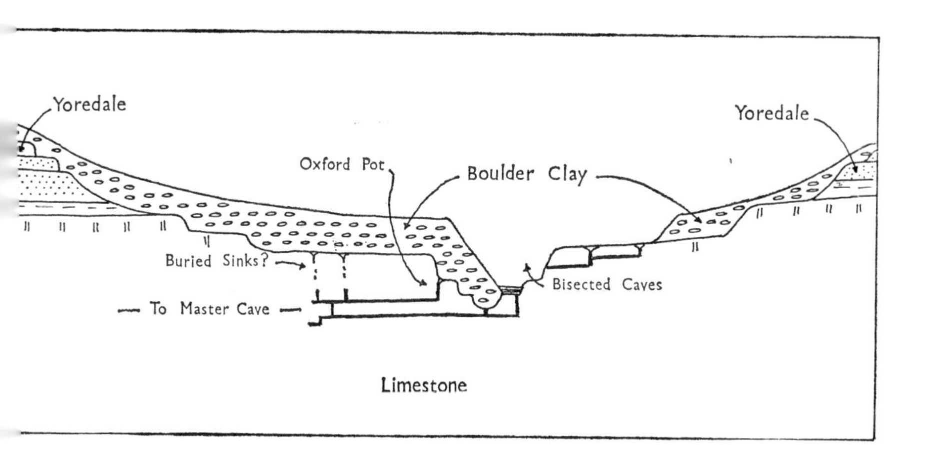

Fig 52 Section through Ease Gill at Oxford Pot, showing glacial fill, the truncation of old cave passages, and the rejuvenation of the active sinks

water eventually finding a more direct horizontal path at the 250m (825ft) level—the Dry Way.

In the Ease Gill caverns, the initial primary feeders from the area near Cow Dubs converged to form a main trunk route carrying the water to Lancaster Hole at the 250m level. This extensive phreatic development at 250m (825ft) suggests an erosion base of considerable duration.

The occurrence of extensive glacial sediments in these fossilised passages support the theory that they were in existence prior to the main glaciation. The sediments themselves show considerable re-sorting and variations due to climatic change; gravels, laminated clays and stalagmite layers occur in alternating layers. Taken together with the existence of two consistent nick points above 250m, both on the surface and underground, this would suggest that this first cycle of cavern development occurred in stages which were climatically controlled. During this long period of time, the limestone/shale boundary was receding, and rejuvenation extended the limestone outcrops upstream beyond the Cow Dubs area, thus extending headwards the cavern development in Ease Gill.

The characteristic feature of vadose development within this first cycle is the development of bedding and joint-determined passages. Wide flat-roofed caverns have developed where roof and wall sections have peeled away; the fallen blocks are commonly overlain by sediments. The phreatic development of low-arched passages are extensively affected by clastic infilling; awaiting discovery must be many more passages of similar morphology which will eventually relate the caves of Casterton Fell to those in adjacent areas.

The second major development cycle was essentially vadose in character. The passages very rarely contain sediments; only a few stream-deposited gravels and flood clays have been formed. The rejuvenation which initiated this second cycle has resulted in either the modification of existing caves or the formation of completely new channels. The youngest of the nick points is still in

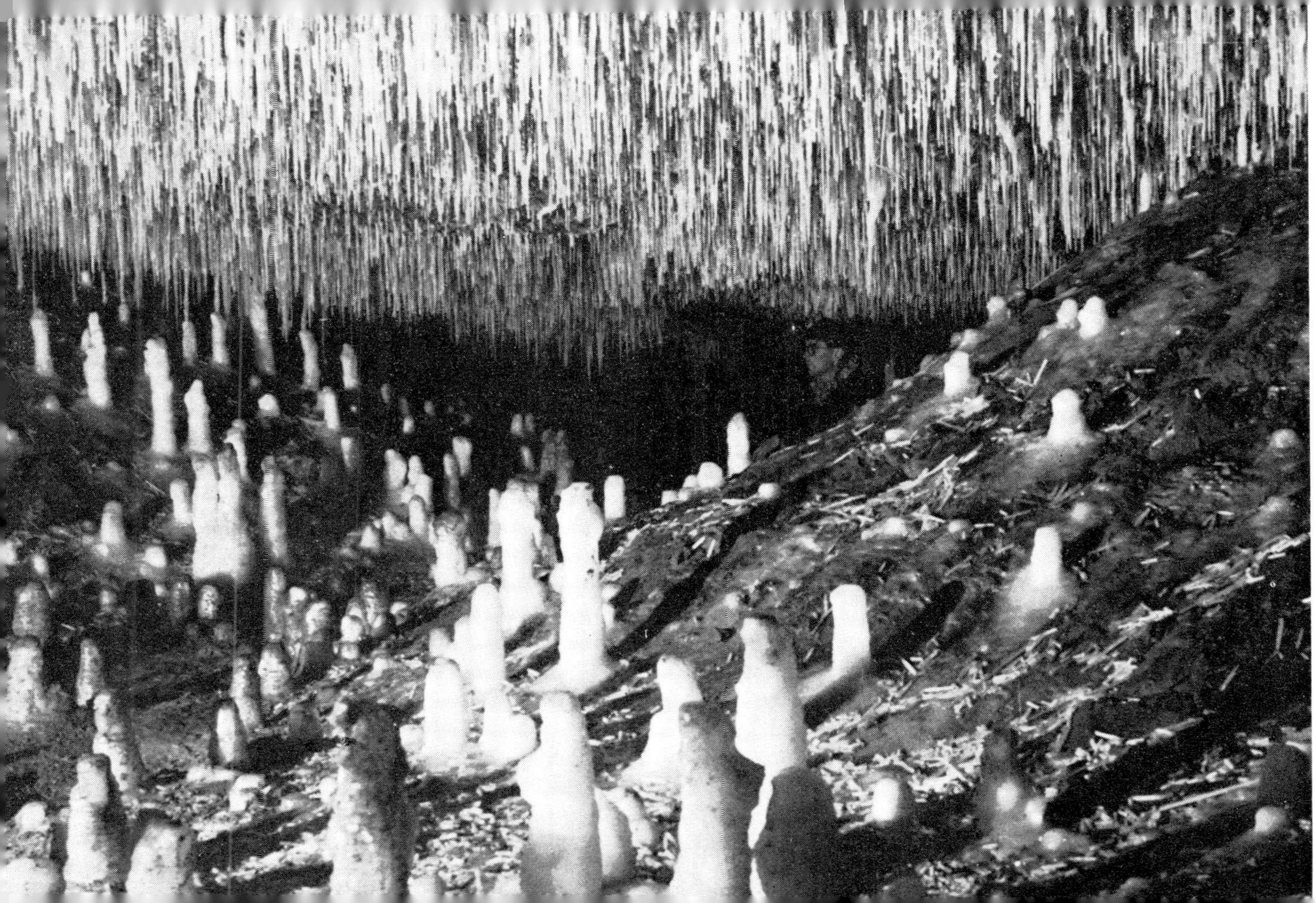

Easter Grotto in Ease Gill Caverns

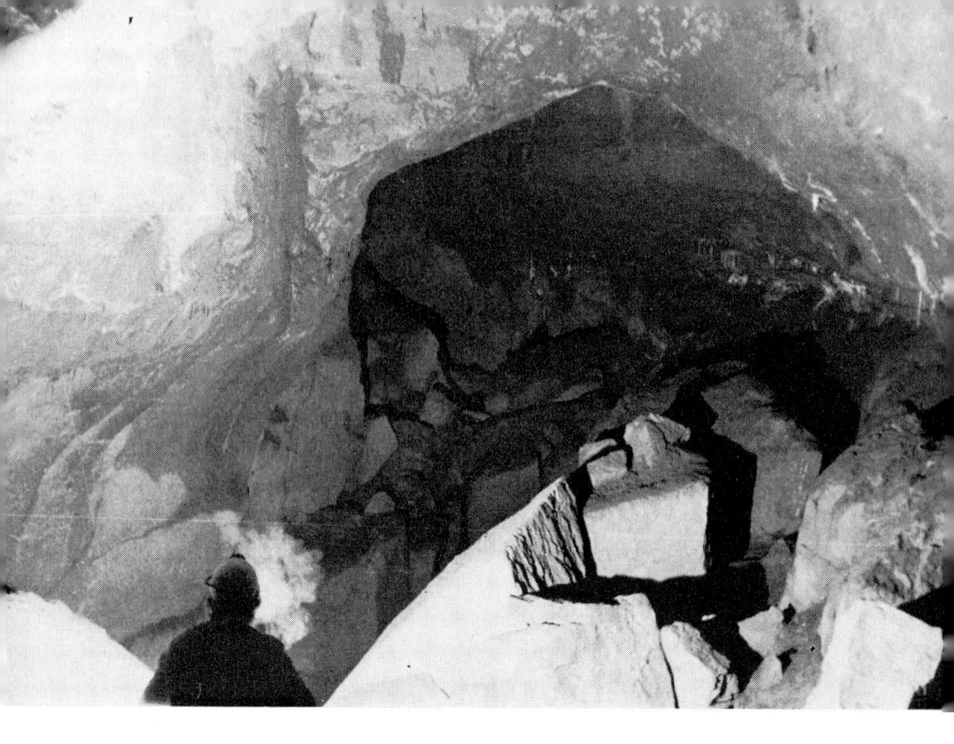

Page 270

(*above*) The main tunnel of Montague East Passage in Lancaster Hole. The phreatic half-tube roof spans a mass of collapsed blocks and even the lower parts of the walls are solutionally cut and not marred by breakdown (*below*) The shallow elliptical opening in the shale bed horizon forming the roof of the New Roof Traverse in Lost Johns system

process of advancing upstream from the Lancaster Hole sump along the various active systems. It has not quite yet reached Bull Pot, as Waterfall Passage takes the bulk of the water from Bull Pot, Wilf Taylor's Passage acting only as an overflow during wet conditions. This rejuvenation is further complicated by deserted oxbows in Waterfall Passage.

The Main Drain in Lancaster Hole carries the Ease Gill waters westwards, and rejuvenation appears to have occurred in three successive stages; this has affected both the surface and the underground. As Upper Ease Gill cut down, the old systems have become abandoned and in some instances, as near County Pot, they have been bisected by the gill itself. New sinks have also developed, the waters seeking fresh routes into the caverns below (see Fig 52).

Prominent nick points within the caves are located at Oxbow Corner, Holbeck Junction and Limerick Junction. Between these points the inlet systems show similar development sequences of the Main Drain (Eyre & Ashmead, 1967). The number of consecutive stages of development within the inlets decreases upstream of the nick points; the active passage above a nick point becomes the fossil passage below. Above Limerick Junction, the passages show two phases, a fossil phase and an active phase, for example, the CRG Extension and the Booth-Eyre Crawl. Between Limerick Junction and Holbeck Junction, two fossil series occur above the active stream passage: Nagasaki Passage lies above Old Limerick Passage, in turn above Far East Passage. Below Holbeck Junction the inlet series show three fossil stages above the present active route: all are present within the Trident Series.

SUMMARY

It has been shown that the Casterton Fell caverns have been developed in two major cycles. The first cycle has developed in stages graded to a water-rest level of 250m (825ft) and this correlates with the 250m erosion surface, which, on the evidence available, is pre-glacial. Phreatic lifting passages in Bull Pot, and temporary phreatic features in the fossil vadose high levels of Lancaster Hole/Ease Gill caverns, provide evidence for the existence of older, higher, base levels. The second cycle commenced after the main glaciation of the district, and included at least three stages of rejuvenation.

The factors controlling cave development have, therefore, been geomorphological, as well as geological.

15

Speleogenesis of the Caves of Leck Fell

A. C. Waltham

Comprising the western flank of Gragareth, Leck Fell contains a magnificent collection of caves and potholes. Almost in the centre of the Fell lie the entrances to a whole group of interrelated caves, of which the Lost Johns system is the most extensive and best known. A kilometre to the north (see Fig 53) the complex system of Pippikin Hole lies under the margin of the Fell, adjacent to the Ease Gill valley. A similar distance in the other direction, Notts Pot and Ireby Cavern underlie Ireby Fell, the southerly continuation of Leck Fell. In many ways these three separate groups of caves are related, and their patterns of development bear distinct similarities to each other, but it is the caves of the central area only with which this chapter is particularly concerned.

In a block of limestone less than 1km square, central Leck Fell contains nearly 10km of known cave passage. Yet there are indications that probably an equal length of cave still awaits discovery. Consequently there are many unsolved problems concerning the origins of even the known caves. All the known passages have been surveyed (see Figs 54, 55) but there are still some doubts about the relative altitudes of the different unconnected caves; there is therefore no accurate vertical scale on Fig 56, though this diagrammatic section does show the general relationships of the known caves.

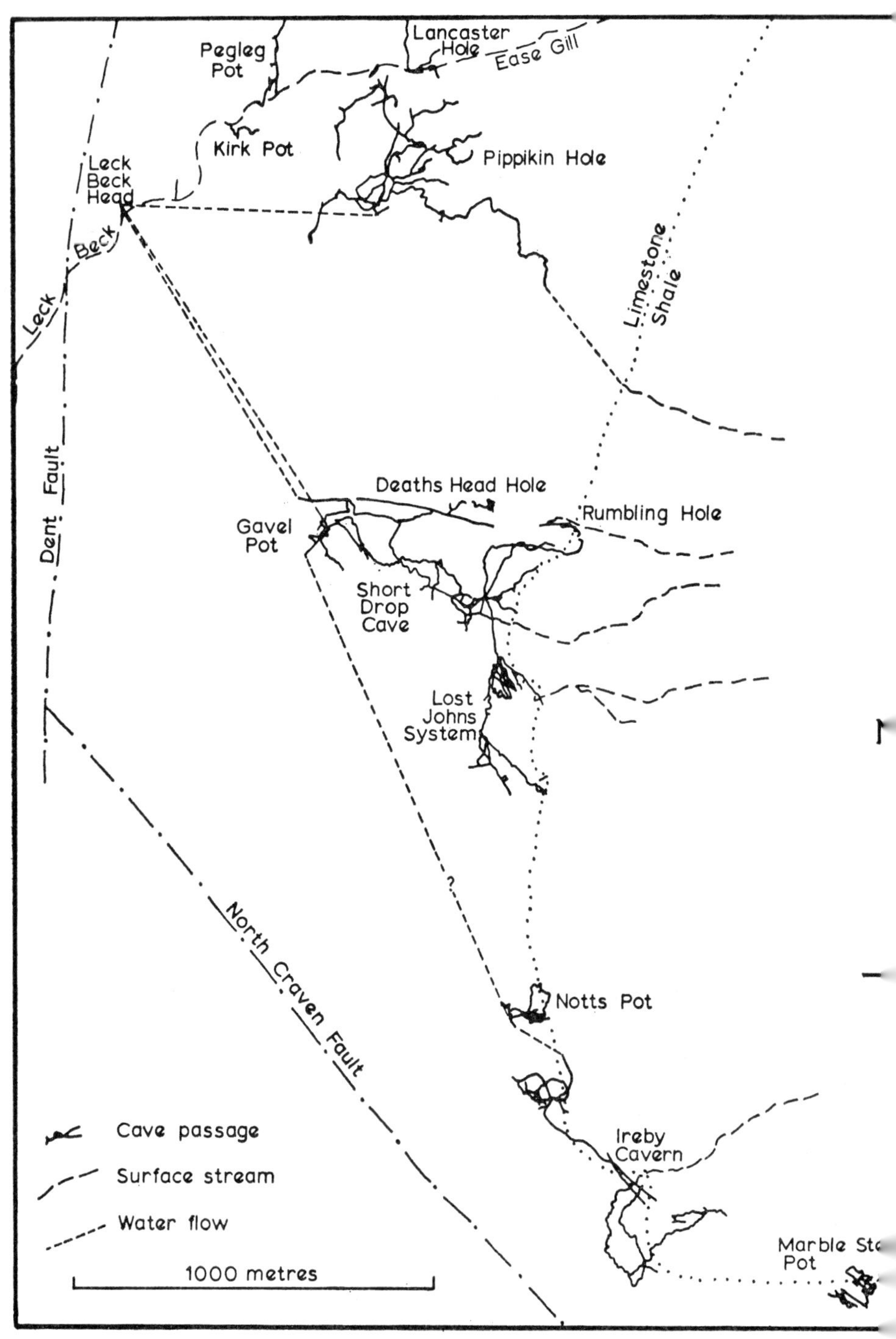

Fig 53 The caves of Leck Fell and Western Gragareth

Leck Fell consists of a gently inclined and slightly undulating limestone bench, nearly 1km wide, covered by an almost complete layer of boulder clay locally up to 10m thick. The eastern margin of the bench lies at an altitude of 360m and is overlooked by the steeper Yoredale Series slopes; west of the bench the limestone edge drops steeply to Leck Beck more than 100m below.

There is practically no surface drainage on the limestone of Leck Fell, though after wet weather the overlying boulder clay supports a vast amount of water and numerous short active stream courses. The series of streams draining the higher shale flanks of Gragareth disappear in a line of sink-holes and open caves along the limestone/shale boundary at about 360m (1,200ft) OD. However, two conspicuous dry valleys can be traced right across the limestone outcrops (see Fig 57) and these appear to have been formed by surface streams flowing during the periglacial conditions of the Pleistocene. Hundreds of closed depressions also break the surface of the Fell, and vary considerably in size. Most are shakeholes incised only in the boulder clay where it overlies the fissured limestone; most are only 1–10m in diameter, and water sinks in the floors of many of them in wet weather. Only a few of the depressions are more than 15m across, and these expose limestone in their lower slopes, indicating that their formation has involved erosion of the limestone. Gavel Pot is the largest example, formed by collapse into a large old cave passage.

TOPOGRAPHY OF THE CAVES

The 10km of known cave passage on central Leck Fell comprise a number of separate cave systems, which are genetically related to form two superimposed drainage systems; at present both are active yet with almost no hydrologic connection (see Fig 56).

Rumbling Beck Cave consists of a short, active, generally small, vadose canyon cave, developed in the uppermost beds of the limestone. It ends at an opening into Rumbling Hole.

Rumbling Hole is well known because of its deep wet entrance shaft fed by the stream of Rumbling Beck Cave. From the foot of

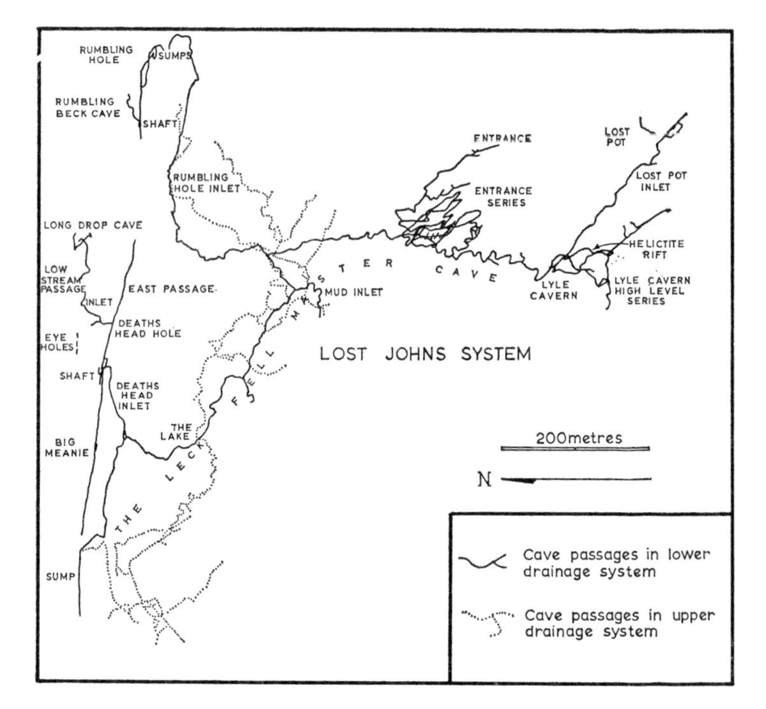

Fig 54 The lower cave systems of Leck Fell

the shaft, a narrow, gently descending, rift passage follows a fault to the east, but after less than 100m turns south to drop down a series of joint-controlled shafts connected by short rifts and bedding plane passages. Below these, the stream soon runs into a very constricted sump, and entering the final chamber is the inlet passage. In the lower reaches the inlet has a phreatic half tube in the roof and a deep vadose canyon in the floor, but neither of these features continues upstream and the passage ends in a bedding plane sump.

Long Drop Cave is cut into the Fell at the end of a prominent dry valley. Inside, a series of vadose shafts and small bedding plane crawlways lead to the high, narrow fault chamber, and, below this, a low bedding plane streamway crosses over twice between the two parallel faults, and is interrupted by a number of high, vadose avens. At the eastern end of the fault chamber, a phreatic bedding plane passage connects with the high rift streamway and upper inlet.

Death's Head Pothole has a round, spectacular 60m deep shaft as its entrance. Its walls are clean and fluted by filmwater, though at the foot of the shaft is an immense pile of boulders and boulder clay sloping down into the main chamber, also with a boulder floor. Perched 25m above the chamber floor is the East Passage, an almost straight and level 5m diameter phreatic passage, partly filled with a variable amount of glacial detritus, and with the roof broken by a series of avens. The western end of the passage has a deep vadose trench in the floor, cut by the stream flowing from the bedding plane inlet on the north side.

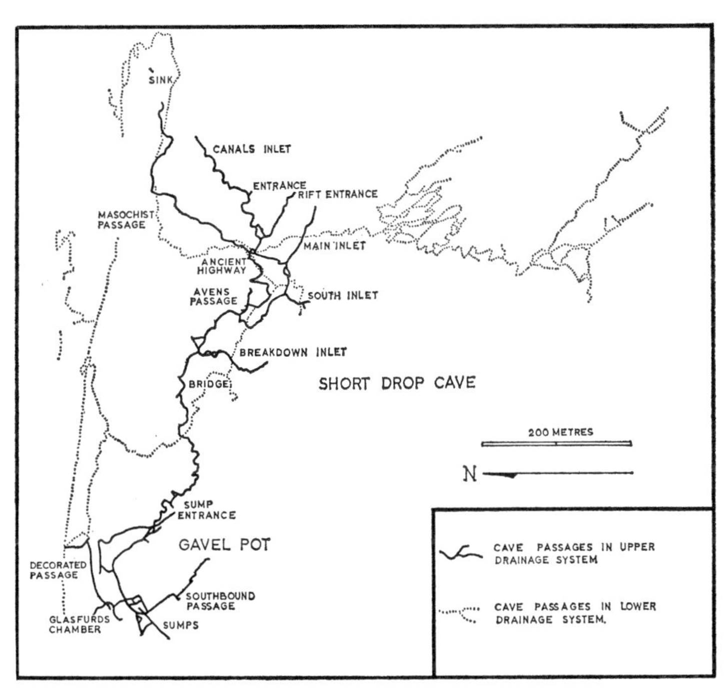

Fig 55 The upper cave systems of Leck Fell

On the opposite side of the Death's Head main chamber from East Passage, a similar ancient, partly sediment-choked, tunnel continues to the west, nearly to Gavel Pot. Part way along, a young 45m shaft drops from a surface doline; the cave was explored via this shaft and is known as Big Meanie.

Short Drop Cave is one of the major cave systems of Leck Fell, though it is less than 40m deep. The upper part of the cave has a nearly dendritic drainage pattern with four important inlet passages. Masochist Passage is a long, tedious, small and very muddy streamway feeding from near Rumbling Hole. Canals Inlet carries a larger stream and is a fine multi-level vadose canyon for most of its length; a hole in its roof is the normal entrance to the system. Rift Entrance Passage is an abandoned, dry, high-level vadose canyon, enterable from the surface, and Main Inlet is an active passage from a boulder choke immediately under the surface sink; the latter is a youthful vadose canyon, now partly dry as the water flows just underneath in a low bedding plane. From the junction of the first three inlets there are two separate downstream passages. The old high-level route is along the aptly named Ancient Highway, in a loop back to the main streamway. The younger active passage is a high narrow rift which joins the Main Inlet near a complex series of high-level oxbows and then leads to a junction with the South Inlet. Though now almost dry the latter is the larger passage at the junction and downstream of here the main passage is of impressive dimensions—up to 15m high and 5m wide—obviously very old, with much collapse and fill, and in places well decorated. Further downstream the main passage is choked, and the stream flows in a tall, narrow, youthful vadose canyon down to the terminal sump. Most of the older and larger high-level passage is choked in this area except for the series of meanders now visible above the sump. Here Gavel Pot has been formed by collapse into a 50m length of this ancient passage.

The huge opening of Gavel Pot is a length of large unroofed meander passage. At the eastern end Short Drop Cave leads in, and the opposite end is nearly blocked by a massive boulder choke. Below the choke the passage soon develops an L-shape cross section, the upright part being mostly choked with loose sediment from Ashtree Hole immediately above, and the horizontal part being a low, wide boulder-strewn streamway. Beyond the first bend the passage is a 5m diameter classical phreatic tube; a similar sediment-strewn tributary enters on the right, and this is the lower end of an old oxbow originating beneath the entrance collapse. Downstream the ancient tunnel continues, also in a downstream direction, along Southbound Passage to end in a choke and a series of small rifts. The stream now takes a separate course in a fine vadose trench, cut beneath a series of smaller half tubes, and then drops down joints to its final sump. North of the streamway, a second large phreatic tube is accessible through a narrow phreatic rift passage. It is choked at both ends, though the eastern limit is very close to Big Meanie.

Lost Johns system contains 5km of passage but has only the one entrance. The main feature of the system is known as the Leck Fell Master Cave, though this name should bear no morphological connotations. The complex entrance series is only a tributary to this important streamway. The principal flow in the upstream part of the Master Cave is from the Lost Pot Inlet, the water falling down a group of huge avens formed on the main fault. After an initial series of canals, the stream flows along a large phreatic passage formed at the intersection of the fault plane and a series of important shale beds. The 'L-section' passage thus formed contains considerable amounts of sediment, and has a small vadose canyon along much of its floor. It ends in a boulder choke underneath Lyle Cavern, with an old oxbow on the north side and, entering from the south, a bedding plane inlet with a well-arched phreatic roof and a sharply corroded floor. In the Lyle Cavern High Level Series, the main passage is an old phreatic trunk route with a cross sectional area generally exceeding 10m^2. Scallops show that the flow originated from a large choked tube part way up the south-eastern terminal aven, though the passages now are dry, and contain extensive fill and abundant calcite formations. At least three youthful vadose passages carry very small streams from the high levels to the lower streamways. Very much older, however, is Helictite Rift, the rift passage which once drained part of the main trunk phreatic route to the top of Lyle Cavern, though it too has suffered some vadose modification. Lyle Cavern itself is a large phreatically opened rift formed at the confluence of a number of faults and joints.

The Leck Fell Master Cave starts from Lyle Cavern and the active streamway can be followed continuously for 1,500m to the north-west. In its upper reaches it varies considerably in size where it cuts across a series of joints, but below Groundsheet Junction it is remarkably uniform. Until 100m beyond Mud Inlet it is a gently descending vadose canyon formed below a single bedding plane, this roof descending from over 6m to 2m high. The floor is covered with closely packed, small, black gritstone cobbles, but the stream is now re-eroding this detritus, incising a channel through it, and locally cutting into the underlying limestone, leaving the cobbles perched on low terraces. Beyond the vadose canyon, erosion forms on the Master Cave roof clearly indicate its phreatic origins. A uniform phreatic half-tube is developed in the roof right to the sump and maintains a constant stratigraphic horizon. Through the Lake the old phreatic tube is complete, but beyond there is a young vadose canyon cut into the floor of the tube, its nick point being at the very end of the

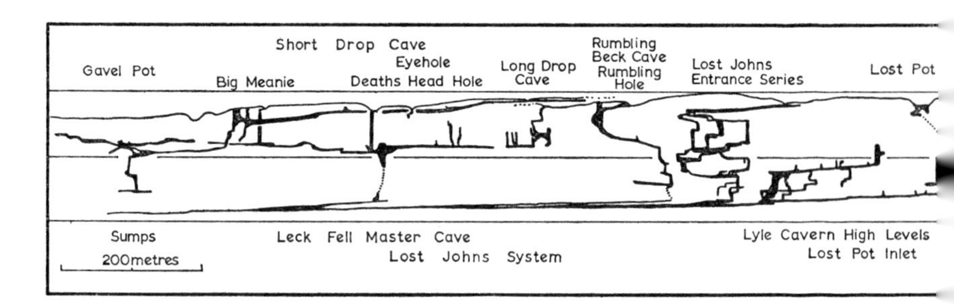

Fig 56 Semi-extended profile of the caves of Leck Fell

Lake. The sump appears to be an unmodified tube continuing down dip with at least one of its air-bells probably formed on the Lyle Cavern–Ashtree Hole Fault.

The Lost Johns Entrance Series carries an important tributary down to the Master Cave. As far as the New Roof Traverse, the entrance and tributary passages are multi-level vadose canyons, but at the Traverse the upper, well-stalactited, canyon level is completely choked, and it is the lower, cleaner, level which continues to subdivide into the further passages. From this point to the Battle-Axe pitch, there is a complex series of joint-controlled rift passages and subordinate vadose canyons developed below bedding planes, ranging over 75m of depth. In each series the uppermost passages are mainly bedding-controlled vadose canyons, and all the major pitches are down joints. Below Battle-Axe pitch the abandoned Maypole Passage is shortly choked, while the single passage leading to the Master Cave is another simple meandering vadose canyon developed below bedding planes. However, the maze of passages in the centre levels show evidence of extensive phreatic enlargement, and the proportion of subsequent vadose modification varies from almost nil to very considerable. In a number of the rift passages scalloped wall niches reveal the position of old phreatic tubes along which the water flowed obliquely up the joints (for example, above Dome Pitch), and a complete tube is still preserved between Dome and Dome junction. Large scallops, up to 25cm across, formed by slow-moving phreatic water, are particularly well developed on the walls of the rifts around Battle-Axe pitch, and contrast sharply with the much smaller scallops being formed today by the rapidly flowing vadose waters. The roof of the presently active streamway also exhibits a variety of typically phreatic erosion forms.

The other tributaries to the Master Cave are of much simpler morphology. The Rumbling Hole Inlet is a very long constricted vadose canyon running roughly down-dip as is the exceptionally muddy Mud Inlet just downstream. In contrast, the Death's Head Inlet is an almost completely drained phreatic tube which can be followed up to a series of choked, also phreatic, avens; the passage is formed along the same bedding plane as the Master Cave roof tube.

Lost Pot is a large open pothole, formed largely by collapse on the fault zone. The stream which enters it drops out of sight in one corner and can be followed only a short distance along a tight boulder-strewn passage.

HYDROLOGY

Two, almost completely independent, hydrologic systems at present drain the centre of Leck Fell, and the two underground trunk streamways are almost directly above and below each other (see Figs 54, 55). The upper drainage system (Short-Drop–Gavel) absorbs the largest single stream flowing off the shale—the Short Drop Cave Main Inlet, with a mean flow of about 70l/sec where it sinks. Being shallow this system picks up numerous small inlets as well as some smaller streams flowing off the shale.

100m below, the Leck Fell Master Cave picks up streams draining from the shales both north and south of the Short Drop Cave catchment area. The three major tributaries, Rumbling Beck, Lost Johns and Lost Pot streams, each have average flows in the order of 50l/sec.

Excepting a few very small outlet streams in Gavel Pot, all the water is last seen in the sumps of Gavel Pot or Lost Johns system. Dye tests have shown that the Lost Johns' water goes to Leck Beck Head, in times varying from about four to thirty days, dependent on stage. The Gavel Pot water has not been tested but almost certainly goes to the same place. Furthermore there is a very large inlet actually in the sump of Gavel Pot. This water is known to come from Notts Pot and Ireby Fell; divers have reported a large underwater passage, and, when finally entered, the open streamways of this system will add considerably to the picture of Leck Fell cave development.

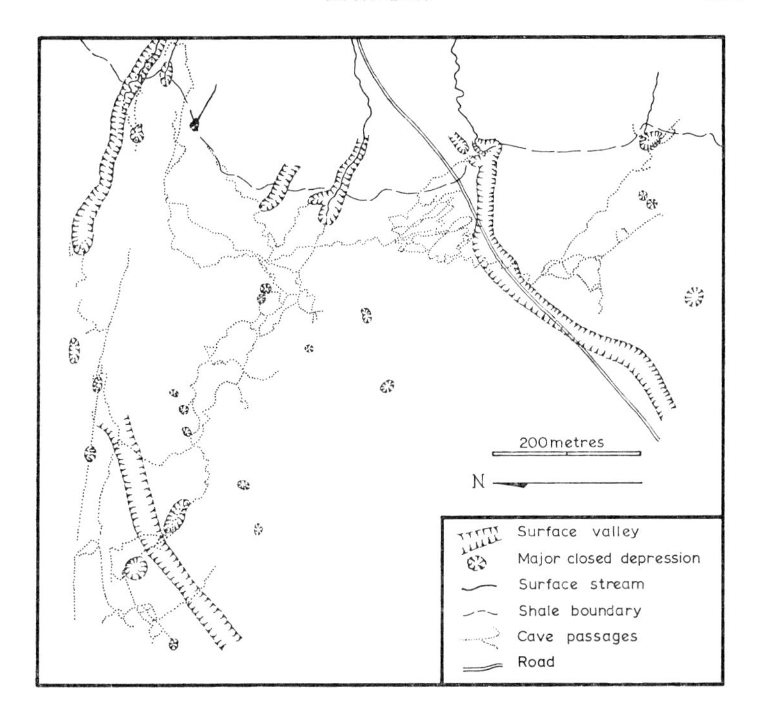

Fig 57 Surface features above the Leck Fell caves

LOCAL GEOLOGY

The Great Scar Limestone of Leck Fell is typical of its form throughout the Yorkshire Dales. However, the base of the limestone is not exposed and the known caves are restricted to a stratigraphic thickness of 145m from the *Girvanella* Band downwards.

The calcareous succession is broken by numerous shale bands ranging in thickness from only millimetres to a maximum of 2m, in the band exposed in Lost Johns Shale Cavern. The shales have been an important influence on cave development, as nearly all the sub-horizontal vadose passages have their roofs on shale beds. (A notable exception to this is the Leck Fell Master Cave, which, along its whole length, originated on a bedding plane with no visible shale.) Furthermore the overall vertical distribution of known cave passages closely reflects the frequency of shale beds through the limestone succession. Particularly prominent is the abundance of vadose canyon-type streamways in the uppermost 20m of the limestone, which contain so many shale beds.

The principal folds, faults and joints in this structurally simple area are shown in Fig 58. Excepting many small and very local fluctuations in the dip, the only important fold is the syncline with its axis oriented just north of west and plunging gently in that direction. The two limbs of the syncline each dip at only a few degrees, but the strong stratigraphic controls of cave development have resulted in the two major vadose cave streams flowing directly down the synclinal axis in the direction of the plunge.

Two important faults crossing the area, though having little

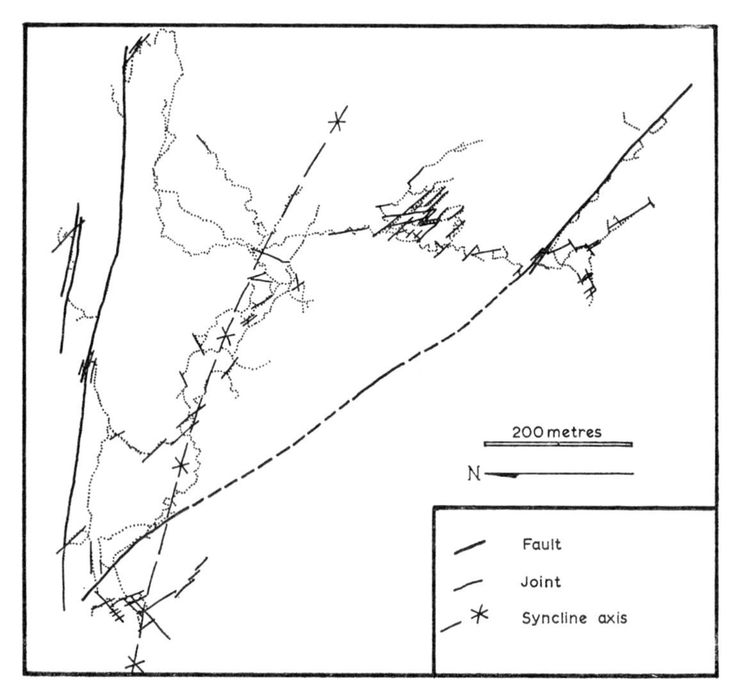

Fig 58 Elements of the structural geology of Leck Fell and its caves

displacement, form wide shear zones and have strongly influenced cave development by creating nodes of preferential water circulation in both vadose and phreatic environments. At least three faults run E–W across the area and have controlled the development of Death's Head Pothole, Long Drop Cave and Rumbling Hole. The other major fault runs NW–SE through Lost Pot, Lyle Cavern (where it is a complex fault zone) and Ashtree Hole. The confluence of the two main fault zones has not yet been reached by the known caves.

The two locally dominant joint systems parallel the two major fault zones, but other conspicuous trends are almost N–S and close to NE–SW. Some joints may be followed horizontally for distances approaching 300m, and others have a known vertical range of over 30m, particularly in the entrance series of Lost Johns system. No one system of joints appears to have had any preferential influence over cave development, though a really detailed analysis of the survey data has not been carried out with this in mind. Fig 58 does not give a realistic impression of joint distribution, as the majority of the cave passages are overridingly controlled by the shale bands, but joints are most abundant close to the fault zones, and there is also a considerable concentration of joints revealed in the Lost Johns entrance series.

SEQUENCE OF CAVE DEVELOPMENT

A series of erosive phases can be recognised within the geomorphic history of the caves of Leck Fell, each phase being characterised by a successively lower resurgence level, giving a greater depth to which vadose erosion has taken place. There are basically three erosion phases separated by two periods of calcite deposition and sediment infilling, the last erosive phase being the present one. Undoubtedly this five-phase history is a simplification (see chapter 24), and several successive events within each individual erosion phase may be discerned, and are noted below, though only the two sub-phases of phase 1 are of a significance sufficient to warrant nomination of this division.

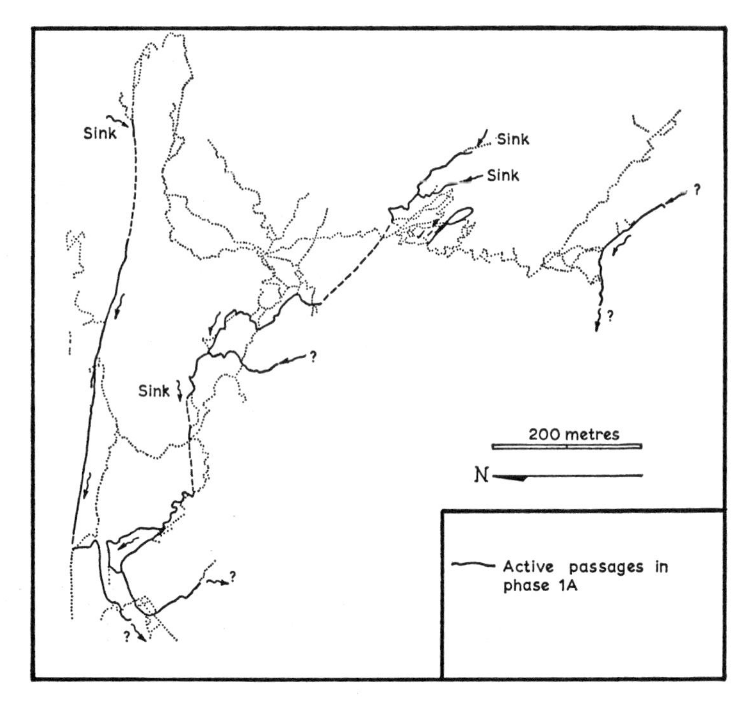

Fig 59 Cave passages of Leck Fell developed during phase 1A

Phase 1a

Of the three elements belonging to this initial stage of cave development perhaps the most prominent feature is a series of large phreatic tubes which were the main trunk routes of the first efficient underground drainage in the area. Also important are the vadose feeders to the phreatic network, and a number of almost isolated joint caves which were opened out by very slow-moving water.

Fig 59 shows the probable state of cave development at the end of this phase, and on it can clearly be seen three independent drainage systems. Furthest to the south is the main passage of the Lyle Cavern High Level series, now choked at both ends. Scallops indicate flow to the west and its original source may be the old *(right)* The 15m waterfall of Wet Pitch in Lost Johns system *(below)* A phreatic half-tube above the vadose canyon in the lower reaches of the Leck Fell Master Cave, Lost Johns system

(*left*) The main streamway of White Scar Cave (*below*) A narrow vadose slot in the floor of the main passage in White Scar Cave

high levels in Notts Pot; its destination is completely unknown.

Gavel Pot contains three large phreatic tubes, though the southerly two are only loops on a single drainage line. They are all of cross sectional area around 10m^2 and generally sub-tubular, but in many places enlarged into significant chambers, mainly at the intersection of joints. The passages reveal many classical phreatic features, including roof niches and reverse gradients, and are all characterised by an extensive fill dating from phase 2. Scallops show the flow directions to have been westward but the continuations of the two outlets, through a mud choke near Glasfurds Chamber, and hidden under the floor of Southbound Passage, are completely unknown. The beheaded remains of the old resurgences, presumably in the Leck Beck valley, are also unknown due to the extensive boulder clay cover.

In the northern phreatic tunnel of Gavel Pot, the inlet end is now choked. But the upstream continuation of the passage is the main tunnel of Big Meanie, and East Passage in Death's Head Hole. Both these sections lie on the fault, and the source of their water was probably Rumbling Hole—a phase 1 sink further along the same fault; the entrance shaft of Rumbling Hole is incongruously larger than the rest of the present active system draining in the other direction.

In contrast, the southern phreatic tube of Gavel Pot, with its oxbow, can be traced upstream to its vadose feeder streamway which is still active over much of its length. The main passage of Short Drop, the South Inlet and Breakdown Inlet, together with the entrance passages of Lost Johns, are all large vadose passages mostly with nearly flat bedding plane roofs, and are characterised by an abundance of stalactite deposits close to their roofs. Fig 59 shows how these passages are the remains of a phase 1 vadose system, now seen as a complete contrast to the much cleaner, younger streamways. Further evidence of the age is provided by the large amount of fill in the Short Drop streamway, so much that the passage has been blocked and abandoned in its lower reaches just before Gavel Pot. The increase in amount of fill in this area suggests there was probably an old entrance just downstream of the boulder bridge. The massive canyon passage continues through Gavel Pot, where it has been unroofed, and enters the ancient phreatic zone where the caver now leaves daylight on his route down Gavel.

Consequently it is clear that during phase 1A times there was a water rest-level, perhaps the resurgence level, at about 310m (1,010ft) OD, nearly 100m above Leck Beck Head; and two of the oldest sinks on the Fell are Rumbling Hole and Lost Johns.

Also dating to this phase is the phreatic, solutional opening of a series of rifts. Those that are known include the Battle-Axe, Centipede and Dome-Crypt rifts in Lost Johns and possibly part of Lyle Cavern. Scallops over 30cm long on the walls of these chambers indicate that they were enlarged by very slow-moving water, which was probably barely able to circulate beyond the confines of these joints. A few upward-trending scallops prove their phreatic origins, and their elevations clearly indicate a phase 1 origin. Their later invasion by vadose streams was merely coincidental.

The phreatic caverns of phase 1A are noticeably concentrated at two levels (see Fig 56). This is due to the sub-horizontal stratigraphic control of cave development, and there is no evidence to show that the passage distribution is related to fossil water-levels.

Phase 1b

No conclusive evidence has yet been recognised of a phase of deposition separating the two erosion phases 1A and 1B, though it is quite possible that one did exist. Furthermore this lack of an intermediate fill stage means that some passages cannot be dated accurately in their respective subsection of phase 1. The subdivision of phase 1 is based mainly on the clear evidence of two important underground stream captures which may indicate a major change in external environment, though there is no evidence of a lowering of resurgence level at this time.

The passage along the fault from the Lost Pot avens to Lyle Cavern exhibits many phreatic features, which ascribe it therefore to this period of development (see Fig 60). The avens at the upstream limit of this passage are clearly of vadose form, and it appears that the water originally entered the passage down small phreatic openings in the fault plane which were only later enlarged to their present state. Lost Pot also dates from this phase, but the present streamway is a youthful feature, and the water must have originally sunk in the floor of the open doline and passed directly down the fault.

Later during the same phase, the waters of the Lyle Cavern High Level trunk route were captured by Helictite Rift developing

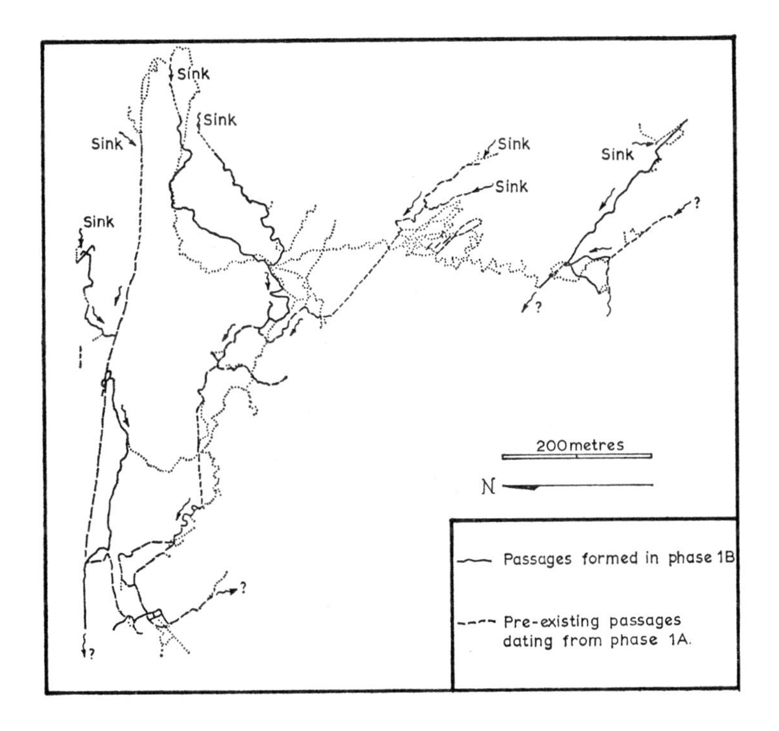

Fig 60 The caves of Leck Fell during phase 1B

headwards from the previously opened Lyle Cavern. However, downstream of Lyle Cavern is only a much younger vadose passage, and the phreatic outlet from the chamber is unknown. A well-scalloped half-tube, plunging at about 70° down the wall of Lyle Cavern at its north-west end, marks the original main hydrologic route but its continuation is invisible in the roof of the younger streamway. This inclined half-tube drops over 20m down the wall, and clearly demonstrates a minimum depth to the active zone of erosion within the phreas (see Fig 61).

Major changes took place at this time in the caves at Death's Head Hole. The main feature was the diversion of the westbound phreatic trunk route at Death's Head Hole Main Chamber into a lower level. At such a depth this new passage was still phreatic, but it ran down the fault and then along a favourable bedding plane, and now forms the Death's Head Inlet and downstream roof tube of the Leck Fell Master Cave. Furthermore, during the

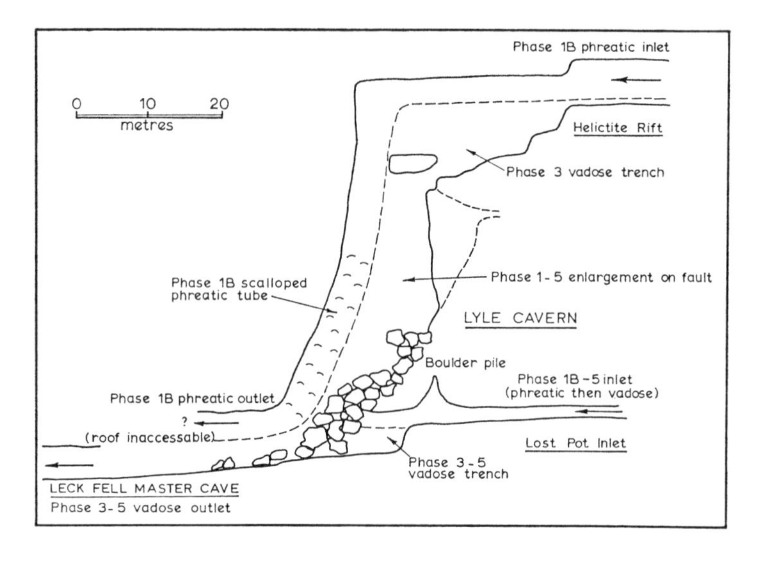

Fig 61 The sequence of phases of development which opened up Lyle Cavern

same phase the Death's Head entrance shaft must have been opened. Though the shaft is well fluted and bears ample evidence of vadose enlargement, a limited ancestor to it must have been present, as the immense quantities of coarse phase 2 boulder clay at the foot of the shaft can only have been washed down it.

The lower parts of Long Drop Cave contain an abundance of phreatic features and their altitude therefore suggests a phase 1 origin. The present entrance passage from the first pitch to the top of Fault Chamber is a youthful, and much later, development. However, the entrance, at the foot of its blind valley, is clearly an old feature, and the original stream route appears to have been through the rubble (washed in during phase 2) at the foot of the first pitch, down to the aven in the High Streamway; thence along this joint passage, and up Crutch Pot under hydrostatic pressure, through the bedding passage to the Fault Chamber and along the present stream route to join the Death's Head trunk passage.

Little is known about the origin of the Eyeholes as they are isolated from all the other caves, but the boulder clay in them and the fact that they are not an integral part of the present topography, proves an origin previous to at least one glaciation, most likely in phase 1, but possibly in phase 3.

The two most northerly inlets (Masochist Passage and Canals Inlet) to Short Drop are still active and yet are very old. Their cross sections are remarkably uniform: a wide clean lower section represents the present vadose erosion, but above this is almost invariably a very high, narrow, vadose channel, liberally filled with stalactites, representing a separate earlier period of development. Their long profiles are graded to the main Short Drop phase 1 passage and they are therefore ascribed to a similar stage of erosion. Just near their confluence they did not utilise the route of the present streamway, but instead their waters flowed through the now abandoned Oxbow and along the Ancient Highway, to join the main stream passage via the upper levels of Avens Passage. These passages, too, are characterised by an abundance of stalactite deposition.

Phase 2

Though not a period of significant erosion inside the caves, the physiographic development which took place during phase 2 was probably the most important in the history of the caves of the area. Within the caves the phase was one of extensive sedimentation, both of inwashed clastic material and also calcite speleothems. The deposits are in many places well and completely preserved, as the phase also marks the time of the change, for many passages, from a phreatic to a vadose environment, due to a substantial lowering of the resurgence level.

Practically all the phase 1 passages, with the notable exception of the isolated rifts in Lost Johns, now contain deposits dating from phase 2. In the case of the stream passages in upper Short Drop and Lost Johns Caves, continued stream erosion means that only stalactites are now preserved, in the upper parts of the canyons. However, the larger phreatic passages and the abandoned parts of the old Short Drop streamway still contain abundant deposits of variously sorted clastic material, mostly of sand grain size. In places the fill completely blocks the passages, for example between Rumbling Hole and Death's Head Hole. The locations of the main entry points for this allogenic clastic material may be deduced from the local downstream concentrations of fill; the major inlets must have been Rumbling Hole, Death's Head Hole, Gavel Pot and a now blocked opening in the roof of the Short Drop streamway near the boulder bridge. Collapse of the streamway roof was initiated during this phase to form the surface opening of Gavel Pot.

Following the main inwashing of clastic material was a period of calcite deposition, with the spectacular results now best displayed in the northern passage of Gavel Pot, where numerous stalagmites rest upon the banks of sand and mud.

The phreatic phase 1 passages are liberally decorated with calcite, some of which was deposited in a vadose environment in the later part of phase 2. Between these two events the rest level of the water must have fallen substantially, and this was almost contemporary with the clastic infilling. To date, no study of the sediments has revealed their precise mode of deposition.

Phase 3

Two completely separate, but superimposed, drainage systems developed during this phase and formed the main outline of the present-day caves. Within the explored area, both systems were dominantly vadose, and hence formed dendritic networks of passages converging on the two main drains, each running down the axis of the gently plunging syncline (see Figs 54, 55).

The passages of this phase are characterised by their dominantly vadose forms, even at the lower altitudes, distinguishing them from most of the phase 1 passages. Also many of these caves contain loose clastic deposits and locally some stalactite decoration, dating from phase 4, which thus distinguish them from the most recently formed, phase 5, passages.

The upper drainage system is almost entirely vadose, where known, and has formed in the central part of the area, centred on the Short Drop streamway, where horizontal development along shale beds has been dominant. The streams flowed down the plunge of the syncline as far as Gavel Pot before reaching joints adequately developed to permit penetration of the waters to lower levels.

In Short Drop Cave itself there was a pre-existing vadose network which was now only modified, extended, and rejuvenated. The Masochist and Canals Inlets were rejuvenated, and the erosion of the wider low parts of the canyons was now initiated. The First Oxbow was abandoned for the low crawl route, and Rift Entrance Passage took the main stream inlet for only a limited time, before being beheaded by the present main inlet. The Ancient Highway was also abandoned, as the waters flowed south, down a clean joint passage, to join the newly formed main inlet, leading to the pre-existing vadose passage downstream from South Inlet. In the Main Streamway, and in Breakdown Inlet, this phase was marked only by re-excavations of the sediments and slight entrenchment of the solid limestone floors.

Below the boulder bridge the fill in the old passage was so complete that the new stream took a fresh course to the south, where it now flows along a clean youthful canyon. High, decorated roof chambers mark the two points where this canyon has locally cut back into the old passage, just below the 5m pitch and just above the sump. The phase 3 passage beneath the Gavel Pot collapse is now a sump due to ponding behind a local accumulation of sediment.

In the upper parts of Gavel Pot the vadose stream has barely modified the drained phreatic passage. This is probably largely due to a continued supply of detritus slumping in through the

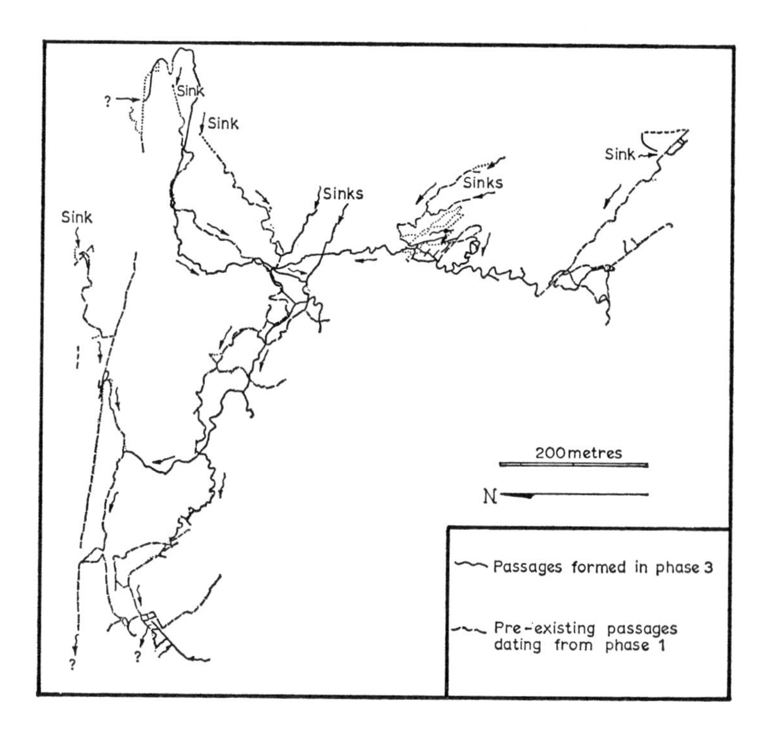

Fig 62 The caves of Leck Fell during phase 3

floor of Ashtree Hole, which the Gavel stream has therefore had to steadily remove. The rubble and mud pile from Ashtree is still visible, on the left side of the streamway, where it almost completely chokes the fault-guided, upright section of the L-shaped passage.

Lower down in Gavel Pot, the phase 3 stream has cut a long vadose canyon in the floor of the network of phreatic passages. The final product of this twofold erosion has been passages with a classical 'keyhole' cross section, the upper tunnel with its stalactites and remnant ledges of sediment adhering to the walls, contrasting with the clean washed, active, vadose canyon in the floor. Even further downstream the water has entirely abandoned the phreatic routes, to utilise a series of joints to drop rapidly to the sump which forms the present limit of exploration.

During phase 3 the Long Drop–Death's Head passages suffered little modification beyond minor vadose entrenchment, except at the Long Drop entrance. From 10m below the entrance the original passage was so choked by phase 2 sediments that the stream opened a new route through to the top of Fault Chamber (the route now used by cavers), though this passage was soon abandoned by its stream sinking upstream of the Long Drop entrance.

Streams sinking into the limestone either side of the Short Drop catchment area intersected joints and faults which permitted easy and rapid descent to lower levels. In phase 3, at least four major streams did this, and all ceased their vertical descents in order to flow along a single bedding plane 130m below the top of the limestone. Once in the bedding the streams flowed down dip and merged in the trough of the syncline to form the Leck Fell Master Cave (see Fig 54). These passages were all dominantly vadose due to the considerable phase 2 lowering of the resurgence level. However, further downstream the Master Cave passage maintained its stratigraphical horizon to sink below the resurgence level and so enter the phreas.

The roof of the Master Cave reveals a distinct morphological

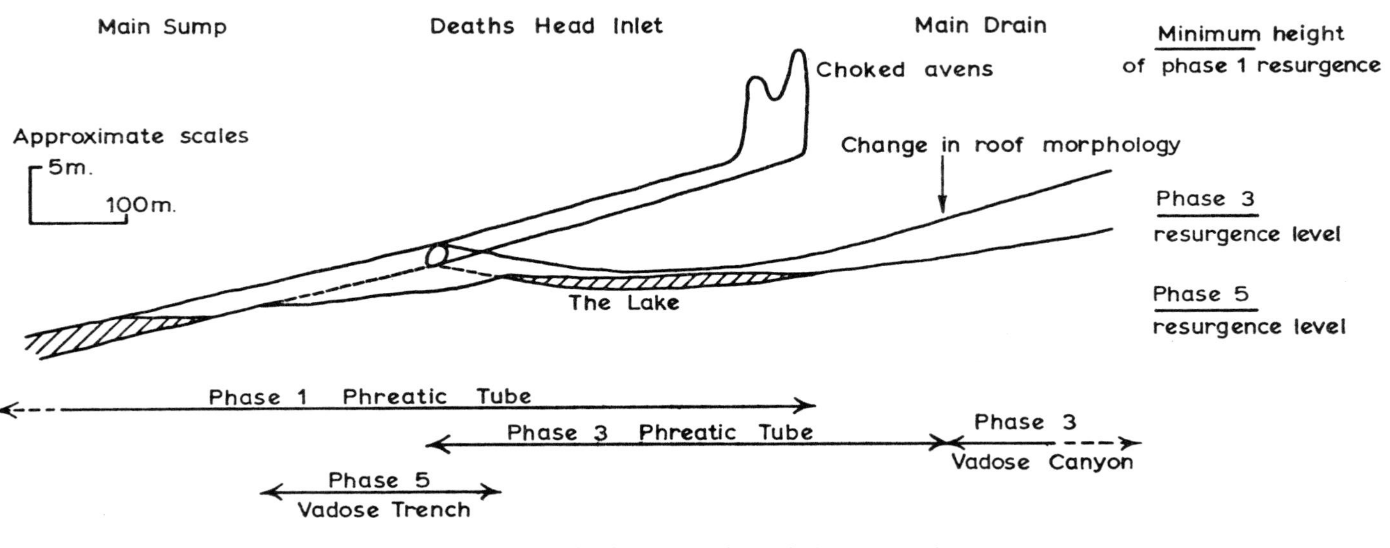

Fig 63 Development of the lower reaches of the Leck Fell Master Cave

change about 100m downstream of the Mud Inlet. Above here, it has an almost uneroded flat bedding roof, indicating vadose initiation, while downstream the roof is a scalloped half-tube, clearly formed in a phreatic environment. The height of this roof change therefore indicates the maximum level of the phase 3 resurgence, about 7m above the present rising (see Fig 63).

The lowest phase 3 section of the Master Cave was developed headwards from the pre-existing Death's Head stream passage, and, within the phreas, this vectored down dip on to the syncline axis, at the lake, and then up the plunge of the syncline. It therefore developed along the shortest and most efficient route connecting the synclinal flow of vadose water to the easiest possible outlet down the open, active tube.

In order upstream, the first inlet to the vadose section of the Master Cave was the Mud Inlet, a simple vadose passage now almost filled with sand and mud. The Rumbling Hole Inlet was initiated in this phase but its source was not the present Rumbling Beck stream; the older inlet is the tributary entering just upstream of the Rumbling Hole sump. The earlier development of this passage is indicated by its later vadose rejuvenation, consequent upon the later rapid downcutting by the Rumbling Hole stream. Though the major length of this inlet to the Master Cave was vadose, the part upstream of the present sump was phreatic due to local updip flow. A roof half-tube marks the position of an efficient drainage passage developing headwards along this stream route, but it does not reach the end of the explorable passage having been fossilised by the phase 5 vadose rejuvenation.

The third inlet to the Master Cave was the Lost Johns Entrance Series stream which comes in at Groundsheet Junction. In phase 3 the Hammer Pot and New Roof Traverse routes were opened in turn, utilising the phase 1 entrance passage and low-level rifts; lower down, the Maypole Passage was formed and then abandoned in favour of the bottom streamway. (Development of this series is further described below.)

Capture of the water in the Lyle Cavern Fault resulted in

development of the top part of the Master Cave. The stream route of the Lost Pot Inlet was already open and remained active, though no major stream was entering from the Lyle Cavern High Levels, there being only slight vadose modification of Helictite Rift. However, small trickles of water did converge in the High Levels and drain down to the main streamway, via at least two vadose passages—Lyle Aven Crawl straight into Lyle Cavern, and the Five Pitches Route down to the head of a phase 1 bedding plane passage. The original route down the fault from Lost Pot was by now blocked by sediment and during this phase the present streamway in Lost Pot must have developed, as it is now choked by phase 4 boulders and cobbles.

Phase 4

A second major phase of deposition terminated the erosive activity of phase 3. However, during phase 4 there was also a second lowering of the resurgence level, indicated by rejuvenation of the streams in phase 5.

In many cases the clastic sediments and calcite deposits of this phase are difficult to distinguish from those of phase 2, which were generally more abundant. However, the age is clear where the sediments occur in phase 3 passages, in many cases sealing them from further exploration, as in the Lost Pot streamway and the Mud Inlet in Lost Johns. Stalactitic deposits of phase 4 are widespread and occur in most of the phase 3 passages; the succeeding change of environment in the modern phase has resulted in some of these being re-eroded while others are still actively forming.

The clearest example of phase 4 deposition is found down the entire length of the Leck Fell Master Cave, where clastic sediments are frequently preserved on ledges a few centimetres above the present stream. Furthermore, the grade of sediment closely reflects its distance from source, as upstream of Mud Inlet banks of imbricately laid gritstone cobbles are dominant, while the only sediment downstream of the lake is fine sand and clay.

Phase 5

Continuing till the present day, phase 5 represents a period of dominantly erosional activity, which, due to only relatively minor changes in the drainage pattern, mainly continued the processes of phase 3. It is also significant to note that at the present time a certain proportion of the calcite deposits are active, indicating contemporaneous erosion and deposition in different parts of the same system; a similar situation must have been in existence during the early phases, though evidence for it is largely unrecognisable at present.

In the upper drainage system, the Short Drop stream continued to maintain its phase 3 route, with consequently unspectacular activity. A number of short, recent by-passes leaving dry oxbows

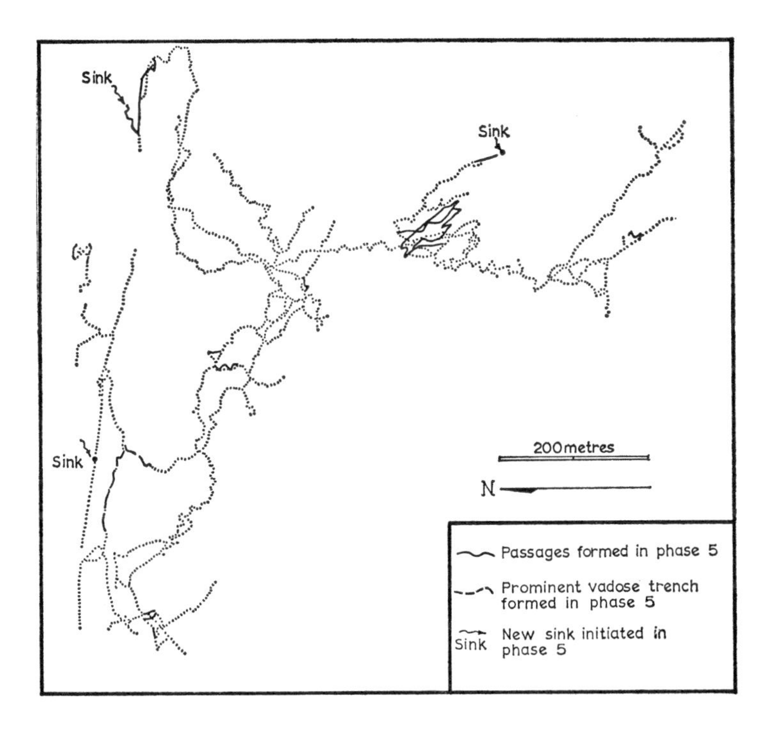

Fig 64 Cave development on Leck Fell during phase 5

can be recognised, notably just above the main pitches in Gavel Pot and in the Main Inlet in Short Drop Cave. At the same time, the Big Meanie entrance shaft was developed while the Death's Head shaft was merely trimmed by filmwater.

The upstream inlet of Long Drop Cave probably originates from this phase, as does the major part of Rumbling Beck Cave and Rumbling Hole. The stream which had previously flowed down Long Drop sank into the limestone at Rumbling Beck Entrance and then flowed downdip to intersect the previously formed Rumbing Hole. However, at the foot of the shaft, the old route to Death's Head Hole was sealed, and the water therefore turned north, along the fault, and then via some major joints, to meet an open passage near the present sump. Rapid erosion by the new stream resulted in rejuvenation of the older inlet.

In Lost Johns further changes took place in the entrance series as a succession of captures first formed Shale Cavern, and then the present streamway (see section below). Upstream in the cave only minor changes affected the drainage of the Lyle Cavern High Levels.

The most important development in the lower levels of the system was the rejuvenation of the Leck Fell Master Cave, due to the finding of a new, lower, resurgence level. Consequently the phreatic tube downstream of the Mud Inlet was partially drained as far as the present sump, and a clean vadose canyon developed, graded to the new sump level, with its present nick point at the downstream end of the lake. The half-tube now forming the roof of this passage is all that remains of the phreatic earlier development, except in the lake where the tube is still complete. However, the lake is only formed by the ponding of the updip flow of water towards the Death's Head Inlet, and it too will soon be drained by continued rejuvenation. Upstream of here the phase 5 erosion has locally cut a trench through the phase 4 sediments and into the limestone floor.

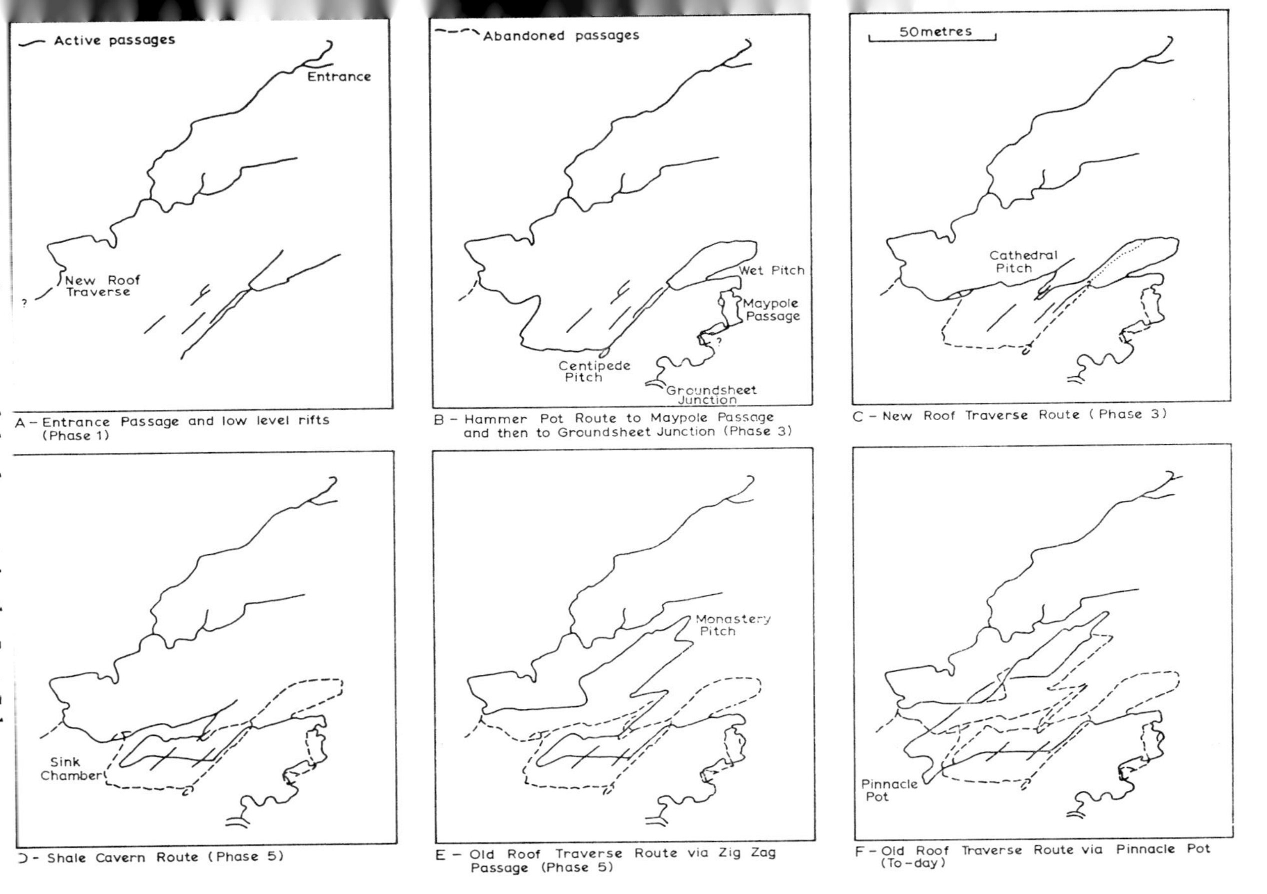

Fig 65 Successive stages of development in the Lost Johns Entrance Series

DEVELOPMENT OF LOST JOHNS ENTRANCE SERIES

Well known for its variety of interconnecting passages, the entrance series of Lost Johns System has been formed by a single stream running off the shales of Gragareth, which, over time, has taken six different routes down through the limestone. This sequence of development is summarised in Fig 65. Much of the detailed morphological evidence is omitted in this description, as it would only be repetitive, the cave development being mainly a series of simple vadose captures.

The vadose entrance passages as far as the New Roof Traverse were formed during phase 1; the stalactite-choked continuation towards Short Drop Cave is visible in the roof of the present streamway. In detail, the entrance passage is very complex; Quicksand Passage with its inlet further out from the shale boundary is clearly earlier and the roof contains a number of old outlets and oxbows. The Battle-Axe rift, Dome Chamber, Centipede rift and the sump air-bells have walls marked by very large scallops and some shallow half-tubes. These low-level rifts were all opened by slowly moving phreatic water, also in phase 1 (see Fig 65A).

The stalactites in the entrance passage roof are the only evidence of phase 2 deposition.

Vadose drainage of the limestone in phase 3 increased the flow rates so that new passages opened, each one developing headwards from earlier low-level caverns. Earliest of the low-level routes was the Maypole Passage, draining the phase 1 rifts out to an unknown destination; but this was soon fossilised by headward erosion from Groundsheet Junction along the present streamway. Upstream, bedding planes were opened to capture the water in Battle-Axe rift; similar development connected this in turn to the Centipede rift and finally a passage opened from the top of this to capture the main stream at the New Roof Traverse, so finally opening the Hammer Pot route to the Master Cave (Fig 65B).

Still in phase 3, a passage next developed headward from Dome Junction to connect with Dome Chamber, first by the pre-existing low phreatic tunnel and secondly by the high vadose rift to the window in the chamber. Then from the top of this high rift, headward sapping opened the Cathedral Pitch and the New Roof Traverse passage, to capture the main stream at Numbers Three, Two and One Holes, in that order (see Fig 65C).

Phase 4 resulted in minor calcite deposits in many of the above passages, which now usefully serve to distinguish them from the later developments.

Early in phase 5, the first new passage to form was from the Battle-Axe and Sump rifts, headward through Sink Chamber and Shale Cavern, to capture the main stream at Dome Chamber forming the present Shale Cavern route (see Fig 65D). At first development in the Chambers was along joints, but later a thick shale bed was washed out to form Shale Cavern. The temporary utilisation of this route by the main stream explains the larger size of the passages to the Vestry compared with those from Dome Chamber to Dome Junction which were abandoned earlier.

After only a short time, however, the zig-zag rifts were opened through to Monastery Pitch and from there development along bedding planes rapidly progressed until the main stream was again captured at the New Roof Traverse (see Fig 65E). For the first time there were now completely dry passages from the New Roof Traverse to Battle-Axe pitch. The final modification to the series was the opening of the present streamway from Sink Chamber mainly along a series of joints to capture the stream near the foot of Monastery Pitch (see Fig 65F). Also, further upstream the Hampstead Heath passage was abandoned as an oxbow.

ADJACENT CAVE SYSTEMS

The multiphase pattern of cave development, so prominent in the Lost Johns area, is similarly evident in the caves on the north and south flanks of Leck Fell.

Pippikin Hole

There are nearly 7km of passages known in Pippikin Hole, but, being under the lower-level benches of northern Leck Fell, their vertical range is only 110m. The main passage is an abandoned phreatic tunnel from the Hobbit to Gour Hall; most of it is 6m in diameter, but it contains extensive clastic fills and some areas of fine calcite formations. It is clearly a phase 1 passage and the source of its water was either the caves of Casterton Fell or an old sink in the Ease Gill valley. An inlet, at Dusty Junction, was fed by the important fossilised sinkhole of Nippikin Hole, and an oxbow contains the very large Hall of the Mountain King. Its destination, beyond Gour Hall, is unknown, but may have been to the same resurgence, in the Lower Leck valley, as was fed by the phreatic drains of Gavel Pot.

Phase 3 in Pippikin saw vadose rejuvenation and a whole series of vadose inlets feeding into and below the main tunnel from the east. Most important of these were Leck Fell Lane and some parts

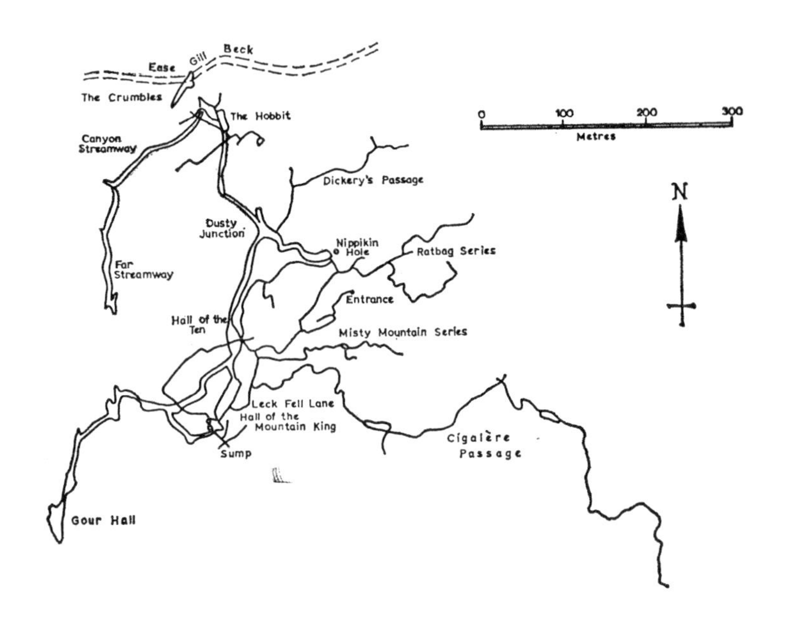

Fig 66 Plan of Pippikin Hole

of the passage from the present entrance; all these passages are characterised by a profusion of stalactites, and fed to a new low level sump—the one still now active. Canyon and Far Streamways probably also formed at this time, beheading any continuing drainage from the direction of Ease Gill.

Up till this time, the shale boundary must have lain very close to the present entrance of Pippikin, for most of the inlets reach no further east. The exposure of the present limestone bench, even though it is drift covered, is a relatively young feature; and it was succeeded by the development of the Cigalère Passage. This, and the new longer route for the Pippikin stream on its way to the sump, were the principal effects of phase 5 erosion. Undercutting and collapse finally connected the old phreatic tunnels to the younger streamways where they passed below. The main stream now flows down Cigalère, and picks up most of the other, very much smaller, streams before flowing into the sump, almost certainly to resurge at Leck Beck Head, which is at the same level.

Caves of Ireby Fell

A 15m deep blind valley, cut into the boulder clay, marks the entrance to Ireby Cavern. This carries a small stream down a series of shafts into a long meandering streamway finally to break out, at a depth of 120m, into a massive ancient phreatic tunnel—Duke Street. The water flows along this to the north-west, though it is clearly an underfit stream, meandering between huge banks of clastic fill. Both upstream (south-east) and downstream, Duke Street is blocked by inwashed debris and collapse; it appears to have carried the Marble Steps water through to the Leck Beck valley during phase 1. The upper part of the Ireby entrance series consists of classic vadose canyons, but lower down, towards Duke Street, a fine phreatic tube is developed in the roof. This inlet was therefore a phase 1 tributary to Duke Street, and its roof change indicates the old resurgence level—at about 300m, and very close to the level indicated by the Short Drop Cave passages.

The pattern is complicated by further old high-level tunnels off both the entrance series and the western end of Duke Street.

The Ireby water leaves Duke Street part way along and follows a younger passage through a sumped connection to Notts Pot. Notts itself consists of an incredible three-dimensional maze with numerous vadose shafts cut at the intersections of a number of joints and faults. Most now contain only drips of water, but one series carries a sizeable stream during wet weather to join the Ireby water just before another sump. From there its course is unexplored, via the sumps of Gavel Pot to Leck Beck Head. The oldest passages in Notts are in a short but complex high-level series; they are now largely choked but may relate to the Lyle Cavern High Levels in Lost Johns.

CORRELATION OF CAVE DEVELOPMENT

The sequence and patterns of cave development outlined above only represent a skeleton of the true picture. Undoubtedly the situation is much more complex and a complete sequence must be divided into dozens of erosive periods. The five phases described embrace the major morphological changes, and further subdivision has not been attempted as it not only makes for repetitive geometric analysis of the caves, but also becomes of increasingly dubious significance. Furthermore it is the very complex systems such as Pippikin Hole which are alone able to provide such detailed evidence of origin and development.

Another difficulty with this morphological correlation of cave passages concerns the occurrence of past water levels within the limestone. A sump does not necessarily occur at the resurgence level. The Great Scar Limestone is not a uniform aquifer, and vagaries of geological structure may commonly result in perched bodies of water—whether they be in individual cave passages or in complete systems. The Ireby Cavern–Notts Pot sump, perched above two 5m cascades, is an ideal example of such a situation. Consequently the changes, within a single passage, from vadose to phreatic features do not necessarily indicate an old resurgence level, though such features repeated through an area at the same altitude do provide reasonably convincing evidence for the elevation of an ancient rising.

Lacking, so far, any absolute dating of the clastic sediments or stalagmite deposits, the chronology of the Leck Fell caves may only be postulated on the basis of morphological correlation with surface features, and even these are not adequately dated themselves. Within the caves, the alternating phases of erosion and deposition can reasonably be related to rapid changes in external environment, and such marked climatic variations can only be ascribed to the glaciations of the Pleistocene. The three erosional phases within the cave represent a high level of stream activity during the warmer or interglacial periods, while the two phases of fill and deposition indicate reduced stream power during periglacial or glacial climates, combined with an abundance of detritus supplied by frost shattering. Then, the major lowering of resurgence levels (within the limits of their validity described above) during phase 2 must correlate with the main glacial excavation of the principal valleys.

Further, more precise, or absolute, chronology (see chapter 24) must at present lie open to debate.

16

Cave Development in Kingsdale

D. Brook

The King's valley is situated close to Yorkshire's western border with Lancashire and just north of the village of Ingleton. Morphologically it is cut into a plateau as a straight trough oriented SSW, 4km long, 400m wide and 120m deep. The plateau is a structural feature dipping gently NNE and is the top of the massive Great Scar Limestone, here 170m thick. Above the limestone benches, which border the trough, rise the ridges of Gragareth and Whernside, composed of 300m of alternating shales, sandstones and limestones of the Yoredale Series. The run-off from the Yoredale ridges gathers into streams and sinks abruptly on reaching the massive limestone. Normally the main Kingsdale valley is dry across the outcrop of the cavernous limestone, as far as the Keld Head rising. The River Doe, which rises here, meanders south across the flat alluvial floor of the valley, cuts through a barrier of limestone and glacial drift 25m high, and drops, at Thornton Force, to meet the contorted Ingletonian rocks. The limestone resting on these basement rocks displays a classic unconformity at the Force.

Downstream the North Craven Fault terminates the Ingletonian and marks the edge of the elevated Askrigg Block which forms the typical structure and scenery of the Craven Dales. Kingsdale has a similar structure and erosion history to other karst areas in the Craven Dales, but the exploration of the caves (Brook, 1971) has revealed more evidence about the past and present subterranean drainage than any comparable region.

THE CAVES

The most southerly caves known to feed Keld Head are in the region around Marble Steps Pot. The latter is a great shaft at the termination of a considerable stream, and takes the form of a gully cut back from a hading fracture plane, which controls the whole complex of active and abandoned entrance routes down to the Main Chamber 60m below. Here the water sinks and is not seen again until Keld Head. The deserted outlet runs south to a long rift (the 240 Feet Rift), sub-parallel with the entrance fracture, before turning south once more and dividing, as it drops quickly to the sump at a depth of 130m. The system has a complex history, with deserted phreatic development at all levels linked by vadose trenches and pitches.

Low Douk Cave, situated in the dry valley below Marble Steps, has a prominent entrance, with a small tube descending to a slope of black cobbles and into a flat-roofed cavern. Downstream, a tortuous vadose canyon ends at Steps Chamber and roomier, phreatic passages of more recent origin lead to the active and static sumps. Between the sumps a crawl enters a hading-shattered rift ascending to ancient choked tunnels.

East of Marble Steps are numerous shallow potholes up to 25m deep, which are vadose shafts choked by debris or becoming too constricted. Outstanding is the Mohole, a pothole 85m deep blocked by boulders of calcite, limestone and sandstone.

Shallow potholes predominate north of the Mohole as far as Swinsto Beck. Turbary Pot, near the flood sink, is a short crawl to a 30m shaft into a long chamber floored by cobbles. The upstream limit of the Turbary Inlet of Swinsto Hole lies beneath the northern end of the rift.

Swinsto Hole lies up the beck on the (true) left bank. The system of essentially vadose passages and pitches is 130m deep, and is at present the highest entry to the Kingsdale Master Cave. A short trench passage and pitch reach the infamous Swinsto

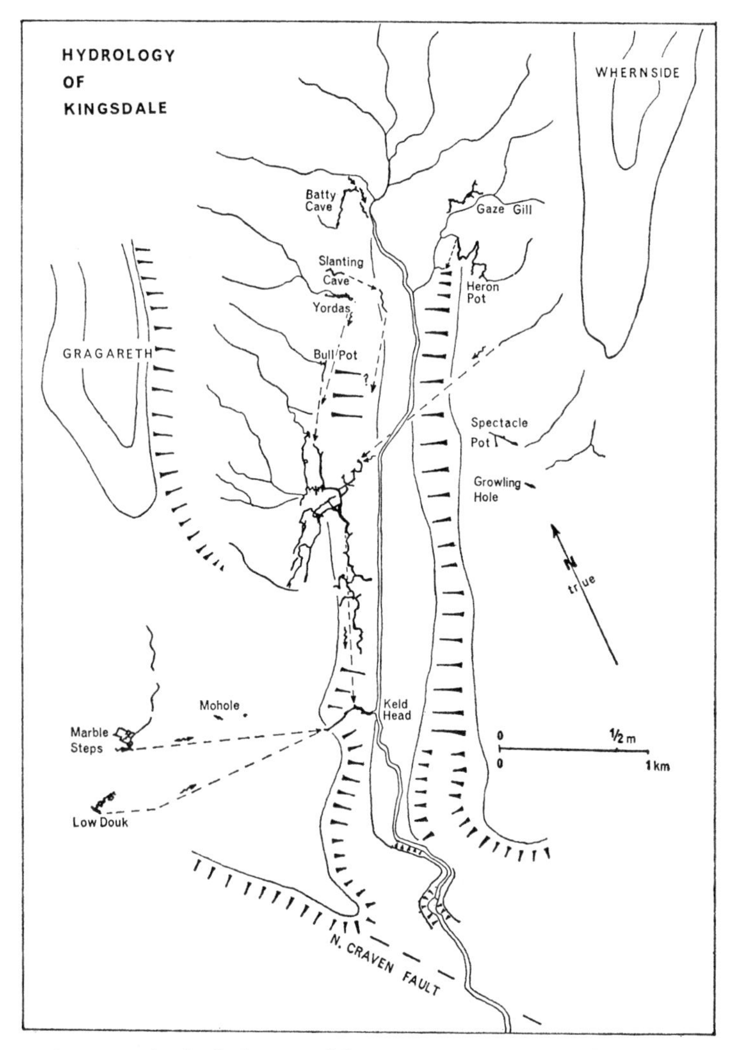

Fig 67 The hydrology and known cave passages of Kingsdale

Long Crawl, but much of its 300m is an awkward stoop with two flat-out sections. More short, sporting pitches then lead to a 30m shaft broken by a stormy ledge, and the continuing streamway enlarges with the entry of Turbary Inlet. The inlet is a long rift displaying fine phreatic features developed before vadose entrenchment. Swinsto continues as a succession of small falls until a large inlet enters via the magnificent aven of Slit Pot. The combined waters then cascade into the Final Chamber to sink into boulders below the Great Aven which towers 45m above.

North of Swinsto are small shafts and caves which drain into the Long Crawl, while just beyond these are Simpson's Caves, a 300m long complex of shallow caves, in which the water finally sumps and a mud crawl emerges at Simpson's Pot entrance.

Simpson's Pot starts as a crawl in a small stream until the main water enters from a slit just above the Five Steps cascades. The stream then drops into the Pit, and hence to Storm Pot, but the caver can descend via dry passages by traversing over the Pit. From Storm Pot a constricted duck suddenly enlarges to a phreatic tube before a succession of small pitches leads onward to Aven and Slit Pots where Simpson's unites with Swinsto Hole. Above Aven Pot a traverse and climb lead to a superb 38m pitch straight down Swinsto Great Aven. Above this pitch another traverse continues to an old passage above the Swinsto streamway.

The most significant feature of subterranean Kingsdale is the Master Cave, whose very existence was doubted for so long. The original entry was the Philosopher's Crawl under the west wall of Swinsto Final Chamber. It follows the Swinsto water, which eventually sumps, and a domed deserted passage meets the main underground river flowing south at the Master Junction. Downstream the canyon cuts down into the floor of the older phreatic tunnel, and, above the inevitable Downstream Sump, the tube continues as the deserted Roof Tunnel for 400m. This terminated in a choke, but has since been dug to break out to the surface as the Valley Entrance. A crawl diverges from the Roof Tunnel into a lengthy series of low phreatic passages, which terminate, in Carrot Passage, within 300m of Keld Head. The Cascade Inlet is a clear-water tributary to this series, and descends via large chambers from a boulder choke.

Returning to the Master Junction, the area is a complex of abandoned phreatic tunnels. One route leads to the east side of Swinsto Final Chamber and provides the easiest entrance to the Master Cave. The complexities of the Master Junction unite upstream in a deep canal to the River Junction. Here the two major tributaries unite. The east branch soon sumps, but can be regained by a dry crawl in the left tributary. This east branch (the Mud River) has been explored beyond two sumps for over 300m to a third about 20m deep (Deep Rising). The west branch (Rowten Passage) also ends in a sump, which may be passed to the final pool of Rowten Pot; but this is merely an airspace, and a further upstream sump emerges in Frakes' Passage, 250m of wide streamway terminated by another large sump. A small tributary also ends in a sump.

Rowten Cave is a large shallow stream cave which reappears 8m below the edge of the great gash of Rowten Pot. The pot is 25m long and 12m wide, and is oriented along a master joint, down which the Rowting Beck makes a precipitous descent to the final pool at a depth of 105m. The caver can avoid the water by utilising dry pitches and oxbows within the system of bunched joints.

Jingling Pot, to the north, is also on a master joint, and provides a fine 30m ladder pitch into a large rift, closing down to a series of choked crawls and fissures. An alternative sink for the stream is Jingling Cave, which comprises 350m of crawling and walking to its junction with Rowten Cave.

The next sink, Bull Pot, is a series of small vadose shafts whose complexities are determined by another suite of bunched joints. The water vanishes into a flooded crack, but above the last pitch a sporting passage continues south-west to drop into a static pool, which is the deepest point and 85m below the entrance. The gradual northward dip of the rocks brings the bench closer to the valley floor until at Yordas Gill, only 45m of limestone is exposed. The stream enters Yordas Cave as a vadose canyon before dropping into the impressive Main Chamber, 55m long,

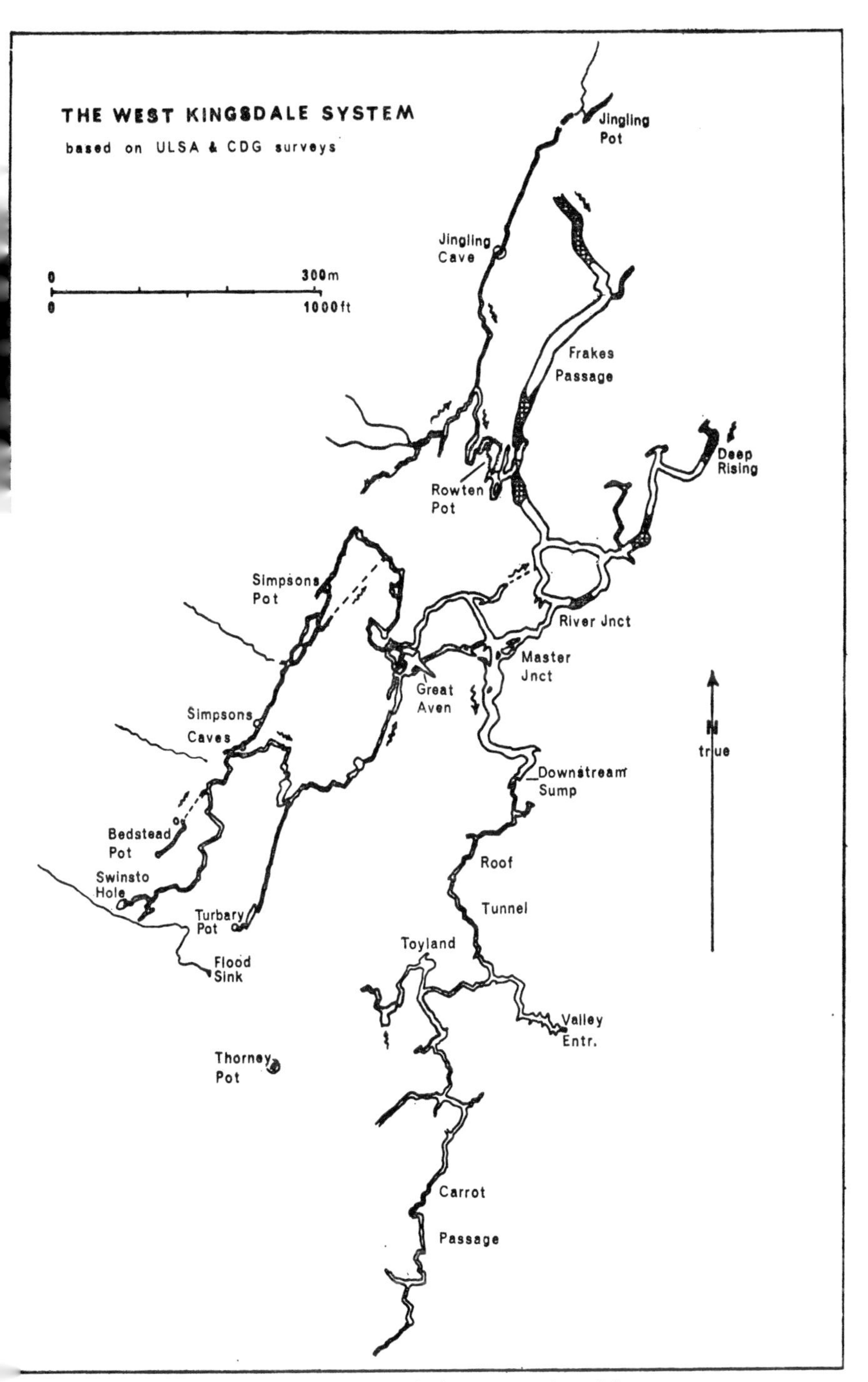

Fig 68 Plan of the caves of the West Kingsdale system

and 15m high and wide. The cave terminates in a shingle choke under a large shakehole.

Nearby is Slanting Cave, a long shallow twisting fissure which ends in a shaft down to a sump, but Batty Cave is a more significant system. It consists largely of lengthy crawls initiated under phreatic conditions and the small stream reappears at Kingsdale Head Cave, a resurgence which comprises 60m of bedding cave to a silted pool.

The remaining streams at Kingsdale Head are loath to sink, and the area is characterised by small caves and shallow choked pots until the eastern tributary, Gaze Gill, is encountered. Gaze Gill Cave appears to have been cut in two by the gill, the remaining passages forming underground oxbows to the surface course.

A major tributary to Gaze Gill is the outlet from Heron Pot, a streamway cave 1,000m long. Most of the water in the cave is derived from Thack Pot Gill and the top entrance is near the sink. The stream course is a high vadose canyon to two short pitches and a further canyon lowering to a canal showing strong phreatic features. Beyond Thack Pot Gill, the bench rises steadily to the area behind Braida Barth Farm which is pitted by potholes. By far the deepest of these are Growling Hole and Spectacle Pot, both just 100m deep and having similar morphology—tributary streams uniting at shallow depth, and dropping quickly to large shafts oriented along fracture zones and choked by boulders.

LOCAL SETTING

The present topography, hydrology and cavern morphology of Kingsdale have been created by the earth movements, and forces of erosion since Carboniferous times. Stresses developed within the limestone by movement of the Craven Faults have been relieved by patterns of joints and small faults. Initially most of the joints were tight cracks, impervious to water movement, but the master fractures were more open and permitted juvenile water to circulate within the limestone. Mineralisation also took place in the deepest and most open fractures during the Amorican earth

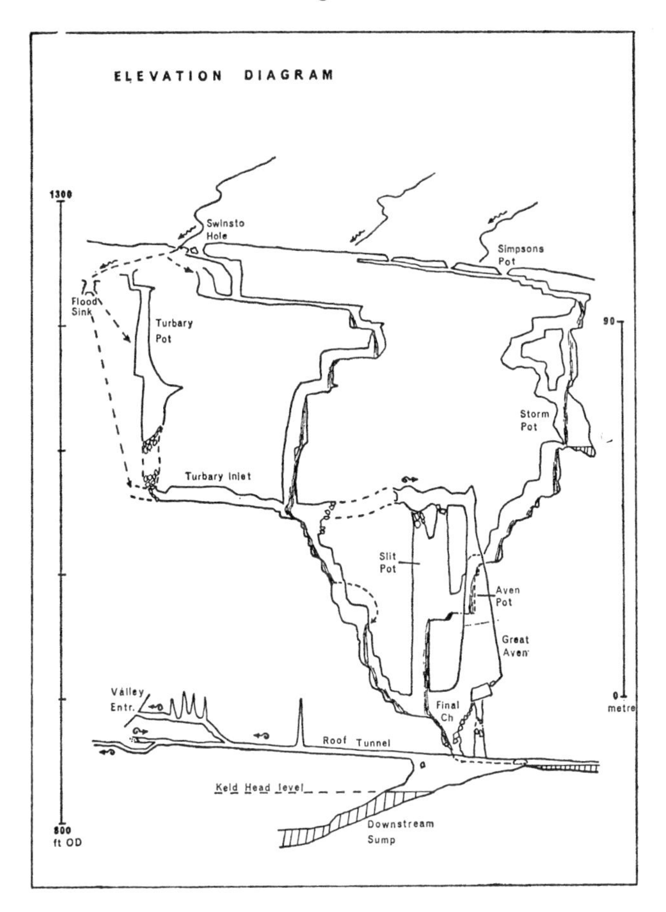

Fig 69 Projected elevation of part of the West Kingsdale system. (The vertical scale is greatly exaggerated.)

movements when calcite, malachite, and barytes were deposited in the north-west oriented veins behind Keld Head. The restricted occurrence of mineralisation indicates that, at this time, circulation of solutions within the limestone was very localised.

It seems unlikely that any enlargement of joints and bedding planes, to allow water movement, could take place until the limestone was exposed by erosion following uplift in Tertiary times. The Kingsdale valley was initiated on impermeable rocks draining south to the lower plain left by movement of the Craven Faults. The linear nature of the valley is almost certainly due to fractures and shatter zones within the rocks, though this trend has been emphasised by subsequent glaciation.

During the Tertiary, the Great Scar Limestone was probably exposed in the floor of the young Kingsdale valley, and water circulation began to open up the tight joints and bedding planes for the first time. Cave formation at this time was unlikely because of the restricted limestone outcrop and slow circulation.

As the climate cooled at the beginning of the Pleistocene, Kingsdale was probably a steep V-section valley with small caves formed by the sinking river. As the valley deepened, so many of the caves would be invaded and destroyed. The small tributaries draining the gritstone plateaux cut shallow valleys in the Yoredale beds, and incised gullies in the side of the main valley. Any caves formed by these tributaries would be shafts with short lengths of horizontal passage linking them with the Kingsdale Beck or its caves.

All these early caves were removed by the erosion of ice and water during the succeeding glaciations, when the Yoredale strata were plucked from the Great Scar Limestone to form the characteristic benches. The Dale was straightened, widened and deepened by successive valley glaciers, and ice sheets plucked blocks from the lines of strong joints and faults to form troughs and gullies. Caves which formed at the inner edges of the benches, have engulfed surface streams and prevented fluvial dissection of the benches.

CAVE ORIGINS

Before the caves could be formed, networks of interconnecting open joints and bedding planes must exist. Aggressive meteoric water entering joints soon became saturated, but mixing of different solutions permitted continued erosion (Bögli, 1971), and gradually produced micro-drainage networks. At this stage only the larger openings contained vadose water, but the scale of this change from the phreatic state depended on the hydraulics of the complex systems created.

The next stage in Kingsdale was the invasion of the micro-networks by large surface streams draining the Yoredale ridges. The invasion flooded the networks with aggressive water, thus causing rapid enlargement under phreatic conditions and allowing still more surface water to be engulfed. The larger fissures permitted faster flow and so increased in size, pirating flow from the tighter networks and creating a simple hydraulic path between sink and resurgence. Continued enlargement produced airspace in the sections of steep hydraulic gradient, creating vadose pitches linking phreatic zones. Entrenchment, by rapids retreating from the pitches, then enlarged the phreatic zones, thus draining them to form vadose streamways.

At this stage great changes took place in the morphology of the Kingsdale caves, since gravity was now the prime controlling force in vadose development. The path taken by subterranean water is governed by two simple laws:

(1) The main flow of water in phreatic passages occupies the path of *overall* least resistance between the inlet and outlet of the flooded zone.
(2) Vadose water, flowing under the influence of gravity, always takes the immediate path of least resistance when presented with an alternative of routes.

These laws explain the morphology of almost all stream caves and should always be kept in mind when attempting to analyse past development and hydrology.

Phenomena associated with these hydrological laws are retreating sinks (Yordas, Rowten, Swinsto, and Marble Steps), vadose capture, and shortcircuiting (Marble Steps, Simpson's Pot, Rowten Pot and Bull Pot). Down-dip bedding planes had only a fleeting phreatic existence before being drained; Swinsto Long Crawl is a fine example of a 300m long bedding plane, which has been enlarged by a series of concurrently retreating vadose trenches, whose upstream limits are marked by small rapids with flat-floored bedding caves upstream. Up-dip bedding passages had a long phreatic development which resulted in tubular-elliptical cross-sections as seen in the roof of the Master Cave.

Relatively horizontal sections of the original networks, formed along open joints, had more erratic profiles, and low points remained flooded long after other sections of the channels had been drained. Examples of this can be seen in the roof of the 240 Feet Rift in Marble Steps, below Storm Pot in Simpson's Pot, and in the Turbary Inlet of Swinsto Hole. Where an upward joint forms part of the stream course, a considerable length of upstream passage remains flooded on the U-tube principle. It remains so until vadose entrenchment removes the whole joint or another route opens up to drain the syphon. The Deep Rising is an active deep phreatic loop due to up-joint flow.

STRUCTURAL CONTROLS

The survey of the West Kingsdale system (see Fig 68, and Brook & Crabtree, 1969) illustrates the joints and bedding planes which have controlled its development. The vadose bedding plane passages are down dip but this trend is often disturbed by strong NW–SE joints which control such features as the Five Steps and Swinsto Great Aven. North–south fractures have also determined the Storm Pot Rift and the huge shaft of Rowten Pot. The latter occupies the core of a very gentle syncline and hence is fed by essentially vadose passages from both north and south. Jingling Caves are oriented to the south because of the $\frac{1}{2}°$ southerly dip of the top bed of limestone in the area. Due to thickening of the beds however this trend is not apparent at depth in the Jingling Pot shale bands. The general dip in Kingsdale is about 3° NE but this varies considerably as the thickness of the limestone units changes laterally. Waltham (1970 *b*) attributes the sudden eastern (*right*) The deeply fluted entrance shaft of Cow Pot on Casterton Fell (*below*) Sand Cavern in the Gaping Gill system

Page 322
The Main Chamber of Gaping Gill

swing of Simpson's Pot to a second shallow syncline, but this structure is better described merely as a local flexure of the dip.

Thick shale beds are both uncommon and limited in extent, in Kingsdale, though thin shales (only 1cm or so thick) persist over large areas. The topmost beds are separated by as many as five, closely spaced, shale beds, but none of these is thick. The thicker shales are more common much lower in the limestone succession, and have a tendency to become thinner to the south. Shales may suddenly thin out, merge, or split up, and although such features are rarely seen in the Kingsdale caves, the underground mapping confirms their existence.

THE PRESENT HYDROLOGY

Due to the topography of the Yoredale ridges, the drainage on to the limestone benches on either side of Kingsdale is not evenly distributed. The Gragareth ridge, although 120m lower in altitude than Whernside, has a high projecting spur to the southwest which acts as a gathering ground for many streams, most of which enter accessible cave systems. Several of these can be directly explored to the Master Cave, others have been dye-tested, and the remainder have been interpreted using survey data, flow rates and guesswork. The subterranean hydrology of Kingsdale is best understood by considering each group of tributaries feeding Keld Head.

The first group includes those which can be followed by the caver into the Master Cave. These comprise Swinsto, Simpson's Pot, Rowten (and part of Jingling Pot) streams. The former pair pick up Bedstead and Double Three drainage, as well as small percolation inlets, eventually entering the sump of Swinsto Water to reappear in Rowten Passage in the Master Cave. Here they join the Rowten Pot water, together with the inlets from Frakes' Passage, the smaller of which may drain Bull Pot, whose southwestern sump pool is 16m above the Frakes' Passage inlet. The main rising in Frakes' Passage is too large, both physically and hydrologically, to be accounted for by nearby Jingling Pot. Yordas

Cave has been dye-tested and pulse tested (Ashton, 1966) to Keld Head and its most probable course is via the main rising in Frakes' Passage. Slanting Cave has also been dye-tested, and shown to resurge in Kingsdale below Yordas Cave, before sinking once more. Its point of entry in the Master Cave has not been proved.

Batty Cave stream picks up water from Backtone Gill, before emerging from Kingsdale Head Cave to join Kingsdale Beck. This is also fed by Cluntering, Fidler and Blackside Gills and nothing is known of the underground course of these headwaters, whose points of engulfment depend upon the stage of the streams. In dry conditions each sinks in its bed, but the normal sink is below Kingsdale Head Cave, although in wetter conditions the sink is opposite Bull Pot on the east side of the Dale. Mild flood brings the final sink forward to the old footbridge on the west side of the valley, but in full flood this is soon overrun and the Beck runs on the surface all the way to Keld Head.

On the eastern bench the first stream of direct speleological interest is Gaze Gill but only a small proportion of the water flows through the cave, most of it continuing along the surface course to join the Kingsdale Beck. A subterranean feeder to Gaze Gill is Thack Pot Gill which is swallowed by Heron Pot, and thus directed to the resurgence cave above Gaze Gill. Part of the water promptly sinks only to emerge at a spring to the south.

Beyond Heron Pot only five notable streams sink on the bench. Two of these enter the deep shafts of Spectacle Pot and Growling Hole and vanish into boulder chokes. Nothing is known of the destination of this water, all tests having proved inconclusive, but a stream to the north has been shown to rise at Keld Head. Its most likely course is via the Deep Rising, and the large stream discharged from this syphon seems to be derived from sinks in East Kingsdale and the main Beck.

All the streams enumerated above unite at the River Junction and flow down the Master Cave into a sump at the same level as Keld Head but 1,100m distant. The direct connection is confirmed, since when Keld Head is lowered by pumping, the Downstream Sump falls in sympathy. The river is joined in its unknown course by the misfit streams of the Roof Series, which sink into floor deposits.

South Western Kingsdale does not drain via the Master Cave but pursues an independent easterly course directly to Keld Head. The largest stream is that of Marble Steps Pot which sinks amongst boulders at a depth of 60m and has been shown to emerge at Keld Head. The sump at the bottom of the pot normally takes only a trickle of water, and is at the same level as Keld Head although no connection has been proved.

Low Douk Cave only takes a small stream and, since the altitude of the sump was in doubt, the resurgence had been problematical since the cave was discovered. The question was recently resolved by the sensitive optical brightening agent technique (Crabtree, 1971), which proved Keld Head to be the outlet.

Very little direct drainage reaches the limestone bench between Marble Steps and Swinsto Hole. The small streams seen in the Mohole, Sheepfold and Buzzer Pots are percolation water or bog drainage, and all larger streams sink into higher limestone beds. Since they do not reappear they must be assumed to drain directly to Keld Head as does the sink to the south of Swinsto Hole.

During heavy rainfall many Kingsdale streams advance further down their normally dry beds and utilise additional surface sinks. Besides a simple increase in stage underground, constrictions cause serious backing up under flood conditions. Such ponding occurs in Marble Steps, where water has been observed to rise 45m above the sump, and in extreme conditions the main stream invades the entrance gully rendering it impassable. Flood debris in Low Douk Cave indicates a ponding level of 15m above the sumps.

Swinsto Hole can flood to the roof in the Long Crawl, but most of the excess water continues along the surface stream course to utilise numerous fissure sinks. Some of these drain to Turbary Pot, and others direct water under the boulder floor of the pot into Turbary Inlet in Swinsto Hole. Parts of most Kingsdale caves become impassable in serious floods; the full fury of a flood is best seen in the Master Cave, which fills to the roof in extreme conditions, since the sump cannot take all the water except under pressure.

Ashton (1965) studied the caves of West Kingsdale and other areas by generating artificial flood pulses at the sinks and noting their travel time and height decrease at the resurgence. From these results he calculated the amount of vadose passage between sink and rising of Keld Head, before the Master Cave was discovered. His prediction for Swinsto was a considerable underestimate, but the unpublished data of a test on Rowten Cave and Pot gave a much more accurate prediction. It would seem that more calibration tests through known systems are needed before the results from Marble Steps and Yordas can be interpreted. Simple calculations predict 2,650m of vadose streamway in the Marble Steps system with another 400m of flood-prone passage, while beyond Yordas Cave there may be 400m of waterway with air-space.

STAGES OF DEVELOPMENT

Low Douk Cave exhibits at least three separate stages of development. The first is represented by the new series above the shattered rift. It is a fragment of an old cave, now plugged by ill-sorted glacial till and re-invaded by small vadose streams, Later, the present complex of roof passages was formed via the present sinks. They are predominantly phreatic, but the roof profile reveals that, at maturity, sections of flooded passage were separated by vadose waterfalls. Finally, the vadose trenches developed, hastened by a fall in resurgence level, and drained all the phreatic passages above the present sumps. A second major infill by glacial debris is indicated by masses of gritstone cobbles ranging up to boulders weighing several tons. Their positions suggest that infill occurred as the vadose stage was beginning. All the passage associated with the present entrance drained to the sump, and hence to Keld Head, but the sink and resurgence of the earliest development are not known.

Marble Steps Pot has an even more complex and intriguing history. The initial stage involved the formation of the western entrance series, whose inlets are now blocked, and later the entrance gully. At this time large, predominantly vadose passages entered the Main Chamber and continued down to a phreatic zone. From here it can be argued that a flooded passage extended westwards into Duke Street in Ireby Fell Cavern (Brook & Crabtree, 1969). The Pot was then completely filled by glacial debris whose remains still choke the floor of the Main Chamber. This stage is reflected in nearby pots such as the Mohole and Kail Pot which contain sections of similar fill in various states of resorting. As the ice retreated, melt-water cut the overflow channel to Low Douk Cave and beyond. Both this channel and that below Rowten Pot are clean cut and unweathered, thus indicating their active development after the last major glaciation of the area, but this does not preclude their initiation during earlier withdrawals of the ice. The stream gradually penetrated the fill, removing the fines, but first finding a new route via the 240 Feet Rift. This was a phreatic passage whose original outlet at its western end is now blocked; it was later partially drained by a lower phreatic passage which continued over Stink Pot to probably reappear as the Glory Holes in Ireby Fell Cavern. The final stages occurred when the choke in the Main Chamber was fully penetrated by the stream, and the choked passage towards Duke Street, and hence to Leck Beck, was abandoned in favour of an easier path to Keld Head. Now the vadose shafts to the sump were opened up under high-stage conditions and the present water route to the Main Chamber was developed. The former are a fine example of vadose shortcircuiting and re-invasion of an abandoned streamway. More exploration will be required to unravel the tangle of Marble Steps' history, since many of the observations are open to differing interpretations at the moment.

The West Kingsdale system is one of the best examples of integrated cave development in Britain, since it has numerous explorable vadose feeders and a main drain which is remarkable for the total absence of collapse. Hence deserted passages have been left intact for detailed study. Two levels of development are apparent in the Master Cave. The now abandoned Roof Tunnel and the active upstream passages were once a phreatic artery feeding an old resurgence at a higher level than Keld Head. The present active phreas begins, of course, at the sump which is at the same level as Keld Head. A vadose trench has cut back from the sump to the Master Junction, but it has not fully drained the phreatic passages upstream of this point. A marked groove in the walls of all the upstream canals, some 60cm above the normal water level, indicates another very steady water level active in the not too distant past. The most likely cause for this would be a cold, very dry spell giving very slow flow with little fluctuation. A full glaciation with ice in Kingsdale would probably cause deep flooding of the Master Cave and violent seasonal, or even daily, water fluctuations.

When separate vadose inlets enter a phreatic network it is difficult to say, on morphological grounds alone, which developed earlier. Absolute dating of sediments is the only certain method of ascertaining the order of formation in such cases, but usually some inference of relative age can be made from the position of the original entrance in relation to present active sinks.

From its position out on the bench, and the absence of an associated dry valley, Thorney Pot appears to be one of the oldest sinks in Kingsdale, perhaps formed by a crevasse sink in an ice sheet over the Swinsto area. The passages from this long-abandoned sink are thought to enter the system via the Cascade Inlet, but the latter has also been enlarged by other sources, as yet unexplored. At that time the sump, at resurgence level, was in the vicinity of Toyland, and from here the phreatic passage ran south along the Milky Way into the present sand choke.

As the ice retreated the Swinsto stream developed, and went underground at Turbary Pot, where a vadose shaft dropped into the Turbary Inlet. Much of the inlet, and its original course to the top of Swinsto Great Aven, was initially a phreatic course along joints, but this level has no counterpart in Rowten Pot and was a structural feature caused by a rise to the top of Swinsto Great Aven. The latter has no visible phreatic roof development and was most probably a vadose shaft, dropping to a sump at resurgence level. The original phreatic outlets were the mud-choked inlets into the Philosopher's Crawl and hence via the Master Junction and Roof Tunnel to the resurgence beyond Valley Entrance. High roof pockets near the present entrance indicate that the water level at the resurgence must have been 280m (910ft) OD, or higher, to enable upward solution to take place. The 280m level corresponds with the height of the Raven Ray Barrier which impounded a succession of lakes in Kingsdale and has greatly influenced cave development in the valley.

The Turbary drainage channel enlarged quickly by vadose entrenchment above the Great Aven, which was shortcircuited when Slit Pot formed, thus beginning the creation of Swinsto Final Chamber. Later, mud was deposited above and below the Great Aven, which suggests a glacial phase causing streams to be loaded with sediment. Silt deposited under these conditions was, in time, removed from developing streamways, but not from ponded back waters which were being abandoned. As glacial drift was eroded so the surface drainage was modified, and Swinsto Hole and Simpson's Pot were initiated where streams sank into new fissures. A lower route to Swinsto Final Chamber, via the Cascades, had been opening up, and now pirated all the Swinsto water, leaving the mud-choked high-level route to be further blocked by boulder falls and calcite. The enlargement of the Cascades caused rapid drainage, and speeded vadose action in the Turbary Inlet by flood water from Swinsto Beck.

Simpson's Pot is an example of phreatic initiation, vadose enlargement with phreatic remnants, and the complexities of vadose shortcircuiting. In the lower reaches the connection between Aven Pot and Swinsto Great Aven has been produced by an incidental roof collapse, and at Slit Pot the stream entered a mature but abandoned vadose shaft.

Upstream of the Master Junction the order of development of stream channels is much more difficult to interpret, since, until recently, the passages have been completely flooded and erosion has continued in all passages unless they were completely blocked by sediments. Roof half-tubes give an indication of passages which transmitted the larger streams under the last stages of the phreas, but most evidence of earlier flow regimes has been destroyed by subsequent phreatic erosion.

The 105m deep shaft of Rowten Pot drops into the main system of phreatic tubes between two flooded zones, and it is difficult to tell whether or not it predates Frakes' Passage. Originally Rowting Beck sank at the Eyehole since this lies at the end of the dry valley on the bench. Dating from the time of a retreating ice-cap, such as initiated Thorney and later Turbary Pot, the first sink pursued a tortuous route down the staggered joints until such features as the Staircase Bypass were truncated by vadose erosion. As the progressive sculpturing of the main shaft was taking place, the erosion of the surface drift and the exposure of fresh limestone caused the sink to retreat; Rowten Caves then began to form. The changing position of the sink eventually produced the network shown on the survey (Fig 68). Undercutting of a shale band, and probably pressure from a later mass of ice, resulted in collapse and the formation of Rowten Pot entrance as it is today.

The whole of the Upstream Series is characterised by an abundance of fluvioglacial floor deposits which themselves have affected subsequent cave development. Their presence prevents any solution or corrasion of the floor of the passages. By a damming action these deposits create sections of slow-moving stream-course where finer particles of sediment settle out, a situation to be found in the Mud River Series where the older and less graded layer of cobbles and sand (such as those in Force Passage and the Linking Crawl) have been overlain by silt and fine mud.

Of relatively recent origin, these finer deposits must result from erosion of older sediments further upstream where the water flows quickly. Bulk transportation must occur only during flood since the Mud River is not normally turbid. The discovery of the system under East Kingsdale, drained by the Deep Rising, is awaited eagerly by both the sporting caver and the geomorphologist.

Frakes' Passage must be the main drain from Yordas Cave and the Kingsdale Head caves, as well as Jingling and Bull Pots. The apparently simple phreatic artery model is complicated by the fact that the sump in the SW Passage of Bull Pot lies 12m higher than the top sumps of Frakes' Passage. Kingsdale Head Cave and Slanting Cave Rising contain phreatic passages restricted below a clearly defined level at 295m (970ft) OD, indicating a standstill in the progressive lowering of the phreas. Similar features in Heron Pot and Gaze Gill Cave suggest a standstill level 30m higher on the eastern side of the Dale. More information is needed before these phreas levels can be attributed to local geological structures, valley floor elevations or ancient lake levels.

As the lake outlet through the Raven Ray Barrier cut down, so the water level in the lake fell and the Roof Tunnel water began to utilise a more favourable hydrological path, via the Milky Way and Carrot Passage, to a presumed lower rising near the contemporary water surface in the vicinity of Keld Head. Strong flow markings in Milky Way and Carrot Passage support this premiss. About this time it is possible that water from Lake Kingsdale sank to flow via Duke Street, in Ireby Cavern, to a lower outlet in Leck Beck, but until the dry series above Keld Head is explored this is mere speculation. Within the Roof Series large passages sealed by sand and shingle indicate another deserted route to the west of Carrot Passage but its role in the past hydrology of West Kingsdale cannot be ascertained until entry has been made. The old resurgence of the Valley Entrance was finally sealed completely by glacial till, which points to a significant ice advance; this sealing prevented water from later utilising this outlet in high stage conditions.

Much later, after the retreat of the glacier, the level of the lake dropped until it fell below the postulated High Keld Head rising, and thereafter a lower phreatic network (which had been gradually developing) rapidly pirated the drainage, which now emerged at Keld Head. As the water level within the system dropped, so collapse of well-jointed, eroded rock occurred in Swinsto Great Aven, Final Chamber and Toyland. Vadose action now modified the shape of the phreatic complex around the Great Aven.

When the lake level fell below 260m (850ft) the upstream end of the new phreas, which was down a joint, formed a small vadose step and began to drain the Master Cave. The canal thus produced enlarged the roof tube laterally before the falling level of the new sump, in sympathy with the lake, created a waterfall in the Sump Chamber. From the fall, vadose shortcircuiting created the bridge in the chamber, before a series of retreating rapids formed the fine vadose trench seen below the Master Junction. These rapids will eventually cut back to the Deep Rising and Frakes' Passage, providing the level of the resurgence is not raised by a return of the ice.

The fluvioglacial deposits mentioned previously in the Upstream Series have influenced cave development in a second sense, in that they have provided the mechanism for the vadose corrasion of the trench passage downstream of the Master Junction. Above the trench lies the old roof half-tube, and along ledges which constitute the remains of its ancient floor there are banks of shingle and cobbles from which reindeer bones (of a species extinct for 8,000 years) have been retrieved. Much shingle has been washed into the Downstream Sump to be dumped in the near parts of the phreas where the flow rate suddenly drops.

Stalactites below water level in Keld Head and vadose trenching inside the entrance indicate that when the last lake was finally drained by the Raven Ray gorge, the valley floor was at a lower level than now and the resurgence was a short vadose cave fed by a shallow phreatic network. Debris must have been spread over the floor of the old lake by the meandering Kingsdale Beck, but this must have occurred after the retreat of the last ice in the valley and the most likely source of debris would be the drift and solifluction deposits at Kingsdale Head. High-stage conditions would be needed to transport the large cobbles present in the fill and cause rapid erosion at Kingsdale Head. Hence it is concluded that the majority of the fluvial beds in Kingsdale were laid down during periods of high rainfall.

At the present stage of exploration the chronology of the Low Douk–Marble Steps drainage to Keld Head cannot be equated with the West Kingsdale stages, but it is clear that the small, shallow phreatic, inlet to Keld Head, which has been explored towards Marble Steps, would have been a vadose streamway during the immediate post-lake, vadose phase of Keld Head.

Without dating of cave deposits, no absolute timetable can be ascribed to cave developments in Kingsdale. The networks were probably being enlarged over one million years ago but development of the known caves in this area is closely related to the present dale morphology and is obviously more recent. A possible exception is a fragment of cave high up the dry valley behind Keld Head.

The succession of Pleistocene ice advances have largely been responsible for both the sculpturing of Kingsdale and the pattern of development of its caves. The stepped topography of the hillsides is a result of erosion by glaciers, and the most prominent feature is undoubtedly the Great Scar Limestone bench. Successive glaciations have carved out the valley, each one lowering the floor by perhaps 20–50m. The majority of the Kingsdale caves began to form under the earlier ice sheets and development continued during the interglacials and into the post-glacial period. The older sinks of West Kingsdale were probably initiated subglacially more than a quarter of a million years ago, and deepening of the valley by successive glaciers has produced the present complex of caves. One further important factor which controlled cave development was the presence of temporary lakes formed as the Kingsdale glaciers retreated.

Glacial till completely choked the Valley Entrance, before it was opened up by cavers, but this plug is above the lateral moraine bordering the valley, and may date from some earlier glaciation. Another patch of glacial till lies almost uneroded between Swinsto Hole and Swinsto Beck, indicating the cave is not completely post-glacial, for during a succeeding glaciation the Swinsto entrance sink was choked with boulder clay. The stream then took its present course as this ice melted.

During each retreat of the later glaciers, melt-water washed masses of debris into the caves, in many cases blocking the older passages. Finally, Keld Head became the resurgence, the vadose canyon upstream of the Master Cave sump began to form, and the caves at last took on their present appearance.

17

The Caves of Chapel-le-Dale and Newby Moss

D. Brook

To the south of Ingleborough's satellite, Little Ingleborough, the Great Scar Limestone reaches its highest elevation west of the Ribble, at 458m, on the drift-obscured bench of Newby Moss. To the west, this bench is bisected by the large valley of Crina Bottom, also much choked by glacial debris, before continuing as Lead Mines Moss to the small fault scarp of Green Edge. Here a huge fracture causes the bench to the north to be lowered abruptly in altitude, before continuing its gentle north-easterly dip towards Ribblehead. The position of the majority of the resurgences in the area is determined by the topography of the pre-Carboniferous basement rocks, and classic examples of this controlling factor are seen at Moses Well, Cat Hole, Skirwith Cave, White Scar Cave, Granite Quarry Risings and God's Bridge.

NEWBY MOSS

Newby Moss is characterised by over thirty shafts ranging in depth from 15m to 157m (at Long Kin West). None has any significant lateral development, and they all terminate in chokes of stream-washed debris derived from the glacial drift. This situation may be attributed to structural control, by deep joints, of predominantly vadose subterranean drainage. The scattered position of the shafts and rarity of well-developed surface drainage or dry valleys above them suggests that many of the pots may have been initiated by melt-water sinking into crevasses or at the edge of a small ice field. On the eastern fringes of Newby Moss there are some potholes which possess more normal cave topography, having sub-horizontal streamways caused by bedding planes guiding the primary hydraulic pathways. Newby Moss Pot (length 300m) and Grey Wife Hole (length 425m) are the best examples of this type, the former being terminated by the typical choke, while the latter ends in an unexpected high-level sump pool.

The P2 sink, close by Newby Moss Pot, has been tested, and resurges at both Clapham Beck Head and Moses Well; thus a link with the Gaping Gill drainage has been demonstrated. However, the Long Kin West stream only feeds Moses Well, though via a sink to rising altitude difference of 250m, the largest entirely within the Great Scar Limestone.

The Moses Well rising has a flood relationship with Cat Hole similar to that of Brants Gill Head to Douk Gill (see chapter 19). During flood Moses Well increases in stage to about 1cumec, but continues to run clear, whilst Cat Hole, 300m distant and 20m higher, suddenly discharges a torrent of turbid water.

LEAD MINE MOSS

The majority of the area consists of a bare limestone tableland with sparse grass, pavements and clint fields in which passable openings are rare. On its borders however a number of streams sink into the limestone. On the southern margin, the Hard Gill stream flows for some distance across the outcrop of the Great Scar Limestone since its valley is floored by boulder clay. It eventually sinks at Greenwood Pot, a tortuous, 51m deep pothole, but in floods it continues across alluvial flats to the impenetrable Rantry Hole sink. This is thought to drain to Skirwith Cave, a now-defunct show cave, with a total of more than 1km of passage terminated by a series of sumps.

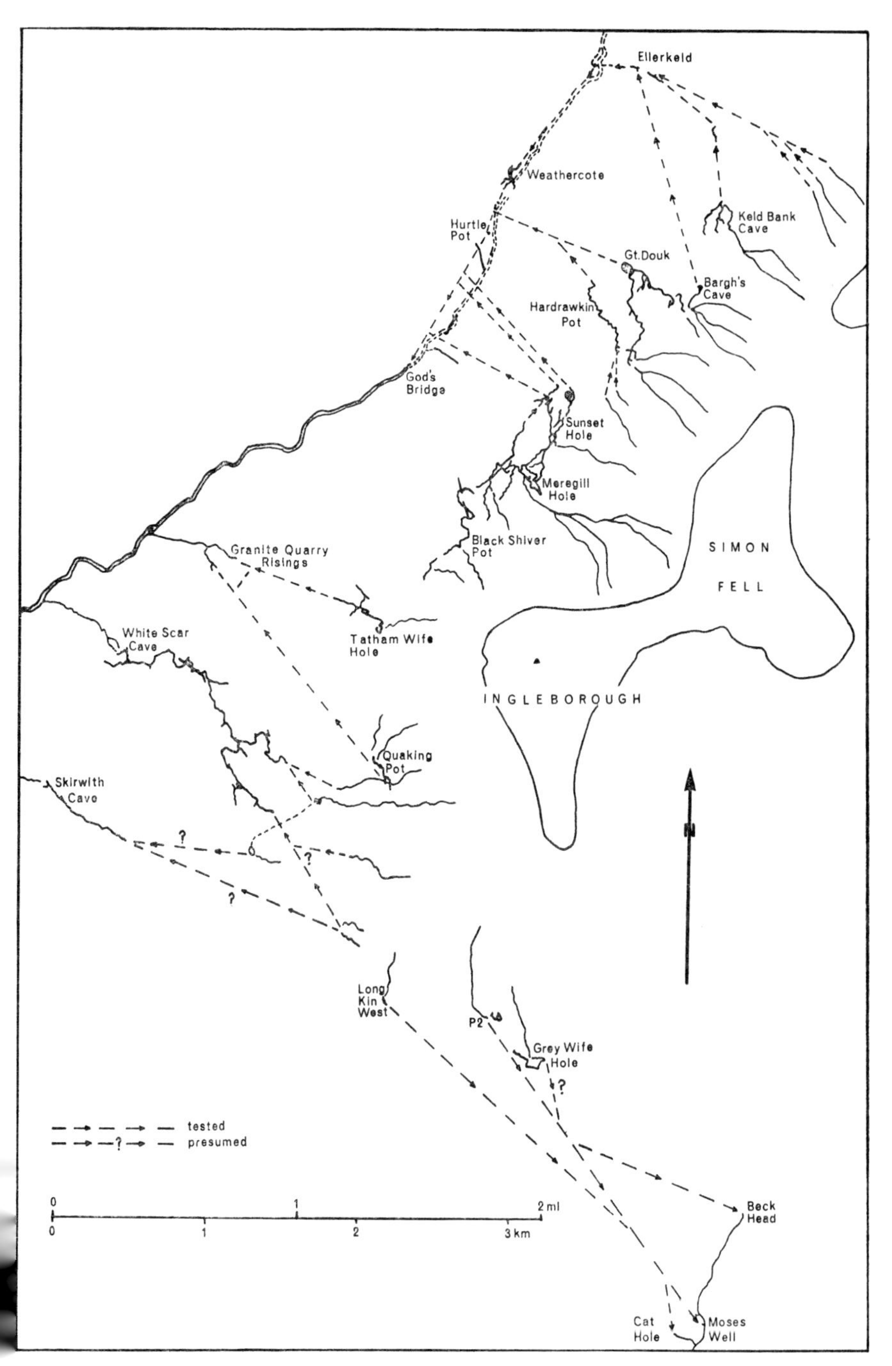

Fig 70 The caves and hydrology of the eastern side of Chapel-le-Dale and the southern slopes of Ingleborough

The water in Greenwood Pot reappears in White Scar Cave, a magnificent streamway over 3km long with several inlets. One of these is very extensive and has an ancient phreatic level about 20m above the main drain; decorated phreatic tunnels at this same level have also been found leading from a huge, 70m long, chamber located above a boulder choke in the main streamway. White Scar Cave passes under Lead Mine Moss but only one other sink close to Greenwood Pot has been shown to feed the large stream in the cave. It is suggested that sinks to the west of Newby Moss, around Boggart's Roaring Hole, drain to White Scar Cave and provide the source of the cave river.

Quaking Pot, 300m north-east of Greenwood Pot, is a series of vadose passages and pitches which eventually become too narrow to explore, but the water next sees the light of day at Granite Quarry Risings. The dividing line between the White Scar and Granite Quarry drainage basins appear to be a long NW–SE fracture prominent on aerial photographs. Between Quaking Pot and Green Edge, the northern boundary of Lead Mines Moss, there are several minor pots with no significant passages. At the Yoredale/Great Scar boundary, on Green Edge, the fault throws 12m down to the north and brings the Hardraw Limestone against the Great Scar Limestone. Here water rises from the Hardraw and promptly sinks into the Great Scar, at Tatham Wife Hole. The cave is 1km long and 155m deep, the principal stream course running along the Green Edge Fault whose plane dips steeply to the north. North inlet is the only tributary passage of any size and originates in inlets on the North Branch Fault which dips steeply to the south. The system is the northern limit of drainage to the Granite Quarry Risings.

CHAPEL-LE-DALE

Beyond Green Edge and the boggy plateau of Tatham Wife Moss are Green Ridge Caves, which comprise essentially a shallow vadose channel occasionally breaking out to the surface. The continuation of this system forms Black Shiver Cave and then the

Page 339

The main shaft in Hurnel Moss Pot

Page 340

(*above*) Roof pendants in a bedding plane passage in the upper part of Bar Pot, Gaping Gill (*below*) Trow Gill looking upwards

pot of the same name. In Black Shiver Pot, the water drops deeper, via a series of short vadose shafts controlled by a small fault, until a north–south fault has caused a shaft 85m deep. Further vadose passages and waterfalls lead to the sump which is back on the fault plane; above this level and also on the fault is a deserted passage choked by silt. The water proceeds via the lower passages of Meregill Hole to the God's Bridge resurgence.

The area around Meregill contains two distinct types of cave development. One is typified by the P101–Hallam Moss Cave system which is a simple bedding plane passage, running with the dip, only 5m below the clints for 1,000m, until it encounters a shatter zone in Roaring Hole and drops into a boulder choke. Meregill Hole, on the other hand, encounters a strong fracture at its point of engulfment and falls quickly via the high and impressive Canyon to a main drain at a depth of 150m; this is also a downdip passage controlled by bedding planes. The Black Shiver inlet (The Torrent) is formed in the same bedding planes and has similar morphology.

Sunset Hole to the north is a mixture of these two extremes; the abrupt descent occurs after 300m of trench passage, but ends in a massive zone of boulder chokes beneath the immense doline of Braithwaite Wife Hole. The latter appears to be bounded by two north–south faults, and these mark the termination of the block draining directly to God's Bridge.

The streams on Southerscales Fell are known to reach God's Bridge, but via Hardrawkin Pot or Great Douk Cave, before all uniting at Hurtle Pot. In the former the shallow passages eventually drop 30m down a joint into a sump, while Great Douk is a huge collapse hole fed by 800m of large vadose passages. These originate at the Middle Washfold Sinks and a similar length of tributary cave carries the water from Southerscales Pot. The latter inlet includes a fine example of perched phreatic development held up by geological structure.

Southerscales Pot takes water from a sink on Fenwick Lot, but here there is a watershed between the subterranean catchments of God's Bridge and Ellerkeld, further north. The latter is a resurgence for Bargh's Cave, Scar Close Cave and Keld Bank Caves which are all shallow bedding plane developments. Bargh's Cave pursues an independent course to the lowest Ellerkeld rising and little is known of its underground path but Scar Close and Keld Bank Caves unite before taking a very shallow course beneath the clints to Ellerkeld top rising.

CHAPEL BECK

Beneath the floor of Chapel-le-Dale the beck plays hide and seek before finally reaching the base of the limestone at God's Bridge. The course is conveniently divided into the two sections above and below Weathercote Cave. The water rising at Ellerkeld soon sinks again, and must join Chapel Beck near its sink at Haws Gill Wheel. From here a shallow and often flooded course resurges at Weathercote Cave to drop dramatically to the level of God's Bridge. Further down the dry river-bed, Jingle Pot and Hurtle Pot are windows into the active phreas, and the water from Great Douk reappears between Weathercote and Hurtle Pot. Nearer God's Bridge, Midge Hole and Joint Hole have revealed, at a depth of 6–8m, two huge, flooded tunnels which seem to be unrelated to each other. Meregill Skit contains a similar tunnel at a depth of 25m, which indicates that the pre-Carboniferous basement must descend steeply behind its outcrop just below God's Bridge, only 200m away.

The west bank of Chapel-le-Dale is undoubtedly cavernous; the endless limestone pavements of Scales Moor overlook a whole series of small risings 150m below. Dry Gill Cave has a large catchment area but can only be explored for 200m to a sump pool. Further down the valley Dale Barn Cave contains some massive phreatic chambers, and an important streamway beyond a sump which therefore awaits further attention. These and the almost complete lack of ancient phreatic development on the Meregill side indicate merely some of the problems and potential of the valley.

18

Cave Development in the Gaping Gill System

R. R. Glover

South-east of the narrow, curved, ridge connecting the summits of Ingleborough and Simon Fell an undulating plateau spreads south for well over 1km at a height of about 400m (1,300ft). In places, wide but low mounds of glacial drift give rise to patches of boggy ground, but elsewhere the surface is grassy, with fragments of limestone pavement showing through. Small streams, rising on the south-east slopes of the Clapham Bent corrie, unite to form Fell Beck, which runs in a trench-like valley cut into the drift. After flowing south for more than 1km, the stream starts to flow over nearly horizontally bedded limestone, and begins to lose water through innumerable cracks and joints. A few hundred metres further on, the stream abruptly disappears down a large oval hole in its bed. Beyond, the valley ends in a semicircular slope of drift over 10m high.

Called Gaping Gill since time immemorial, this is possibly the most widely known pothole in the British Isles. In flood, Fell Beck hurls itself over the lip of the shaft, falling 110m to the floor of the enormous Main Chamber, which under these conditions becomes filled with flying spray, driven by winds of hurricane force. It constitutes one of the highest waterfalls in Britain.

Today, over one hundred years after the first attempted descent by John Birkbeck, and the scene of some of the earliest speleological explorations in the country, Gaping Gill continues to exert a fascination for sightseer and potholer alike. In view of this long history of interest, exploration and discovery, it is surprising that theories of the origin and development of this magnificent system have hardly advanced at all over the years.

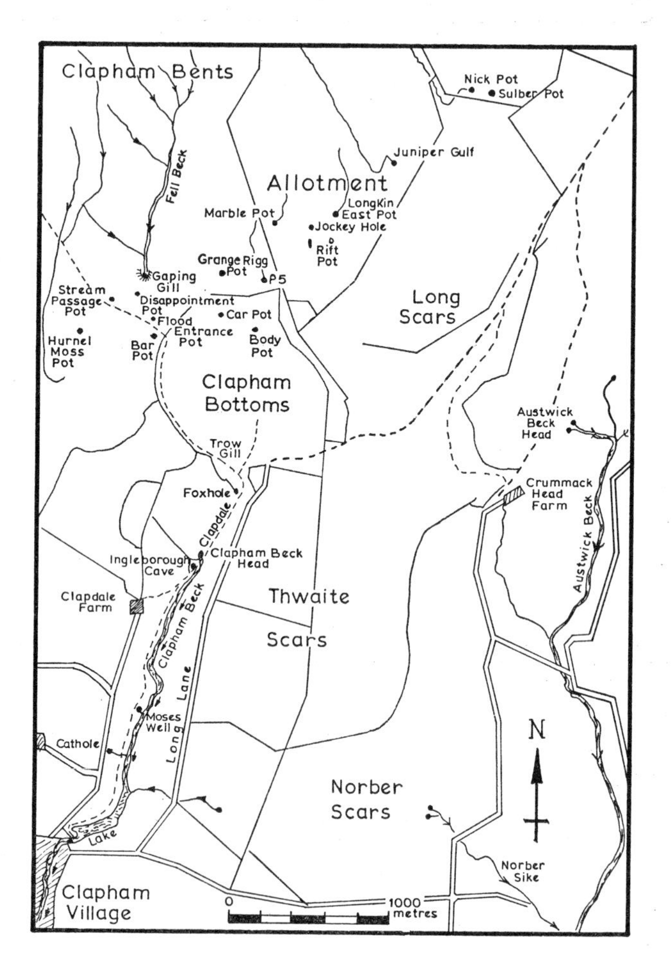

Fig 71 The Gaping Gill–Ingleborough Cave region

GEOLOGY OF GAPING GILL

The Great Scar Limestone in the area is up to 180m thick and its upper boundary is usually taken to be the *Girvanella* Band, which here outcrops in the bed of Fell Beck above Gaping Gill. The Porcellanous Band, a dense blue-white limestone, marks the boundary between the S_2 and D_1 zones. Clearly traceable at the surface in Crummack Dale and in Chapel-le-Dale, it is not seen in the area except in parts of the Gaping Gill system, notably in the Main Chamber. The limestones lie unconformably upon lower Palaeozoic slates and grits which locally show considerable relief, and which can be seen at the 200m (650ft) level near the head of the lake, where the path crosses the normally dry stream bed below Cat Hole. The presence of a ridge in the basement, lying to the north-east of Gaping Gill, has been suggested in order to explain the drainage anomalies of the area (Carter *et al,* 1904).

This relatively undisturbed thickness of limestone abruptly ends to the south-west along the line of the North Craven Fault, which passes through the head of the lake, but is not well exposed. North-east of the fault line, a related series of fault/fracture zones can be clearly traced which have to a large extent controlled the initial penetration of surface waters (see Fig 72).

SURFACE LANDFORMS OF THE AREA

South of Gaping Gill, the 400m (1,300ft) plateau forms a series of bare limestone pavements dissected by faults and joints, and culminating in Clapdale Scar, overlooking the Craven Lowlands. To the east, a great expanse of pavement, Long Scar, sweeps round the east flank of Simon Fell, below the Allotment, and ends in the block-faulted ridge of Thwaite Scar and Norber, overlooking the village of Austwick. Crummack Dale lies to the east of the ridge, while to the west Clapham Beck flows in a deep, narrow ravine incised in the floor of a wider shallow valley.

The mouth of Ingleborough Cave opens at the foot of a small limestone cliff just below the point where Clapham Beck rises

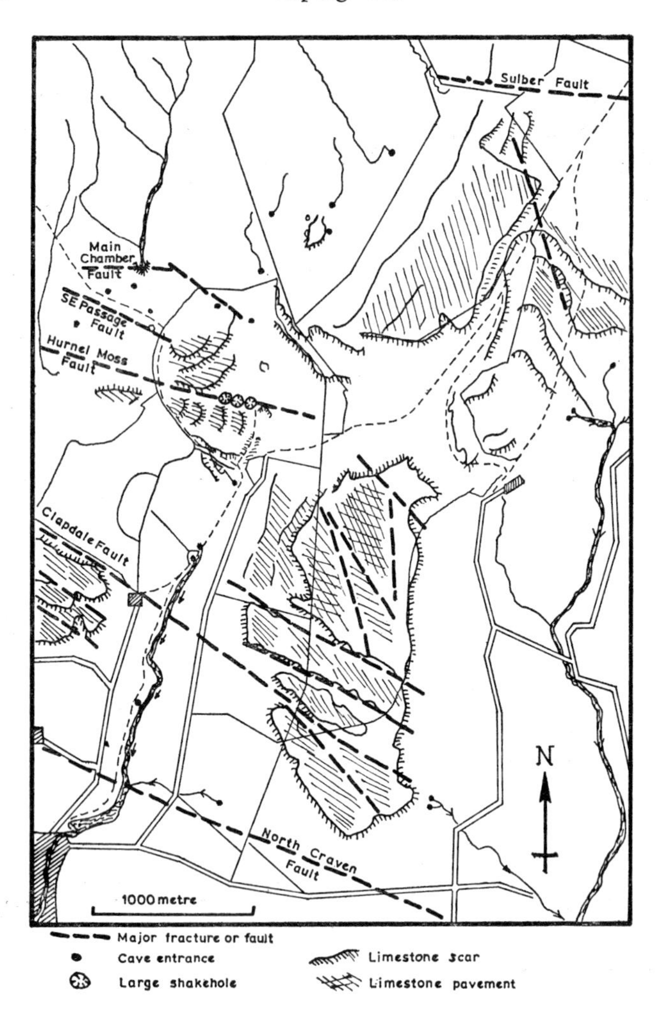

Fig 72 Main features of the surface morphology and geological structure in the Gaping Gill region

in a low bedding cave at an altitude of 275m (825ft). Above Beck Head, the dry valley of Clapdale rises steadily in a northerly direction for 500m, to the junction of the impressive dry ravine of

TABLE 12 *History of exploration in Gaping Gill and Ingleborough Cave*

Date	Events in Gaping Gill	Events in Ingleborough Cave	Reference
1837	—	Old Cave opened and Lake Avernus reached	Hill, 1913
1842	J. Birkbeck descends Main Shaft to ledge	—	—
1892	Descent planned by Yorkshire Ramblers Club	—	—
1895	E. Martel makes first descent	—	—
1896	First British descent, East Passage discovered	—	Calvert, 1899–1900
1900–1902	Yorkshire Geological Society's water tracing on Ingleborough	—	Carter *et al*, 1904
1905	Discovery of South West and South East Passages	—	—
1909	Discovery of Flood Entrance Pot	—	Greenwood, 1910
1914	—	Foxholes excavated	Broderick, 1924
1935	Rathole first descended	—	—
1937	Hensler's Passage discovered	—	Simpson, 1937–8
1943	Grange Rigg Pot explored	—	Heap, 1964
1944	Disappointment Pot discovered	—	Gemmel & Myers, 1952
1948	Car Pot opened	—	Brindle, 1949
1949	Stream Passage Pot opened	—	—
1949	Bar Pot opened	—	Atkinson, 1952
1951	—	Beck Head Cave discovered	—
1953	—	Inauguration Cavern reached	—
1957	P5 discovered	—	—
1963	Bar Pot–Hensler's Passage link opened	—	—
1968	Whitsun Series and Far Country discovered	—	Brook, 1968
1970	Hurnel Moss Pot discovered	—	—
1970	—	Terminal Lake Extension explored	—
1971	Far Waters Entered	—	—

Trow Gill and the narrow dry gully leading to Clapham Bottoms.

Trow Gill runs for over 300m in a north-westerly direction from the gate at its foot, and forms one of the best known and most striking features of the landscape in the area. It consists of a narrowing, steep rock-walled gorge cut in solid rock along the line of the major joints in this vicinity. The floor rises a total of 50m, climbing more steeply over boulders at the head of the gorge, where the towering walls converge to within 2–3m of each other, forming a deep narrow slot over the last 50m. On both sides, development of low, shallow, bedding caves has occurred along a conspicuous bedding plane, giving the impression that the gorge originated as a cave, whose roof has subsequently collapsed. However, the wide mouth of Trow Gill is unlike the dimensions of any caves and therefore inconsistent with an origin of cavern-collapse; there is also a marked lack of debris which is characteristic of the floor of collapsed caves elsewhere in the world. Instead, Trow Gill is just a surface gorge incised into the limestone. It is comparable, though on a rather more spectacular scale, to a number of other gorges in the area, some still active, where cave formation and collapse plays only a subordinate role in the excavation of the streambed. Trow Gill almost certainly dates from a periglacial environment when the main caves were sealed by ice.

Above the narrow slot at the head of the Gill, the dry valley continues climbing to the north-west, then to the north-east, until plateau level is reached near Bar Pot. Halfway between Trow Gill and Bar Pot, at the bend in the valley, a shallow dry tributary from the north-west hangs above the main valley.

Clapham Bottoms consists of a wide flat-floored valley lying at an altitude of 300m, surrounded by limestone scars, and with tributaries entering from the north-west, north, north-east and east. In places, bedrock shows through the grassy floor in the form of small semicircular limestone coves, backed by small areas of deeply dissected pavement. In the south-western corner, three very large dolines occur on the valley side, in a line running east–west. It is clear that Clapham Bottoms once formed the head of the older, wider, valley now containing Clapdale, and has been isolated from it by the high moraine barrier visible at the bend in the path below the Trow Gill gate. The gully now giving access to Clapham Bottoms may represent a glacial melt-water channel breaching the barrier and possibly draining a temporary glacial lake occupying the Bottoms.

A few metres below the bend in the path, at the foot of Trow Gill, a low cliff face on the left overhangs the entrance of a small cave known as Foxholes. The cliff above Foxholes lies at the foot of a curious minor gorge running very close to and almost parallel with Trow Gill. Halfway down the length of the gorge there is a very large choked pothole, larger even than Bar Pot. The upper reaches of this gorge are beheaded by the dry valley above Trow Gill, and therefore the gorge itself must predate the latter.

The pavements of Clapdale Scar, Thwaite Scar and Long Scar show the characteristic clint and grike formation, being broken into blocks averaging 2m long by 1m wide, by two main sets of joints. The more persistent set, the major joints, trends NW–SE, parallel to the nearby North Craven Fault, but swinging further north–south away from the fault (Wager, 1931). The less persistent set of joints lies approximately at right angles to the major set, and an intermediate set is occasionally found bisecting the angle between the two dominant sets, but these are rare and impersistent.

Abrupt changes in joint spacing and direction occur at the fault/fracture zones, mentioned above. These zones are in places over 1m in width, and the joint spacing within the zones is reduced to the order of centimetres, parallel with the line of the fault. Over much of the area of the pavements, the beds dip uniformly 1°–2° NNE, but close to the fault zones the dip may increase rapidly by 5°–10°. This may be seen along the southern edge of the east–west fault which runs from Clapham Bottoms to Hurnel Moss, especially just west of Trow Gill. North of this line, the dip appears to have a slight westerly component, especially between Gaping Gill and Clapham Bottoms.

HYDROLOGY OF THE AREA

Fell Beck is by far the largest (mean flow 40–50l/sec) of the many streams that sink at the inner edge of the plateau and it had always been thought to reappear at Clapham Beck Head, but not until 1904 was this belief confirmed.

During the classic series of water-tracing experiments carried out on Ingleborough by the Yorkshire Geological Society, 1 ton

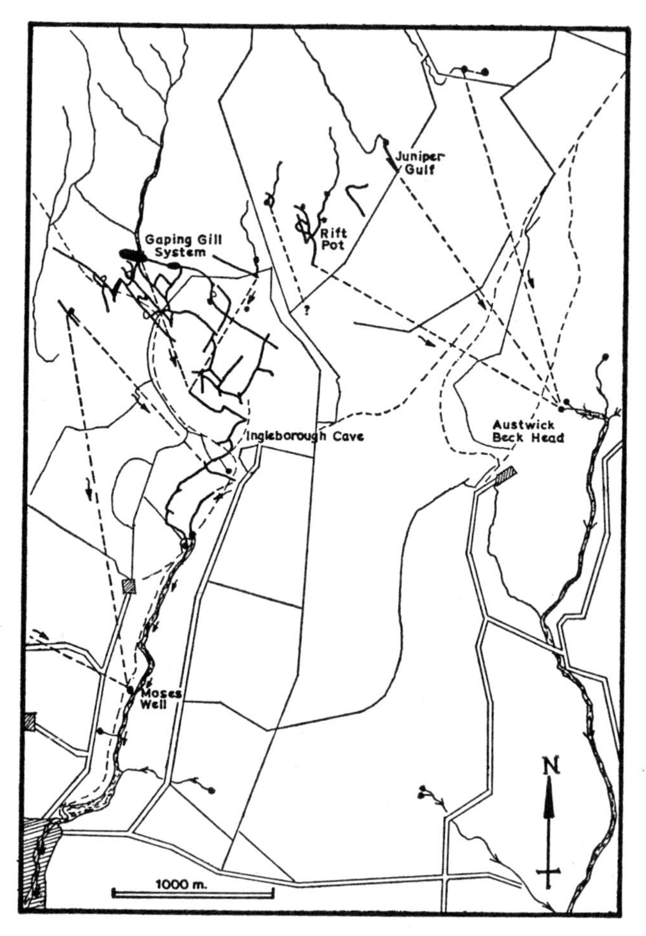

Fig 73 The caves and hydrology of the Gaping Gill–Allotment region

of common salt was added to Fell Beck a few metres upstream of Gaping Gill (Carter *et al*, 1904). Daily samples of the water from Clapham Beck Head were tested for chloride, as were samples from the only two other springs of any size in the area—Austwick Beck Head and Moses Well. The tracer was first detected in the Clapham Beck Head water eleven days later, reaching a concentration peak after a further three days. It was not detected in water from the other two springs. Subsequently, the same group showed that Austwick Beck Head was fed by streams sinking in the several deep potholes in the Allotment. The presence of the underground watershed mentioned above was deduced on the basis of these results.

Moses Well, a small spring (5–10l/sec) lying at about the 200m level close to Clapham Beck near the head of the lake, has been shown to be fed by streams sinking in the Newby Moss area, in particular that entering Long Kin West. Cat Hole, lying to the south of and higher than Moses Well, has never been tested since it flows so infrequently. It lies on the basal unconformity and probably only functions as a high-level flood overflow for the system behind Moses Well. When in flood the flow is many times that from Moses Well.

Of the many other small streams that sink on reaching the limestone between Newby Moss and the Allotment, only those entering Hurnel Moss Pot and P5 have been tested successfully. Results appear to show that, in mild flood conditions, the water from Hurnel Moss reappears at both Clapham Beck Head and Moses Well. Flood water sinking in P5, on the edge of the Allotment, has been proved to rise at Clapham Beck Head. Marble Pot, in the Allotment, close to P5, has been tested on many occasions without success.

It is interesting to note that although water appears to traverse the Gaping Gill system very slowly, taking 11–14 days to travel 1·5km under low flow conditions, the system responds to natural flood pulses in a matter of hours. Moreover, the low $CaCO_3$ content of the water reappearing at Clapham Beck Head at the peak of a flood suggests that it has not spent long underground. Recent dye tests show that the water sinking in the floor of the Main Chamber of Gaping Gill is seen again in the static sump at the foot of South East Pot, nearly 50m below the level of Main Chamber on its route to Ingleborough Cave. It would appear that the normal channels for low flow are at depth, and probably with a considerable reservoir of flooded passage, while under flood conditions, higher-level overflow channels convey the flood water with much greater rapidity as far as Ingleborough Cave.

Many more water-tracing, water-sampling and stream-gauging experiments will have to be carried out under a wide range of flow conditions before the full details of the hydrology of the area are known.

THE GAPING GILL SYSTEM

Presently totalling over 10km in length, with five subsidiary entrances (a sixth, Car Pot, providing visual connection only via a narrow bedding plane), the ramifications of the system of passages leading from the Main Chamber of Gaping Gill make it one of the most interesting cave systems in the country. Among the more distinctive features are vertical systems apparently independent of horizontal development, great accumulations of clastic sediments, the absence of major active streamways and, above all, the pattern of passages radiating outward from the Main Chamber, but under joint and fault control.

Much of the system consists of a maze of low, sub-horizontal, rock-floored tunnels at three or four different levels at around a depth of 110m below the surface of the moor. Most of these passages are normally dry, but parts of the Far Country and Far Waters contain long pools with little airspace. Most of the remaining passages at this depth consist of rifts of irregular width and height and whose floors consist mainly of mud and boulders. The only extensive passages developed above this horizon are the Stream Passage Pot and Disappointment Pot inlet systems.

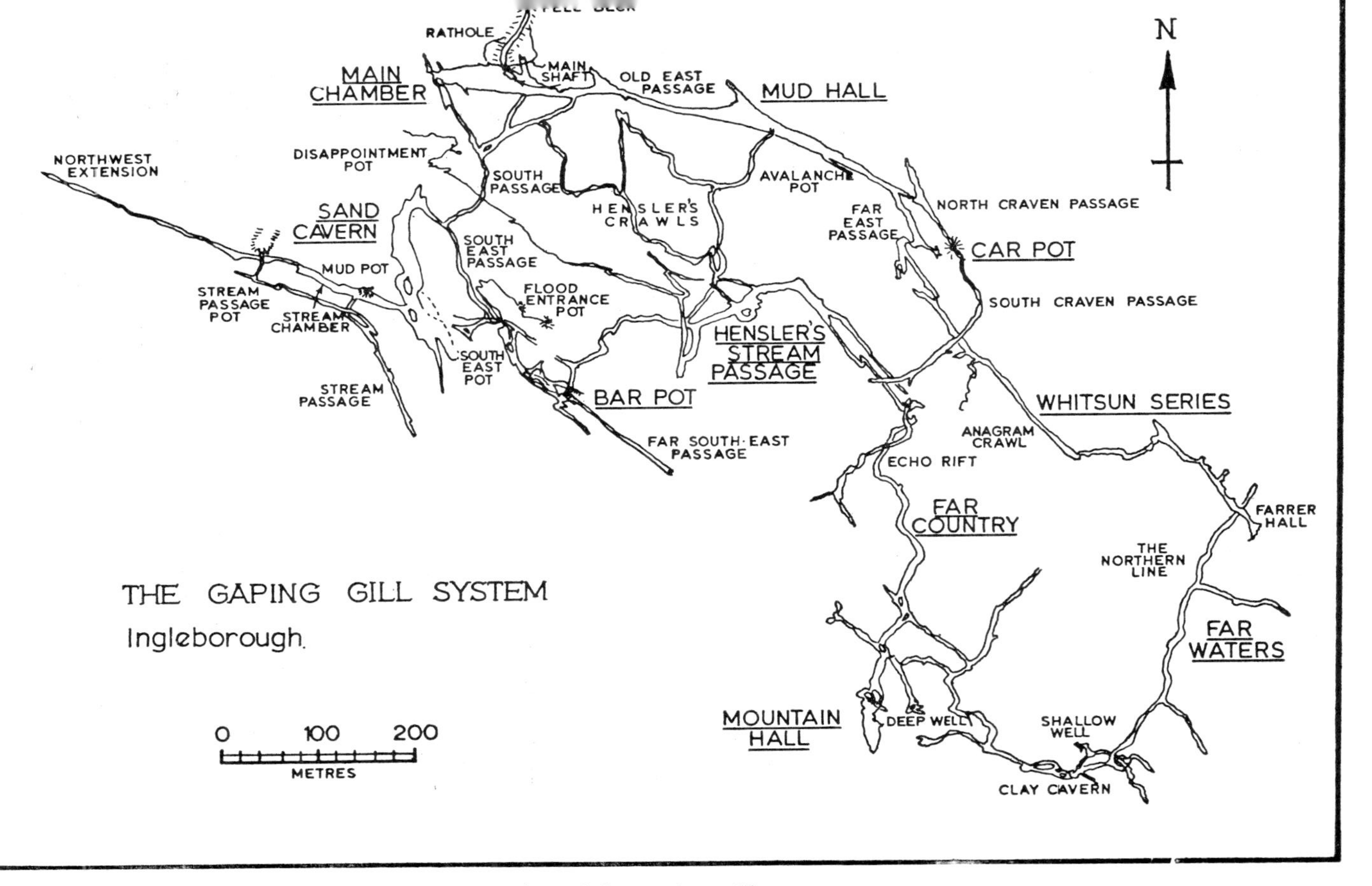

Fig 74 Plan of the Gaping Gill system

This 'ground-floor' system is intersected at many points by shafts which drop from the surface (the Main Shaft) or form high avens (Bar Pot and Mud Hall Avens). Some shafts extend below this level (South East Pot). A series of chambers are found at the main level at various points in the system with the Main Chamber (a problem in itself) as the most striking example. The true shape of many of the chambers is concealed by the extensive deposits of mud and boulders. Active streamways are few, as most streams drop rapidly to sumps or, as in the Main Chamber, sink through the floor of boulders. Hensler's Passage, which is the main exception, is largely fed by the Disappointment Pot inlet.

The apparent complexity of cave passages in the Gaping Gill system is considerably simplified by separating them into five broad morphological types. Most individual passages can easily be ascribed to their correct type by quick examination of their major features. Furthermore such a subdivision is valuable in that it also relates to the patterns and chronology of development of the cave systems as a whole, and this in turn relates to the evolution of the surface landforms of the area.

Vadose inlets

Included in this category are the shafts, potholes and sinkholes which allow the entry of surface streams, or have done so in the past. In nearly all known cases the water leaves the surface down a joint or fault plane. The extent of sub-horizontal cave development below is variable, but wherever formed these passages have the characteristic vadose T-shaped cross section, meandering plan patterns and consistently descending stream routes. The largest of all the active inlets is the Main Shaft—exceptional in having no associated horizontal development.

The Main Shaft is an oval, joint controlled, vadose shaft formed in the line of a fault zone (hereafter referred to as the Main Chamber fault) which has played a major part in controlling the development of the Main Chamber. A breccia-band, 1m thick, is

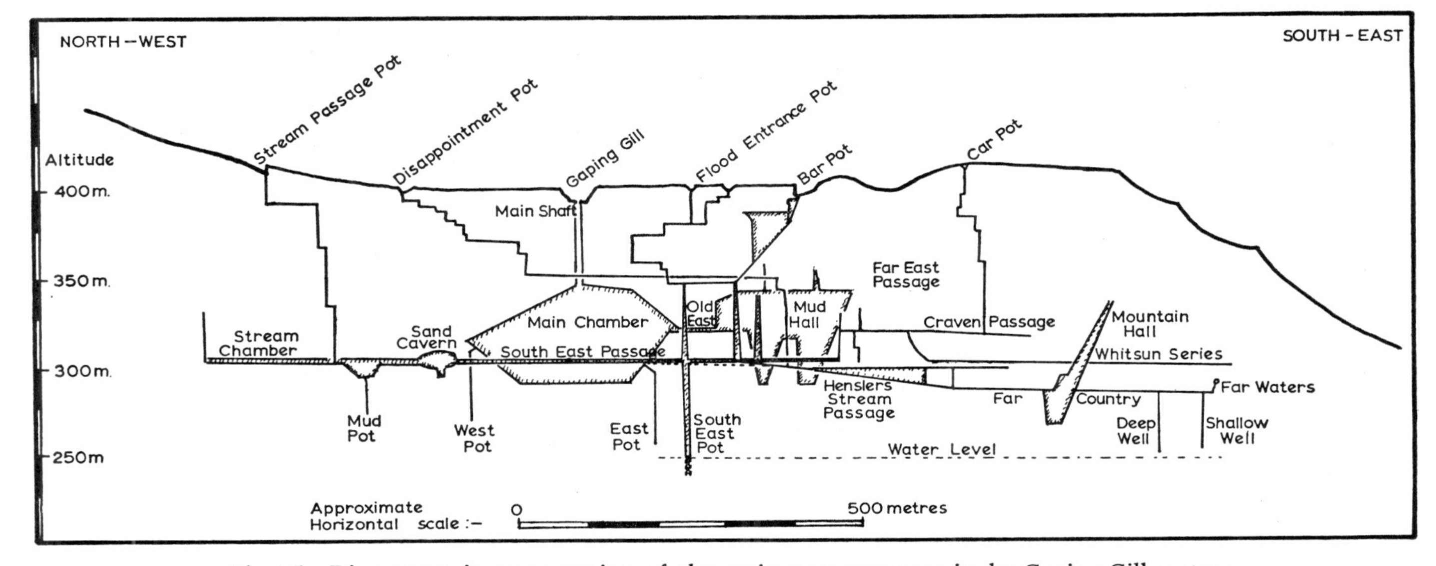

Fig 75 Diagrammatic cross-section of the main cave passages in the Gaping Gill system

visible at the south-eastern end of the oval shaft for much of the upper 60m. Vertical displacement at the shaft top is no more than 15cm. The north-west wall of the shaft opens 20m down into a parallel shaft, the Lateral Shaft, down which the waters from the Rathole, Jib and Spout Tunnel systems pour. This upper portion of the shaft, down to Birkbeck's Ledge at a depth of 60m, shows vadose features in common with many of the deep joint and fault-controlled potholes of Ingleborough and elsewhere. Birkbeck's Ledge represents the last vestige of the floor of the vadose rift which formerly conveyed the waters of Fell Beck from the foot of the shaft in a north-westerly direction, prior to the development of the great void of the Main Chamber below.

The vadose inlets of Stream Passage Pot and Disappointment Pot are similar in form to many of the potholes found elsewhere in Craven, and are relatively new (probably interglacial), having developed from earlier bedding systems by the action of bog and drift drainage streams. OBJ Pot and Frustration Pot possibly represent less well-developed forms of the same general age. Many more systems now hidden under drift may be found in the future by surface digs or climbs from below. Adjacent systems of this type show strong correlation in joint and bedding control, as shown by the fact that the major horizontal development occurring in Stream Passage Pot, Disappointment Pot, Flood Entrance Pot and in parts of Bar Pot is found at the same three horizons (see Fig 75).

In Disappointment Pot, the influence of joints and bedding in control of development is very well demonstrated. The main passage between the acute bend near the entrance to below the third pitch runs for over 300m along one main joint. In addition, the roof of the passage between the second and third pitches lies on one shale band, with the stream occupying the floor of a narrow meandering canyon. Extensive bedding developments above and beyond the first and third pitches suggest at least two distinct stages in the evolution of this system.

In contrast with these essentially vadose inlets, the tight,

Page 357

(*above*) Fell Beck disappearing into Gaping Gill Hole, as seen from the air. A large number of shakeholes—subsidence dolines—break the monotony of the drift-covered plateau (*below*) A low bedding-controlled stream passage in Out Sleets Beck Pot

Page 358
The open shaft of Alum Pot

joint-controlled, passages of Flood Entrance Pot appear, from their triangular cross section, to have developed under phreatic conditions and have suffered little vadose modification by the bog drainage inlets now occupying the system. It is possible that these passages form part of an extensive high-level phreatic network along the line of the South East Passage Fault which includes the upper series in Bar Pot, and which connect, almost by accident, with the lower series via a series of pre-existing vertical fault shafts.

Both surface and underground evidence suggest that there exist at least two further fault zones to the south of that responsible for the Main Shaft. Shear movements have taken place along each fault and have resulted in the production of fault breccia bands over 1m in width, which penetrate the limestone over almost its whole thickness. These fracture zones have undoubtedly played a major role in the development of the system by facilitating the entry of major surface streams.

The most clearly defined of these two contains Far South East passage from the aven at the end, South East Pot, the southern end of Sand Caverns, and Stream Chamber as far as the high aven at the end of the recently discovered extension. This will be referred to as the South East Passage Fault. Lateral displacements of the order of 5cm of sections of solution tubes in the roof of Far South East Passage are evidence of movement along this fault since the formation of the passage. The enormous quantity of fill occupying the southern end of Sand Caverns, and the Stalagmite and Stalactite Chamber series still further to the south, may represent the foot of a giant talus cone filling a major shaft from the surface. The top of the shaft could now be hidden by surface drift, but it may well have swallowed Fell Beck in earlier times.

To the south of South East Passage Fault, the shallow dry valley running from below Hurnel Moss Pot to upper Trow Gill may lie on another fracture zone (Hurnel Moss Fault), which in turn may have allowed Fell Beck to sink underground during an even earlier phase of development.

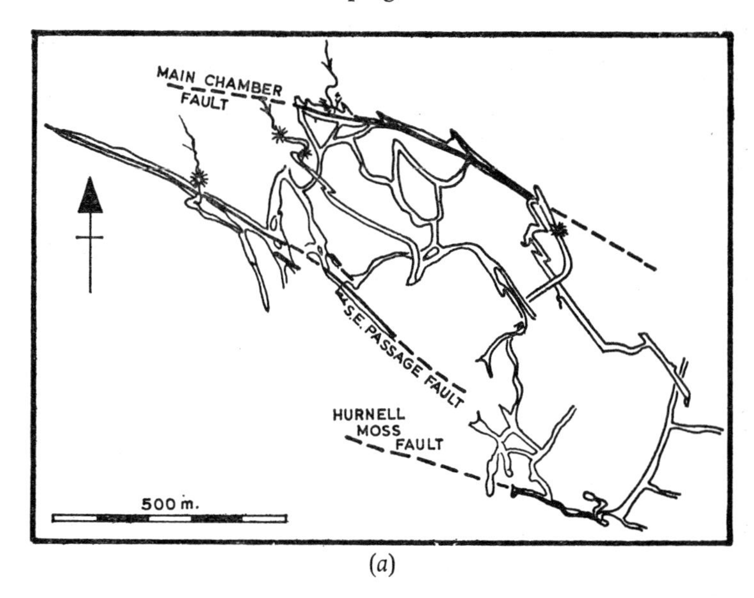

(*a*)

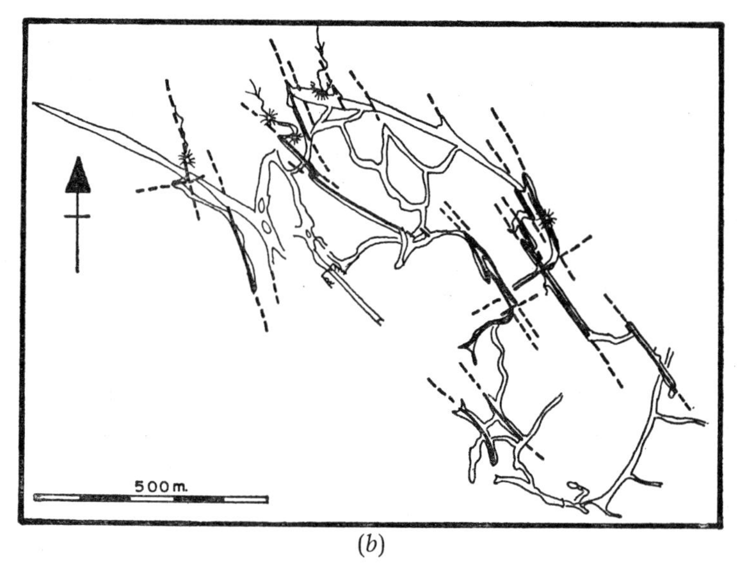

(*b*)

Fig 76 Geological structure of the Gaping Gill caves: (*a*) major faults; (*b*) principal joints

In Disappointment Pot, the south wall of the last pitch consists of a vertical face of sand and gravel which rises out of sight. This may represent the foot of yet another large choked shaft, briefly occupied by a major stream which was responsible for the vadose development now seen in Hensler's Master Cave.

Aven systems

In many parts of the system, high avens are found, contributing small inlet streams which are slowly clearing mud and sand fill from the floor of the joint and fault-controlled passages below. The majority of the avens owe their origin to such inlets, and are merely widened joints.

The three large avens in South East Passage, below Flood Entrance and Bar Pots, are clearly in a different class to those mentioned above. They occur in close proximity, along the line of the South East Passage Fault, under an area of moor with little or no drift cover, and are linked at a depth of 70m by an extensive series of bedding planes. Collapse on a massive scale into the chambers formed at the intersection of the bedding caves with the deep fault rifts has resulted in the formation of Bar Pot. In addition a large, old, phreatic tube links all three avens some 10m above the present floor level. South East Pot, extending some 50m below the present, false, floor level of Far South East Passage, demonstrates the depth of vertical development that has taken place in the fault plane. It is clear that these three avens, while owing part of their present form to vadose inlets, have a long and complicated history and may have either originated or acted as phreatic risers associated with a resurgence in the vicinity of the semicircular cove at the head of Trow Gill.

Vadose trunk routes

Hensler's Stream Passage is the only clearly identifiable vadose main stream passage in the lower levels of Gaping Gill. Blockfall along the major faults has largely obliterated evidence in other parts of the system. The streams in Stream Passage Chamber, Far

East Passage and in Hensler's Stream Passage are underfits in normal weather conditions, although overspill from the Main Chamber via the various bedding routes mentioned below must reactivate vadose erosion in Hensler's Passage during peak flood periods.

Major caverns

In at least five places in the Gaping Gill system, passage enlargement by blockfall along fault, joint and bedding planes has resulted in the development of very large chambers in striking contrast with the network of low tunnels which link them. Much the most impressive is the Main Chamber of Gaping Gill; it measures some 145m, by a maximum of 25m wide along an axis which runs WNW–ESE, and reaches a maximum height of over 35m above the floor close to the north wall. The Main Shaft enters close to this wall near the eastern end of the chamber, and allows enough daylight to enter to enable the rough dimensions to be appreciated without the aid of lamps. The chamber floor is composed of a level layer of water-worn boulders with large sandbanks rising 1m or so in places, although the exact shape and size of these vary due to the effects of heavy floods. Under normal weather conditions, nearly all the water enters in the form of two waterfalls, via the Main and Lateral Shafts, and sinks among the boulders on the floor. Under flood conditions, as many as eight separate waterfalls have been seen, not including a major inlet which has been reported entering high up on the south wall. The floor of the chamber is occupied by a large lake, up to 6m deep, under these conditions.

The north wall of the chamber rises vertically over its full height, but the south wall is vertical for the first 10m only. Above this height it angles sharply over at nearly 55° to meet the other wall. This hanging roof is very shattered, due to the heavily jointed nature of the rock being attacked by high-pressure jets of spray driven by the gale force winds which roar round the Main Chamber when Fell Beck is in flood. Frost action may also be significant with such open access to the surface.

A clue to the origin of the Main Chamber is given by the Porcellanous Band. This dense blue-white micritic limestone which marks the S_2–D_1 boundary is clearly visible in section along both walls of the Main Chamber some 6–8m above floor level. Careful examination of the levels at which this bed is seen in each wall reveals a relative displacement across the chamber, which increases from zero at the west end (where the bed is clearly continuous across the west slope) to over 4m at the east end, down-throwing to the south.

The Main Chamber Fault appears to be a rotational or wrench fault. The absence of any discernible displacement at the top of the Main Shaft, together with the 55° hanging roof feature of the Main Chamber (which can be traced along most of Old East Passage to beyond Mud Hall), suggests that the fault plane is hading, parallel with the hanging roof, and cuts across the Main Shaft at the Birkbeck's Ledge level.

It is likely that the precursors of Fell Beck found a series of widened joints aligned en echelon along the upper part of the Main Chamber Fault. These afforded easy access for the stream down the joints and down the steep southerly dip of the fault plane itself.

The Main Chamber subsequently developed by vadose erosion and extensive collapse of the rock mass lying between the joint shafts and below the oblique fault plane.

The Porcellanous Band, in addition to being a useful marker horizon, has itself acted as an important control over the development of parts of the rest of the Gaping Gill system by limiting the down-joint percolation of water. It lacks joints by comparison with the coarser-grained micrites and sparites found both above and below it, and acts as an impervious layer in a similar manner to shale bands (Waltham, 1970 *b*) but without the mechanical incompetence of shale. Thus many of the passages radiating outward from the Main Chamber are developed on or just above this horizon. Wherever the Porcellanous Band has been penetrated by fault action, vertical development below this horizon is facilitated. Main Chamber itself forms a good example of this effect, as do East, West and South East Pots.

Mud Hall consists of a cavern almost comparable in size with the Main Chamber, but the absence of daylight, the liberal coating of mud on walls and floor, and the curious ridge which partially divides it make it difficult to appreciate its true shape and size. It owes its origin to progressive collapse at the junction of the Main Chamber Fault zone, clearly seen extending east in the roof of East Passage, with a second fault system aligned with the joints and trending more to the south. Vertical development has occurred both upward and downward along the length of both faults, and in the area of intersection, and has resulted in twin roof avens, each with their respective collapse pits in the floor. The mud-covered rock ridge separating the two halves of the chamber has suffered considerable denudation in recent times, especially between the first and second visits by the original explorers (Cuttriss, 1908), and it can be seen that the whole chamber is slowly being cleared of mud by the action of recent bog drainage inlets entering via a number of roof avens. This is particularly apparent near the top of the 40m high boulder slope forming the eastern wall. Here the action of twin aven inlets has cleared the mud from between the boulders and allowed access to a bedding plane tributary to Hensler's Master Cave. A little further on, the same process has revealed (via Avalanche Pot) the extensive vertical development of East Passage.

Although usually regarded as one large chamber, Sand Cavern has at least two, quite distinct, sections. The lower, northern, end is in part a former continuation of the low phreatic tunnel which enters from the T-junction. The southern end is formed in a much larger phreatic tunnel, the floor of which is occupied by a large vadose trench now partially blocked by enormous quantities of mud, laminated sand and clay. The roof of this section is laced by a network of fine, calcite filled, fissures indicating partial brecciation. This is due to the proximity of the chamber to the fault zone running through both Far South East Passage and Stream Chamber. As suggested above, it is possible that the huge pile of sand now almost completely filling the southern end of Sand Cavern was carried in by a precursor of Fell Beck, via a now choked shaft lying on the South East Passage Fault.

The whole of Stream Chamber, between the avens at the end of the North West Extension and the junction with Sand Cavern, appears to have been formed as a result of the vadose invasion of an extension of the South East Passage Fault system, followed by collapse in the fault plane itself, aided perhaps by water from the joint-controlled, vadose, Stream Passage system.

In the south-eastern part of the system, vertical development along the line of the Hurnel Moss Fault has resulted in the formation of the collapse chambers of Mountain Hall and Clay Cavern. In the former, the collapse zone extends over a vertical range of nearly 100m, from the floor of the pit to the roof above the boulder pile. The three large dolines on the west flank of Clapham Bottoms lie further to the south-east on the same fault, almost directly over Clay Cavern, and their origin may be due to collapse in the plane of the fault, possibly associated with former phreatic rising systems feeding resurgences lying on the fault line in Clapham Bottoms.

Phreatic trunk routes

By far the greatest proportion of passages in the Gaping Gill system at or near Main Chamber floor level are of this type. In many cases these appear to have developed on the Porcellanous Band, and owe their origin to the up-dip flow of water, escaping, perhaps under pressure, from around the periphery of a glacial choke almost filling the Main Chamber. It is clear, however, that they are of differing ages, and some at least appear to be related to water escaping from the other large chambers, in particular Sand Cavern and Mud Hall. Many of these phreatic tubes contain traces of one or more fill stages, which in places consists of sand, in other clay or mud, and yet elsewhere a threefold layer

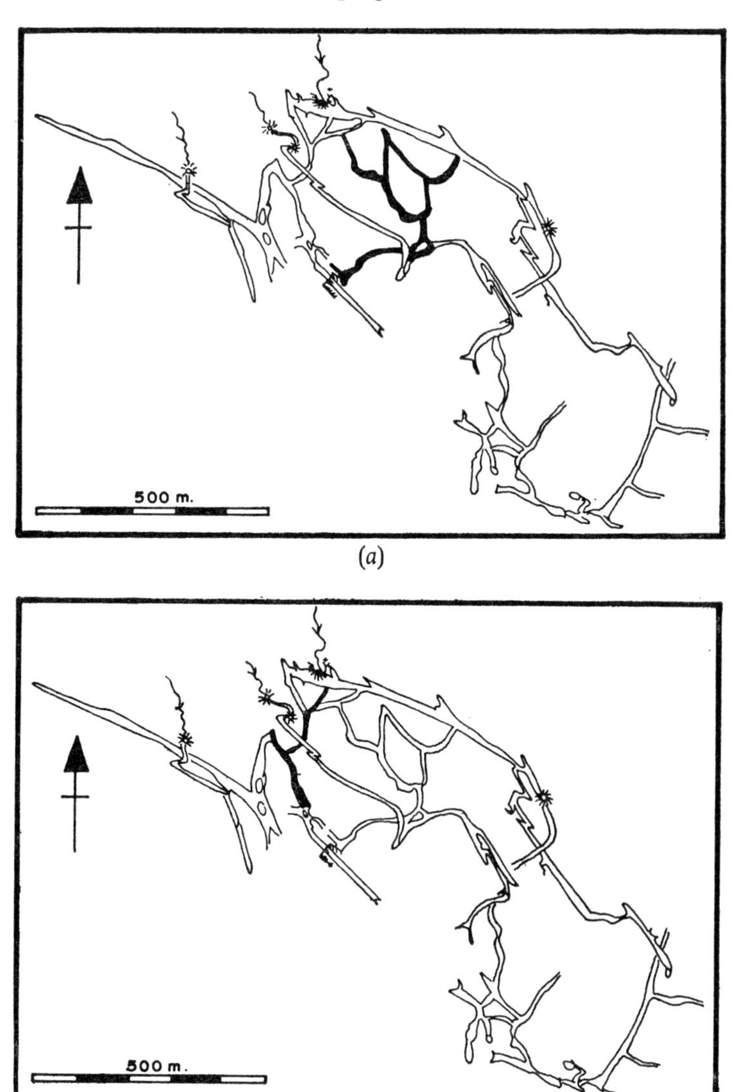

Fig 77 Examples of different passage types in the Gaping Gill system: (*a*) type 1 phreatic tubes of Hensler's System; (*b*) type 2 phreatic tubes of South Passage and its branches

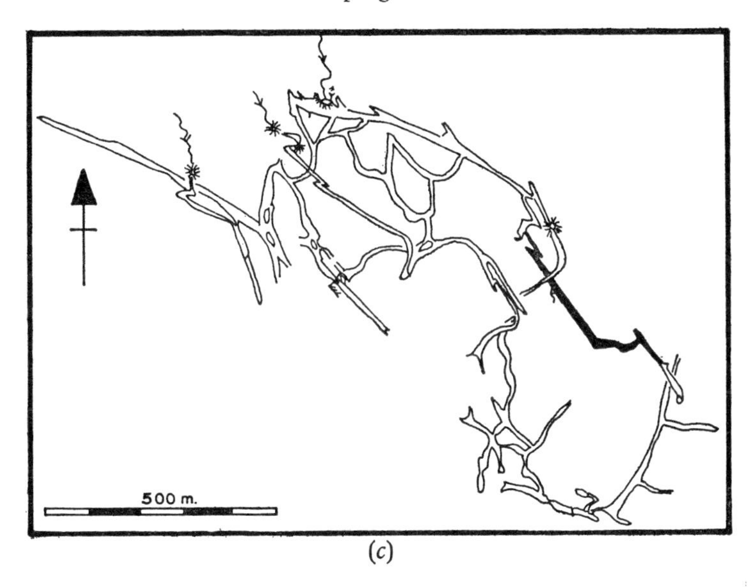

(*c*)

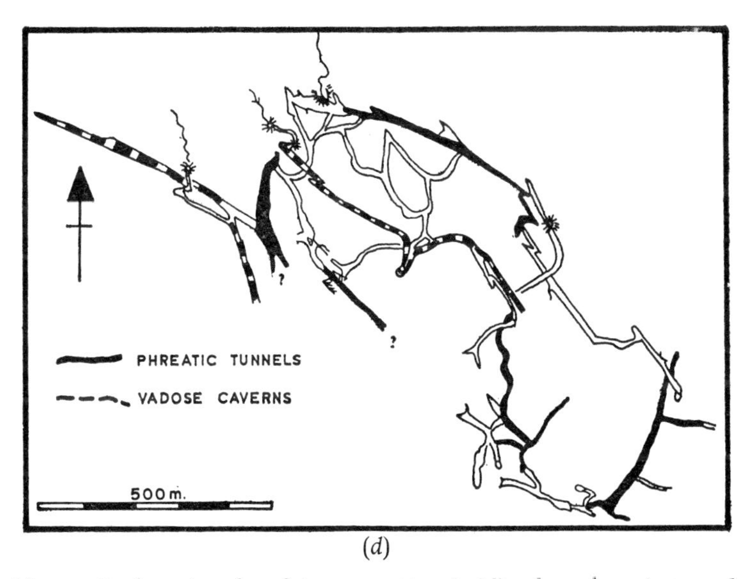

(*d*)

(*c*) type 3 phreatic tube of Anagram Crawl; (*d*) other phreatic tunnels and major vadose passages of boulders, pebbles and mud. The degree of clearance varies from point to point and area to area.

There are three main types of phreatic tube. The first is developed either directly on top of, or in the thickness of, the Porcellanous Band. Where this feathers out to the south-east, the tube is developed in the equivalent bed. They are low tubes, or bedding caves, with small flow-marks and little fill. All appear to be strike or up-dip tributaries of Hensler's Stream Passage, and are perhaps the youngest of these three types of phreatic tunnel. Examples of passage of this type include Hensler's Crawl and the western end of the Booth-Parsons Crawl complex. Mud Hensler's Crawl (from Mud Hall) and South Hensler's Crawl (from Bar Pot) also appear to be of this type.

Phreatic passages of the second type are also controlled by the Porcellanous Band and also flow up-dip or along the strike. They differ from the first type in being larger, with medium-sized flow-marks and contain some fill with the occasional stalagmite development on top. Examples include South Passage, South East Passage as far as South East Pot and South West Passage as far as Sand Cavern. They are probably older than the phreatic tunnels of the Hensler's system. Even within this group, the passages do not all date from the same phase. Wall features indicate an original flow from Sand Cavern to South East Pot, but there was a later reversal of flow when South Passage developed and its water flowed from T-junction back to Sand Cavern and out of its northern end.

Even larger than these is a third type, containing greater amount of fill and extensive calcite deposits. The scallops are big, and development may occur above the Porcellanous Band. Anagram Crawl in the Whitsun series is the best example of this type, which appears even older than both other types of phreatic tube.

Different again is the extensive phreatic development found at higher levels in Sand Cavern, Far South East, Old East and Far East Passages. Much of this development is fault controlled, and later, lower, development along the same lines has partially destroyed the original passages, the remnants being seen at high level in the rifts. Flow appears to have occurred down-dip in South East passage, towards South East Pot, but in Far South East Passage the water flowed in a north-westerly direction towards South East Pot. Old East Passage is one of the oldest in the system. At one place it contains a stalagmited fill, incised by a stream channel and then re-covered by stalagmite, indicating a long and complex Pleistocene history. Further evidence is required before any classification scheme can be extended to cover these passages.

Much of the extensive complex of passages recently discovered at the extreme south-eastern end of the system appears to be phreatic tubes of the same general type as the others discussed above. The greater part of the Far Country series consists of low tunnels partly filled with water or sand and conveying the Hensler's system water toward the Hurnel Moss Fault, and eventually to Ingleborough Cave via the Deep Well. Northgate may represent a former tributary from the Whitsun series. The Far Waters, entered via a complex of passages at several levels beyond the choke in Clay Cavern, contains both vadose and phreatic passages of several differing ages. The greater part of this series, the Northern Line and the Styx, appears, from flow-mark evidence, to have carried water in a southerly direction, possibly toward the Shallow Well, and possibly represents an old phreatic trunk route from P5 or Marble Pot, since it extends beyond and below the far end of the Whitsun series.

The whole of this part of the Gaping Gill system may represent a development level about 17m below the Porcellanous Band, intermediate in age between the present (as yet undiscovered) active phreatic systems and the older developments of South East Passage and the Whitsun series, and forming part of yet another phase of development (Brook, 1968). These passages are of particular interest because, besides providing the point of nearest approach to Ingleborough Cave, they are developed probably less than 50m below Clapham Bottoms and may have supplied resurgences in this area when Ingleborough Cave, either through youthfulness or extensive fill, could not carry all the water entering the system.

Associated potholes on the plateau

Car Pot entrance lies 200m ESE of Gaping Gill, at the northern end of the limestone plateau overlooking Clapham Bottoms. It lies at the foot of the 10m high northern wall of one of the largest of a number of shakeholes. Four short shafts connected by very narrow rifts, lead south-east along the major joints to the head of a wider 40m shaft. This shaft drops into a large, much older, mud-filled passage running also NW–SE and which lies at the 300m (1,000ft) level. This is the Craven Passage, and it appears to be a former extension of East Passage beyond Mud Hall. Visual connection between the two has been established. North Craven Passage has been partially cleared of its mud and sand fill by minor inlets, including the Car Pot stream, but South Craven Passage contains much of its former fill. Glissade Pot may be a younger vadose outlet connecting with the Whitsun series via Windy Aven; the choked end of the main tunnel lies nearly 50m above the sumps at the end of Hensler's Stream Passage.

At the head of Clapham Bottoms, close to the Allotment wall, a small stream sinks at the foot of a low, shattered, cliff of limestone which terminates a short but deep valley. This sink, usually referred to as P5, has been explored for some 60m through collapsed bedding planes and down three shattered, joint-controlled, pitches to a depth of 40m. The cave heads SSE toward Clapham Bottoms. Dye introduced here some years ago was carried through overnight after a heavy thunderstorm, and was seen in the lake above Clapham, having emerged at Clapham Beck Head.

The unstable entrance to Grange Rigg Pot, a relatively young system, lies to the bottom of a large shallow depression some 350m east of Gaping Gill. Partly artificial bog drains collect a small stream which rises rapidly in wet weather. The system starts as a series of low bedding caves and joint rifts, dropping nearly 50m down three pitches, leading to a larger, probably older passage with extensive collapse. The Christmas Pot inlet enters here, having once been a much more important tributary (possibly P5 stream itself) than at present. Lower down, narrow vadose rifts lead to the sump and the large inlet passage from P5. The sump is perched at 350m, and is only a local constriction in the important P5–Ingleborough Cave drainage system.

A few hundred metres due south of P5, on a moraine-covered ridge between two of the shallow dry valleys leading down to Clapham Bottoms, Body Pot lies at the bottom of a deep shake-hole. A steeply descending stream passage with an unstable boulder floor leads to a small chamber at a depth of 25m. The apparent size of the passage suggests it is a former sink of a large stream, but being now situated on a ridge between two valleys, no stream reaches it today. It may prove to be a very old sink for streams flowing down the south-east flank of Ingleborough.

South-west of Stream Passage Pot on the other side of the Gaping Gill system is Hurnel Moss Pot: A very deep shakehole is cut down to bedrock where the water drops straight down a 10m shaft followed by a 55m shaft. These lie on a major fault and drop straight into a very large old phreatic tunnel on the same fracture. It is completely choked with fill after 200m, and the vadose floor slot of the present stream sumps beneath the fill. It is not yet clear how this massive pre-glacial tunnel relates to the similar caverns in Gaping Gill.

INGLEBOROUGH CAVE

The low, arched, entrance to Ingleborough Cave is situated on the west side of the valley of Clapham Beck, nearly equidistant between the village of Clapham and the main sink of Fell Beck at Gaping Gill. A few metres further up the valley, the waters of Fell Beck emerge from a low bedding cave, Clapham Beck Head, at an altitude of 275m (825ft) above sea level. The two cave entrances give rise to a series of passages totalling over 2km in length running under the western flank of Clapdale, parallel to the track, as far as the lower end of Clapham Bottoms. Over half this total length of passage contains the main stream, while the outer half of the cave is normally dry and has been a show cave for over 100 years.

Terminal Lake, at the inner end of the Ingleborough Cave system, is formed in a 100m long, low, phreatic, joint-controlled passage, with the main stream emerging from a syphon at the north-western end. Recent exploration by divers has shown that the underwater passage continues along the joint for a further 150–200m. Several tributary passages were observed on both sides, one of which brings in the main flow of water. At the far end a series of rift chambers above water level give access to a further 250m of higher-level, dry, joint-controlled passage which ends in a boulder choke. These features, together with the aven development in shattered rock in Rimstone East, an old, choked, inlet from the north near the waterfall below Terminal Lake, suggest the proximity of a major fracture zone, probably associated with the Hurnel Moss Fault.

The main stream emerges from Terminal Lake down a 1m high waterfall into Inauguration Cavern. This consists of a mature vadose stream passage incised in the floor of a bedding cave, the remains of which form high-level shelves on both sides. Extensive calcite deposits high on both walls testify to the age of this portion of streamway, but the striking development of large rock flakes projecting horizontally on both sides demonstrates the frequency with which phreatic conditions return during floods.

After 100m the character of the passage changes abruptly, becoming low, wide and nearly water filled. Overhead at this point, a shaft gives access to the Upper Inauguration Series consisting of a series of tight joint and bedding tubes connecting the bases of a series of high avens, and running north-west under the lower part of Trow Gill.

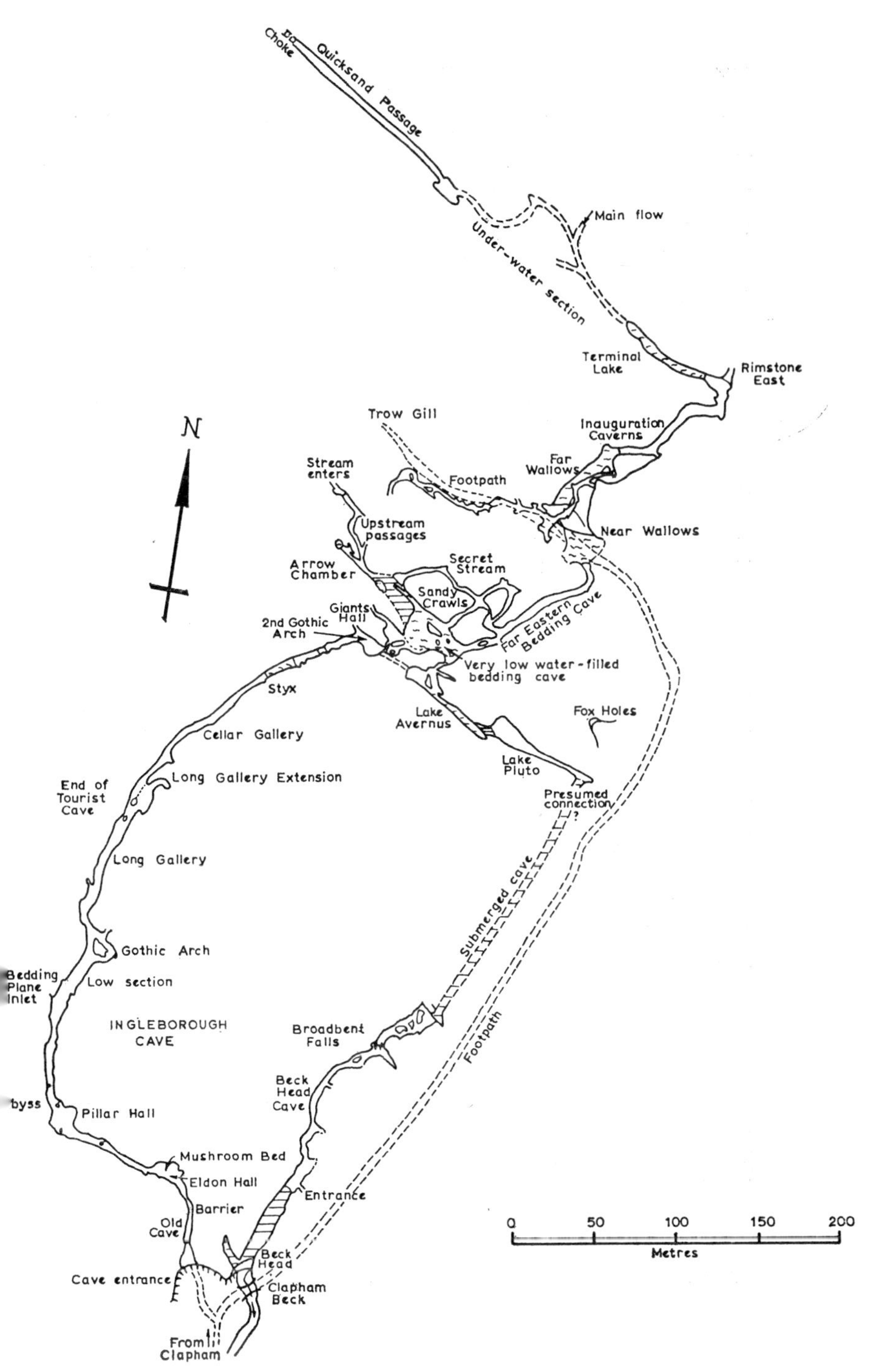

Fig 78 Plan of Ingleborough Cave

For the next 300m the main stream occupies a series of wide, water-logged and partially sand-filled bedding-plane passages and can only be followed through the Far and Near Wallows, before the roof meets the water. Across the Near Wallows, the Far Eastern Bedding Passage bypasses the flooded portion of the main stream passage.

At the end of this series of low, wide crawls, the stream is again encountered, flowing from north-west to south-east in a series of en echelon joint tubes, connected by low bedding plane caves. These include the Upstream Passages, Arrow Chamber and Lakes Avernus and Pluto. The depth of water increases downstream until the roof meets the water at the end of Lake Pluto. Here the main stream exits via a large flooded bedding tube some 10m below water level, and is next seen in Beck Head Stream Cave.

Across the stream-filled bedding plane cave at the outer end of the Far Eastern Crawl, a series of low, phreatic, rising tubes lead to an inclined rift rising some 7m into a higher-level chamber. Named Giant's Hall, this chamber is developed along a joint lying parallel to those seen in Lakes Avernus and Pluto. 10m long by 5m wide, the fissured roof rises 10–15m above a floor of calcite-cemented sand and gravel. High in the roof, stalagmite-cemented boulders indicate traces of a former fill and a possible connection with the surface.

A low arch over a downward-sloping sandbank leads into a further chamber, known as the Second Gothic Arch Passage, which lies on the same joint as Lake Avernus. To the south-east, a low crawl leads back to the lake, while at the north-western end, the passage leaves the joint and enters the large phreatic passage known as Cellar Gallery. At this point a 2m deep hole in the floor gives access to a tight flooded lower passage along the joint. Under normal conditions the water appears static, but in flood, a powerful stream can be seen flowing towards Lake Avernus.

Cellar Gallery consists of an oval phreatic tube, approximately 2m in diameter, running up-dip along the bedding for 150m in

Page 375

Fools Paradise in Gingling Hole, Fountains Fell. The passage has a classic keyhole section and both the phreatic tube and the vadose canyon are liberally decorated with dripstone

Malham Cove from the air. The Watlowes dry valley is in the background and the River Aire flows away to the left

a south-westerly direction. It rises some 3m over its length, intersecting several cross-joints where mixing corrosion has resulted in extensive upward solution. Large scallops on the walls show that it has been formed by the steady, outward, flow of a large volume of water over a long period. The floor is largely sand covered, and a 1m deep pool, the Styx, has developed among the sandbanks at one point. The passage becomes lower towards the end, and a low bedding plane extension on the left connects with a much larger passage 2m above a pool at floor level.

From this point, the end of the tourist part of the cave, this large passage, Long Gallery, runs south-west for a further 150m. The upstream, north-eastern, section of this passage can be seen, across the pool in the floor, choked to the roof by a graded fill of rounded gritstone pebbles in a coarse sand matrix. The pebbles show an imbrication structure indicating that deposition occurred from within and the top 25cm of the fill grades through fine sand to laminated clay in contact with the roof.

Long Gallery itself forms a classic example of a mature, multi-phase, main stream passage. Evidence in the form of a flat roof, ledges on both sides partly covered with calcite-cemented pebbles and stalagmite formations, oxbows, and scalloped rock walls with traces of a former pebble fill cemented in place, points to a long, complex, history of development, fill and re-excavation. It is clear that the main stream occupied this portion of the cave for a very long time.

At the end of Long Gallery, the passage swings left along yet another major joint into the First Gothic Arch Passage, but immediately regains its former direction along a wide bedding passage some 3m below that forming the roof of Long Gallery. The roof maintains this horizon for the next 150m but the passage height gradually increases as a series of minor inlets entering along cross joints combine to cut a deepening vadose trench in the floor. The stream finally drops down the Abyss, a small, 3m deep, pot in the floor, into a series of tight joint rifts 10m below the floor of the next section of the cave

At this point, both floor and roof step up some 3m into Pillar Hall and the passage beyond swings left to run south-east, at a higher level, along the same set of joints as the Abyss stream utilises below. The next 100m of passage, as far as the Mushroom Beds, consists of a large high, old vadose canyon containing many fine calcite formations, including one large stalagmite dam. Finally the roof steps down to its former horizon, and the last 100m of passage consists of a very old vadose channel, flat-roofed, and now largely choked by massive calcite deposits on the walls and floor. The remains of no less than nine stalagmite dams can be seen, behind which pools reached within a few centimetres of the roof, prior to the destruction of the dams in 1837.

Associated low-level caves

The main stream, not seen again in Ingleborough Cave beyond Lake Pluto, is next encountered at the far end of Beck Head Stream Cave. It emerges, from a depth of 10m, in a narrow cross rift sump, some 250m to the south-west of the end of Lake Pluto. For the next 50m the stream flows through a labyrinth of small vadose channels cut in the floor of a wide phreatic bedding cave. The next major cross joint to the south-west allows the stream to drop some 3m to the bed below, and results in a fine cascade, the Broadbent Falls.

From this point out to the resurgence at Clapham Beck Head, 250m downstream, the cave consists of a low, constricted, vadose channel cut in the floor of a much wider, very low phreatic bedding plane passage. Each cross-joint gives rise to a deep pool at floor level and a corresponding solution hollow in the roof above. The absence of fill and calcite formations, together with the sharp-edged scallops and chert flakes, provide evidence of frequent return to phreatic conditions in period of flood. The last 100m of passage is partly blocked by massive roof collapse.

Foxholes lies at the foot of the small cliff on the left of the main path, a few metres below the bend leading to Trow Gill. It now consists of a steeply descending arched cave passage floored with loose boulders which soon fill the passage. Prior to the archaeological excavation carried out in 1914 (Broderick, 1924), the passage was concealed by a thick layer of sand and clay, covered by scree derived from the cliff overhang. The excavators sunk a 10m shaft through the floor and reached a smooth water-worn rock-floored passage choked with sand and gravel. Faint flow markings are visible on the roof and walls of the upper, accessible, part of this passage, and appear to indicate that water flowed in. Foxholes may therefore be the upper end of the choked extension of Long Gallery.

Cat Hole lies on the unconformity at the base of the limestone a few metres south of the lane, halfway between Clapham and Clapdale Farm. It consists of a very low bedding cave running west for some 150m. In wet weather a very large stream emerges from the cave, but the exact source of the water is not known. The cave must lie very close to the line of the North Craven Fault and it is of interest to note that at Cat Hole the unconformity is found at a height of 230m, but at Moses Well, a few hundred metres to the north, it is much lower. Cat Hole most probably acts as a high-level overflow outlet for the Moses Well system, which drains the Newby Moss area.

DISCUSSION

The development of a complex underground drainage network must necessarily evolve as a result of the interaction of surface conditions and processes with the structure and lithology of the rock in which it occurs. The underground environment, unlike that of the surface, often retains or protects earlier portions of the developing system from the destructive action of later processes. Many of the passages in the Gaping Gill–Ingleborough Cave system appear to have undergone several cycles of erosion, infill and re-excavation, which immediately suggests that parts of the system are old enough to have suffered some of the cyclic climatic variations which characterise the Pleistocene period. Each successive glacial period has, in this area, been of sufficient severity to have largely destroyed, at least on the surface, the evidence of earlier glaciations. Consequently the full sequence of events occurring in the area during the Pleistocene is not known in any detail, and discussion of the development of the cave systems is effectively limited to consideration of the main morphological problems.

In Gaping Gill, these are largely concerned with the relationship of the Main Chamber to the other passages which connect with it. For example, the existence of chambers of the size of Mud Hall, Stream Chamber, Sand Cavern and Mountain Hall, having no apparent surface connection, suggests the possibility that the Main Chamber existed prior to the development of the Main and Lateral Shafts, and that the principal effect of the vadose invasion by Fell Beck has been the partial clearance of the fill which the other large chambers still contain. The origin of large blind cavities of this kind is not well understood, but the fact of their existence is indisputable, particularly in the limestone areas of North Wales and Derbyshire, where they are often associated with mineralised faults. At Gaping Gill, however, absence of any evidence of extensive mineralisation in the area, together with the problems posed by the nature and quantity of the fill in these large chambers, renders theories of origin of this sort less satisfactory than the simpler, vadose, origins suggested in an earlier section.

In support of vadose theories of this kind, a number of other factors appear to have played important roles in controlling the nature and degree of development. In the first instance, the position of Gaping Gill, at the inner edge of a plateau, whose surface corresponds to the top of the Great Scar Limestone, and which lies below a south-facing corrie, appears to be significant. The shoulder of Simon Fell has protected much of the plateau from the direct erosion of southward-moving ice, and appreciable accumulations of drift were deposited instead, which aided the integration of the surface drainage of the corrie. Thus a single, large, stream was able to form on the drift and Yoredale rocks before running on to the limestone. Equally significant is the fact that the axis of the catchment area lies approximately perpendicular to both the major joint set and to the three major faults. Thus the precursors of Fell Beck were able to enter the limestone at a number of discrete locations, which were subject to reoccupation on a number of occasions. The gradual retreat of the shale/limestone boundary would imply that, in general, water entered, and development occurred, firstly along the Hurnel Moss Fault, secondly along South East Passage Fault, and only recently, geologically speaking, along the Main Chamber Fault. This is not the complete story, since it cannot account for the evident age of parts of Old and Far East Passage, which are developed on or near the latter fault. The failure to find any major part of the system associated with the Hurnel Moss Fault, despite many years of exploration in the area, poses another problem in this respect, although the recent discoveries in Hurnel Moss Pot itself may soon supply further evidence of development in this area.

The lithology of the Great Scar Limestone appears to have influenced the nature and degree of development of the system in depth. The existence of at least four horizons, at which extensive horizontal development is found, implies the existence, at these levels, of relatively impervious beds (in particular the Porcellanous Band) which have limited the downward movement of aggressive waters. In addition, the type and pattern of passage developed on the Porcellanous Band, converging towards the south-east, suggest that, between the Main Chamber and South East Passage Faults, the limestone exhibits a shallow fold structure whose east–west axis plunges west at a low angle. This structure is therefore probably responsible for the nature of the long, rising, phreatic tube passages which formerly carried water up-dip from the main entry shafts in the west toward former resurgences in Clapham Bottoms.

Another problem posed by the Gaping Gill System concerns the depth to which development has occurred. Apart from the long phreatic tube systems in the Far Country and Far Water series,

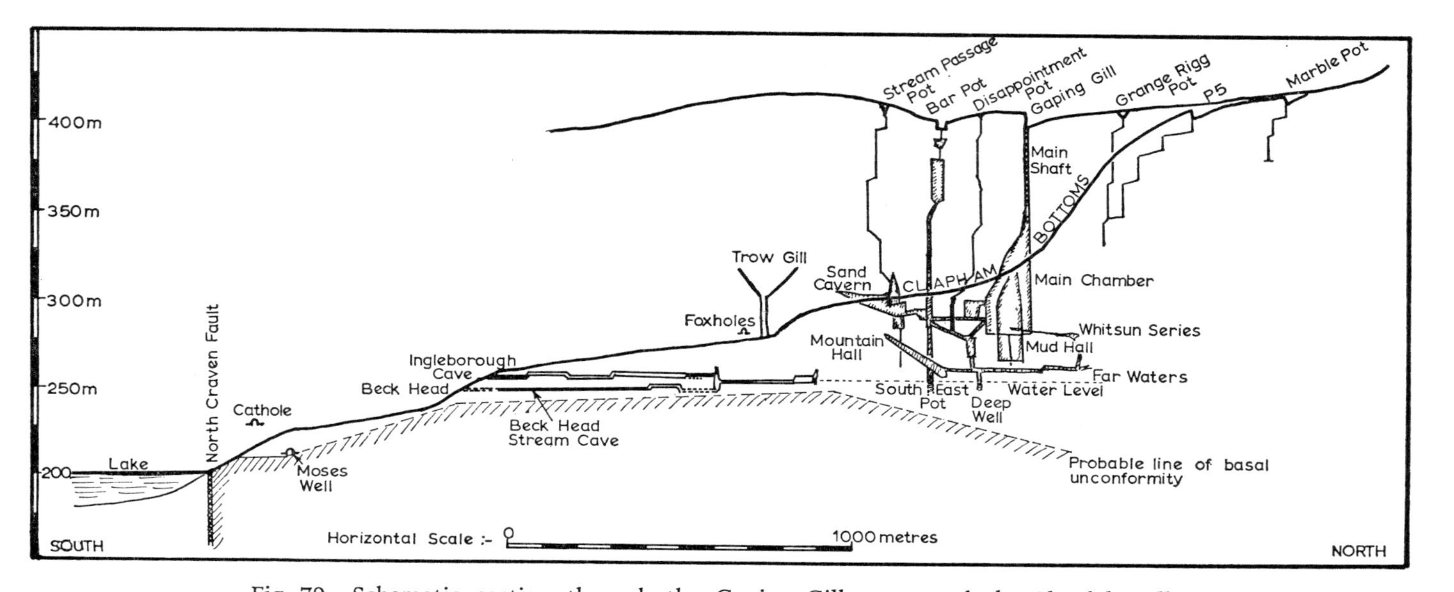

Fig 79 Schematic section through the Gaping Gill caves and the Clapdale valley

there is virtually no accessible development in the whole system below the 300m horizon, whereas the main flow of water appears to take place at least 30m below this level. In view of the known relief of the basal unconformity in this area, which appears to act as an underground watershed dividing the Allotment water from that of the Gaping Gill area, the existence of another buried ridge, running north-west under Clapham Bottoms along the line of the Hurnel Moss Fault, would appear to be the most likely explanation for the existence of the deep undrained phreatic system under Gaping Gill.

In contrast with Gaping Gill, Ingleborough Cave is basically a linear, up-dip, phreatic system developed along probably no more than two bedding planes. These are occupied alternately by the stream (or form part of its former course), the passage rising or falling vertically along strike joints. Vadose trenching has taken place in most sections but is now active only at the outlet of Terminal Lake and over a short section of the stream passage above Lake Avernus. Cellar Gallery forms a second phase by-pass to the choked portion of Long Gallery. The inner end of this choked passage is not seen anywhere else in the cave, and it may connect to Foxholes. Beck Head Stream Cave developed along a third, lower, bedding horizon and is much younger than Ingleborough Cave from which it has captured the stream.

It is not known whether all of Fell Beck normally reappears in Terminal Lake, or whether additional feeders exist in the Wallows area. Nor is the point of entry known of the important Grange Rigg/P5 tributary.

One of the biggest problems of the area concerns the position of Ingleborough Cave, at a height of 275m (825ft), where there is possibly a further 70m of limestone below this level, before the unconformity is reached. The above-mentioned hypothesis regarding the existence of a basement ridge under Clapham Bottoms will, with only minor modification, serve to explain this problem.

Of the many unusual surface landforms in the area, the twin

dry limestone gorges of Trow Gill, and the deep, V-form, dry valley above, in striking contrast with the wider, gentler, form of Clapham Bottoms, and suggesting very different modes and periods of origin, are possibly the most important clues to the history of the area. For the moment, however, these features provide questions rather than answers, and it is possible that the caves themselves, and their contents, will furnish the vital evidence of surface events that has long since vanished from the surface itself.

19

The Caves of Ribblesdale

A. C. Waltham

For almost the whole way from Ribblehead down to Stainforth, the valley of the River Ribble is flanked by high limestone scars, beneath which is a proportionately large series of deep cave systems (see Figs 80 and 81). Below Horton the valley is floored by impermeable pre-Carboniferous rocks, but, even above this point, the Ribble maintains a purely surface flow. Near Horton the river meanders on widespread drift deposits, though the higher parts of the Dale, nearer Ribblehead, are adorned with a particularly fine drumlin field.

South-west of Horton, the Moughton Scars are covered in bare limestone pavement with no caves large enough to enter, and the likely resurgence for most of Moughton's percolation water is the similarly tight Blindbeck Cave, unexplorable after 150m of passage.

However, north-west of Moughton is a group of deep potholes fed by the streams of Simon Fell. Marble Pot, Long Kin East Cave and Juniper Gulf each absorb streams with mean flows of about 30l/sec. The former is choked less than 50m down and the destination of its waters is unknown. From an adjacent shakehole, the constricted passages of Marble Sink lead down to a series of chambers at a depth of about 90m, but how these relate to any fossil passages of Marble Pot is uncertain.

Long Kin East Cave consists of a series of fine, meandering, vadose canyon passages close below the surface, carrying the stream southwards to the Rift Pot Fault. Rift Pot itself is formed almost entirely in the breccia and shear zone of a major fault (though this has little displacement), and is known as a series of large loose chambers leading to a depth of 115m. The Long Kin stream enters part way down and at the bottom runs into a short passage leading to a sump. A long series of bedding plane inlets are formed in the unfaulted limestone, and the main water comes from Jockey Hole, a choked 65m shaft further along the fault.

Juniper Gulf is formed on another fault, but in this case a clean fracture with little brecciation. Consequently the stream flows in deep narrow slots, and high-level niches excavated in shale beds give the explorer the only access between the much larger vadose shafts. The cave sumps at a depth of 125m, only a few metres above resurgence level.

Nick Pot and Hangman's Hole are both formed on a third fault. Each is over 100m deep with little horizontal development and the former contains a 100m shaft where its, normally small, stream descends the fault.

Though overlooking Ribblesdale, the caves between Rift Pot and Hangman's Hole are hydrologically unrelated to the Ribble as their waters all reappear in Crummockdale. Austwick Beck Head, the resurgence, has a mean flow of about 150l/sec. The water runs out on the unconformity at the base of the limestone, and the flooded passages are not yet known beyond some low bedding planes a few metres inside.

A complex series of sinks and risings mark the location of the Fell Close Caves where a little-known series of shallow bedding plane passages guide the drainage northwards, but the next major cave system is that of Alum Pot (Milner, 1972). 3km of shallow vadose passages, all developed in bedding planes near the top of the limestone, transmit the drainage eastwards. The caves are all near the surface—the main stream (mean flow about 60l/sec) even flows above ground between the Borrins Moor Cave and Long Churn Cave systems—until a major joint is reached,

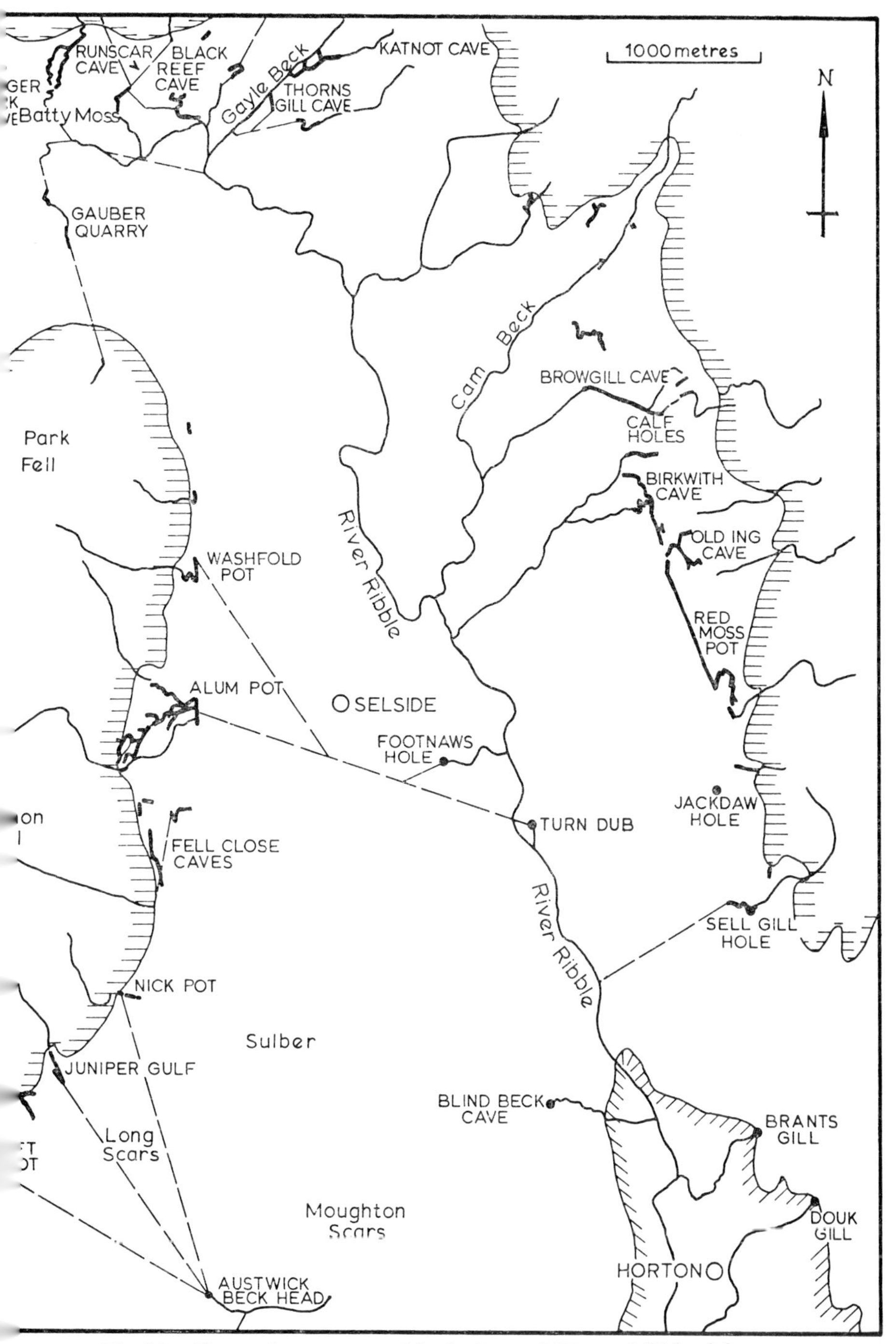

Fig 80 The major caves and hydrology of Upper Ribblesdale
(Key as in Fig 81)

oriented north–south. The water then drops 100m in Diccan Pot, though before this the main stream is joined by a number of tributaries, some in quite extensive vadose passages of their own. A smaller stream also falls down the impressive open shaft of Alum Pot, and the combined Diccan and Alum waters sump in a lower large chamber on the same major joint. The resurgence is only a few metres lower, in the pool of Turn Dub on the east bank of the Ribble—indicating karst drainage beneath and across the main surface river.

Washfold Pot, just to the north, is somewhat similar, in that a vadose bedding initiated passage leads eastwards to a 40m shaft down a main joint. The water then flows north along the joint, even though its eventual destination is again Turn Dub. The other small streams sinking on this flank of Park Fell have not formed deep potholes, but only run in shallow caves down the northerly dip.

At Ribblehead a broad drift-covered col separates the valleys of the Ribble and Greta. Astride the col is Batty Moss, a flat alluvial plain which probably originated in a post-glacial lake, for laminated silts and clays were encountered in the excavations for the Ribblehead viaduct foundations. Bordering the Moss are a number of small resurgences, all at the 295m (975ft) level. From the south, Batty Wife Cave drains the northernmost clintfields of Park Fell. On the northern side, Roger Kirk and Runscar Caves are formed in the top beds of the Great Scar Limestone; these are shallow but mature vadose caves containing some surprisingly well-decorated chambers no more than 5m below the surface. Each of the streams draining off the Yoredale slopes sinks underground for a short distance before resurging to flow into the Ribble headwaters.

The main headstream of the Ribble is Gayle Beck, with an extensive surface catchment area to the north-west. Except in flood conditions its course is normally dry for a stretch of 350m. The entire flow sinks into Thorns Gill Cave, a short series of phreatic rifts beyond which the watercourse is sumped. Some water is also lost into Katnot Cave. South-east of Thorns Gill, drumlins mask the limestone, though a few short caves are associated with the incised course of Cam Beck.

Even further south, tracts of limestone are exposed between the drumlins, and Birkwith Moor contains some more extensive cave systems. Calf Holes–Browgill Cave carries a stream of mean flow in the order of 50l/sec. It is oriented down the dip and the resurgence, at Browgill Cave, and is only 30m below the top of the limestone; from there a surface flow is maintained right down to the River Ribble.

Only a little further south, Birkwith Cave is a similar perched resurgence, from which pours an even larger stream. Though the cave passages behind Birkwith are closely related to the joint patterns (see Fig 80) the overall drainage is down-dip, so that the remotest inlet is well to the south at Red Moss Pot. The streams sinking here and into the Canal Cavern inlet flow through nearly 2km of spacious decorated stream passage, though dropping to a depth of less than 40m. From the sump the water flows into the sump of Dismal Hill Cave and then along the same joint, through another sump, to Birkwith Cave. The first flooded section picks up the water from Old Ing Cave, just to the east.

The fell south of Birkwith Moor contains many small caves and potholes, Most are young and still active, but some, such as Jackdaw Hole, are clearly very old. Sell Gill Hole contains active and fossil passages descending rapidly down a series of joints to a depth of 50m where they converge in a large rift chamber. However, the streamway beyond is very short and sumps at a level about 40m above the rising in the banks of the Ribble.

Between Horton village and Penyghent Hill lies the massive hydrological system of Brants Gill Head. The resurgence with this name lies just north of Horton on the limestone/basement contact. It is sumped only a few metres inside and has an almost constant maximum flow of about 200l/sec. This is because the drainage system has a flood outlet at Douk Gill Cave 500m to the south-east. Similar in that it is also on the base of the limestone and flooded only 100m inside, Douk Gill is normally dry but in wet weather may produce a flow of 1,000l/sec. There are numerous feeders to these risings, but to date none of the influent cave systems has been explored to anywhere near their resurgences (Simpson, 1949).

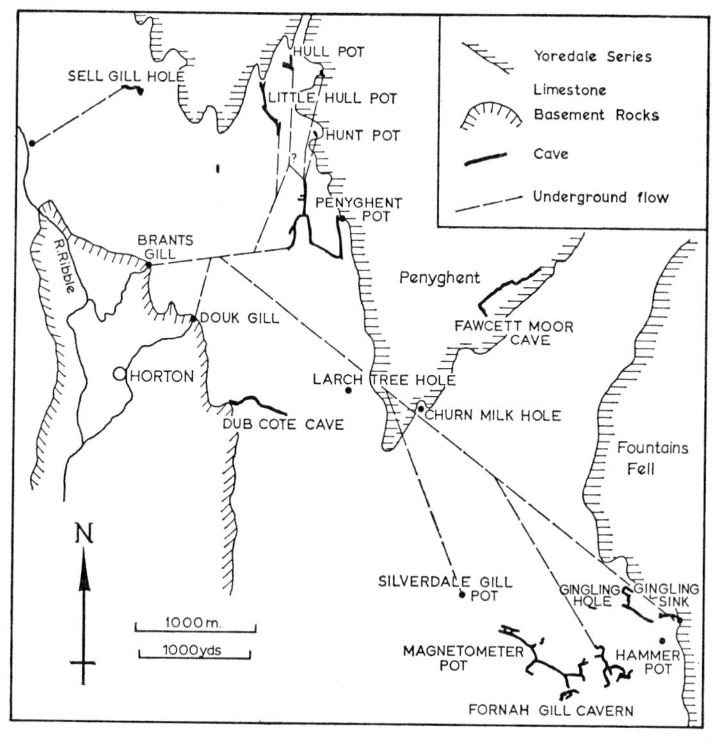

Fig 81 The known caves and hydrology of the Brants Gill drainage system

Hull Pot is the largest single feeder to the system. Unfortunately rubble in the great quarry-like pothole (100m long, 20m wide and 20m deep) has obscured any further passages, though an average flow of about 150l/sec sinks in its floor. Nearby, High Hull and Hunt Pots absorb much smaller streams; both consist of joint located shafts blocked by breakdown at depths of 60m.

Almost certainly, most of the water from these three sinks flows through the lower reaches of Penyghent Pot, on its journey to Brants Gill or Douk Gill. Penyghent Pot itself contains nearly 2km of passage over a vertical range of 150m. In its upper reaches bedding plane passages with and without vadose canyons carry only a small stream first south and then west. This stream then collects a great number of small inlets, where it steeply descends a major joint in a northerly direction. At its junction with the Hunt Pot water it has a mean flow of about 40l/sec; the passages in these lower reaches, though now being actively modified by vadose action, show evidence of an important early phreatic phase. The sump is at the resurgence level, though the submerged downstream passage is at a depth of 15m below the water surface.

Little Hull Pot is an active vadose system where bedding-controlled passages lead to a shaft and continuing rift passages along joints. The low level of its sump indicates that it misses the lower series of Penyghent Pot and flows on a separate route to Brants Gill.

The flanks of Penyghent also contain a number of other small choked shafts, though none at present leads to significant lengths of passage. The largest sink, that of Silverdale Gill, is similarly choked by glacial debris. Two longer caves have been explored—Lanteshop Cave (in the upper reaches of the Hull Pot catchment) and Fawcett Moor Cave. Both consist of small active stream passages in the Yoredale limestones and the waters reappear on the surface before sinking into the Great Scar Limestone. There is also Dub Cote Cave, a small resurgence at the base of the limestone. Nearly 500m of passage, mostly flooded, is known here; it is uncertain how its hydrology relates to the Brants Gill system, though its rapid flooding does suggest some connection.

Brants Gill is also the resurgence for a number of sinks over 5km away on the western slopes of Fountains Fell. These underground streams do not take the obvious westerly course to the Ribble valley, because of a hydrological barrier under the lower Silverdale valley, provided by a marked local rise in the level of the base of the limestone. Drainage is not towards Penyghent Gill due to a lower hydraulic gradient in this direction, and probably also because the greater extent of limestone outcrop would provide a more efficiently integrated network of karst fissures in the direction of Brants Gill.

Gingling Sink engulfs the largest single flow on this side of Fountains Fell—averaging about 50l/sec. But after flowing through a short, large, old passage the streamway cannot be followed due to sediment choking. The adjacent Gingling Hole carries only a very small stream. Narrow vadose canyons, about 300m long, lead into a large well-decorated phreatic tube. The stream has cut a slot in the floor of this old passage, but further on it falls down a series of rifts and youthful shafts to sumps at a depth of nearly 170m.

The passages known in Hammer Pot consist of very narrow meandering vadose canyons leading down to a series of bedding plane passages with more marked phreatic origins. The main stream enters these, and has incised a newer trench in parts, before dropping down a joint shaft to its sump. Mean flow in the streamway is about 50l/sec, and this originates in a zone of youthful and shattered shafts on Out Fell. In Magnetometer Pot, the main passages are large phreatic tubes only 30m below the surface. These normally contain standing water, but in flood periods carry a very heavy flow—probably acting as an overflow channel for the drainage from Out Fell, which normally enters the lower reaches of Hammer Pot through a very constricted passage.

Ribblesdale therefore contains some distinctive contrasts in cave morphology. The Ingleborough side is typified by deep fault-guided shafts, Ribblehead by shallow perched caves and Fountains Fell by its long underground drainage routes.

(*above*) Attermire Scar seen from the south (*below*) One of the large shallow closed depressions on High Mark, Malham

Page 394

Gordale Scar from the air with the limestone plateau in the background

20

Karst Features of Malham and the Craven Fault Zone

J. O'Connor, D. S. F. Williams and G. M. Davies

Along the southern edge of the Great Scar Limestone outcrops lies the Craven Fault zone. The pattern of faulting is complex and has impressed a distinctive pattern on the local geology. Furthermore both the stratigraphy and uplifting action of the fault zone have resulted in a karst topography in marked contrast to that elsewhere on the Askrigg Block.

The limestone features at Malham, particularly Malham Cove and Gordale Scar, are dramatic and well-known tourist attractions. They lie in an area of karst containing a number of landforms almost unique in north-west England. Much of this distinction is due to the area's situation immediately north of the fault zone.

Just west of the Ribble, Giggleswick Scar is a magnificent fault-line scarp, formed on the South Craven Fault, where the Great Scar Limestone overlooks the downfaulted mass of Millstone Grits and shales. A number of small cave remnants are known in the Scar but none is of any significance. Even further west the caves and karst of the fault zone bear only subtle differences from those in the main blocks of Ingleborough and Gragareth.

East of Malham, however, the Craven Faults continue to exert their influence on local karst development. Two particularly distinctive areas are the reef-knolls just south of Grassington and the remarkably cavernous limestones of Greenhow Hill. Each is described below.

THE MALHAM AREA

Geology

Structurally the area is dominated by the Craven Fault system, which marks the southern edge of the stable Askrigg Block. The Craven Faults are primarily wrench faults of post-Carboniferous age, which have moved in phases of the Armorican orogeny and in subsequent earth movements. They appear to mark a very ancient line of weakness. The Mid-Craven Fault, however, appears to have been active during the mid-Carboniferous as well, when it formed the boundary between the sediments being deposited on the stable block to the north and those in the less stable, basin, area to the south.

The tectonic forces which activated the faults are generally held to have caused the joints in the limestone, though the mechanism for their formation is still in dispute (Moseley & Ahmed, 1967; Doughty, 1968). In the area around Malham two sets of joints are generally developed almost at right angles to one another, though often one set may be better developed than the other. In 1931 Wager examined the direction of the jointing in the Malham area and interpreted the changes near the faults as an indication that the North Craven Fault was sinistral and the Mid-Craven Fault dextral. Away from the faults the joint trends are NW/SE and NE/SW. As the North Craven Fault is approached from the north the joints tend to swing anticlockwise, so that those in the north-west sector, swing to a more E/W trend. Less marked, but nevertheless the same, is the anticlockwise deflection found close to the south of the fault. In contrast, the joints close to the Mid-Craven Fault show a clockwise deflection, and joints in the north-west sector tend to swing to a more northerly direction.

Between the two faults are a number of cross faults, most of them trending NW/SE. Their throw is small, sometimes almost negligible, and they are probably wrench faults formed by the stresses and strains set up by the lateral movement of the North and Mid-Craven Faults.

The oldest rocks in the area are the Silurian strata under the Tarn, the continuation of those in the Ribblesdale inlier. The exposures in the Tarn inlier are poor, but the fine-grained, impervious, slaty rocks of the Horton group can be seen in Gordale, just north of Mastiles lane and in Cowside Beck, east of Cowside farm.

The almost horizontal rocks of the Carboniferous lie unconformably on the Lower Palaeozoics. In the Malham area the Lower Carboniferous is mainly represented by limestones. To the north of the Mid-Craven Fault the thick, remarkably pure, bioclastic limestones known as the Great Scar Limestone facies were laid down. To the south there is an intermediate facies consisting of bedded limestones and 'reef' limestones, which separates the Great Scar Limestones from the more muddy, limestone and shale, sequence found in the basin facies. The latter were sediments laid down away from the stabilising influence of the Askrigg Block; the term does not imply any great depth of water.

The so-called 'reef' limestones are quite distinctive. They are composed of very fine-grained, largely unbedded limestone, supporting a characteristic and locally abundant fauna. The origin of these limestones has been a matter of dispute since the last century. One school of thought regards them as wave-resistant structures similar to modern reefs, but with algae as the most likely reef binders or builders (Tiddeman, 1892; Black, 1954; Lees, 1961). At the other extreme they can be regarded as totally inorganic limebanks (Earp *et al*, 1961). Mixed up with the arguments over the origin of these Carboniferous 'reefs' has been the controversy concerned with their topography. At Malham, it has been argued, the knoll-like hills of Wedber, Cawdon and Butter Haw owe their shape to erosion (primarily in Mid-Carboniferous times) and tectonic activity, as is more easily demonstrable in the reef belt between Settle and Kirkby Fell (Hudson, 1930,

1949). On the other hand their shape is said (Tiddeman, 1892; Black, 1954) to represent the original reef structure, that of knoll-reefs or bioherms. The extent to which the topography of the area to the south of the Mid-Craven Fault is ascribed to erosion and tectonics naturally depends to some extent on the authors' opinions about the origin of 'reef' limestone.

Following a period of non-deposition and erosion, the Upper Bowland Shales overlapped the limestone from the south, and subsequently the soft shales have been themselves eroded to reveal the fossil Mid-Carboniferous landscape.

Mineralisation, associated with the Armorican earth movements, took place at the end of the Carboniferous. The mineralisation of the Askrigg Block was described by Dunham (1959), who showed that the veins generally occupy fault fissures of small throw. Around Malham (Raistrick, 1938) the veins, generally thin strings of lead ore (galena), occur in the upper parts of the Great Scar Limestone on High Mark and in the Pikedaw area. They are short, economically poor and follow the major joint directions, which can now be traced by the lines of old shaft hollows.

Little evidence remains of the epochs between the Carboniferous and the Pleistocene, but Sweeting (1950) postulated the existence of several erosion surfaces. Of these the 395m (1,300ft) platform is the best developed around Malham, for it forms the flat area around the Tarn. The others are much less extensive.

Finally, glacial deposits, and erosional features due to ice, are found throughout the area (Clark, 1967; Raistrick, 1930). During the last major glaciation the area was covered by ice and any evidence of previous glaciations can no longer be distinguished. An examination of the deposits shows them to be of local origin with limestone predominating. Ground moraine, till and solifluxion deposits cover much of the area. Fluvioglacial deposits have been recognised by Clark (1967) and are particularly extensive around the Tarn from the east of Black Hill to Gordale. Thus the presence of ice and melt-water has had a profound effect on the present landscape.

The karst depressions

The karst geomorphology of the Malham area is in many ways typical of the Great Scar Limestone, but, more importantly, also provides a number of marked contrasts with the Ingleborough and Wharfedale areas. The Malham karst is typical in that it includes fine limestone pavements, particularly just above the Cove, and also numerous shakeholes and sinkholes on the drift-covered benches. But it is unusual, firstly, in its almost complete lack of known caves, and, secondly, in the presence of the massive features of Malham Cove and Gordale Scar and the largest closed depressions in the country.

The karst hollows are best developed on the high limestone outcrops of Parson's Pulpit and High Mark. There, a series of large depressions, or dolines, up to 750m across, have been the subject of much discussion (Moisley, 1955; Clayton, 1966). They show no signs of marginal faulting or collapse. Indeed the Great Scar Limestone seems so massive and strong that it is unreasonable to postulate collapse as responsible for any major features in the area.

Moisley counted seventeen 'centres of closed depressions' around Parson's Pulpit, and twelve on Proctor High Mark. Some of the larger hollows have two such low points. They appear to be areas of removal by gradual solution initiated by water percolating through from overlying layers of shale and other sediments, none of which remain except for a small outcrop of the Dirt Pot Grit, a Yoredale sandstone, around the summit of Parson's Pulpit. The hollows probably started to form in the late Tertiary, and continued enlarging during the interglacials. Thus their development must have been affected by changing climates. When a limestone rim appeared it retreated by subaerial erosion processes, widening the hollow, while solution continued to deepen it. There is a great variety in depths and widths. Frost shattering may still operate on the rims, while, within the hollows, the deepening continues beneath the glacial sediments and accumulations of loess, soil and vegetation—Clayton's 'acid sponge'. Clayton has also suggested that the same type of process was responsible for the smaller and more scattered group of closed depressions which occurs in the Grizedales area to the south-west of the Tarn.

It is tempting to try to identify more youthful stages of these landforms. The conditions postulated exist at the top of the Great Scar Limestone on Fountains Fell. Water collecting on the Yoredales is percolating into the limestone, producing holes which, if left to subaerial erosion, would widen by retreat of the faces. The largest are the lowest Cockpits, on the banks of the upper reaches of Darnbrook Beck (NGR 886713).

Hydrology

Around Malham Tarn, and north of the North Craven Fault, a number of very small springs drain from the lowest beds of the limestone. Surface flow is then maintained across the impermeable basement rocks of the Tarn inlier as far as the Mid-Craven Fault. South of here the hydrology is complex, varied and distinctive.

Goredale Beck normally maintains a surface flow right across the limestone on its course through Gordale Scar. Springs at the base of the Scar have been proved to be fed from seepage losses in the higher parts of the beck; during the drought of 1899 this underground course absorbed the entire drainage, leaving dry the surface gorge and waterfalls of Gordale.

The outflow from Malham Tarn normally goes underground in a series of boulder-choked sinks just south of the Mid-Craven Fault. In wet weather extra sinks are utilised further downstream; in most conditions all the water has disappeared within a few hundred metres of the first sink, so leaving the Watlowes valley dry down to Malham Cove. However, springtime flooding at least once a decade is adequate to send a surface flow right down to a normally dry waterfall near the head of the Watlowes valley, but rarely further.

At the foot of the Cove a large stream resurges from a wide,

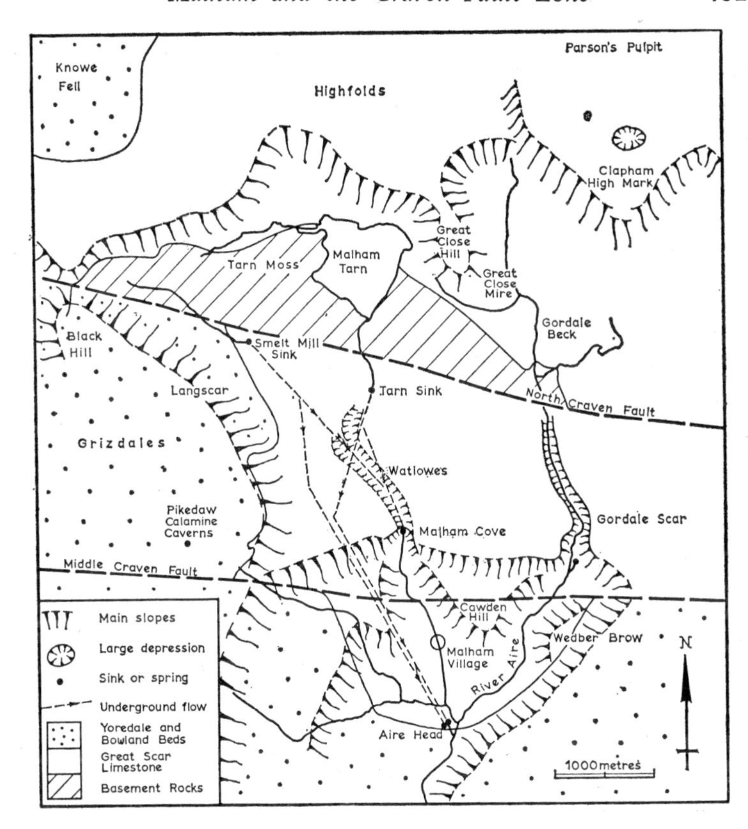

Fig 82 Morphology and hydrology of the Malham Tarn–Aire Head region

low slot below water level. As early as the end of the last century (Howarth *et al*, 1900), it was found that this was not the same water that sank just below the Tarn. Most of the Tarn water was found to flow to Aire Head, a pair of large springs in the west bank of the river just below Malham village (see Fig 82). The rising at the Cove was found to mainly originate at the Smelt Mill sink, just west of the Tarn sinks, and this therefore constituted one of the first-known classic cases of crossing karst drainage routes.

April 1972 saw some more refined techniques applied to the hydrology of the Malham Tarn–Cove area, and the results showed further complexities in the drainage patterns (Atkinson & Smith, in the press). Different coloured lycopodium spores, measured quantities of Rhodamine WT and an artificial pulse wave generated at the Tarn have helped to provide a picture of this underground drainage, where no significant known cave passages have yet offered scope for direct observation.

Water from the Tarn sinks arrived at both of the Aire Head springs simultaneously with a flow-through time between 13·5 and 24hr. These springs accounted for about 97 per cent of the Tarn drainage, and the other 3 per cent went to the Cove rising in a time between 24 and 28·5hr. The spores placed in the Smelt Mill sink arrived at the Cove rising in between 2 and 6·5hr; some of the Smelt Mill spores also went to the southern only of the two Aire Head springs in a time of 6·5–10hr. The flood pulse generated at the Tarn arrived at Aire Head in only 90min. No pulse was observed at the Cove; the same situation occurred in 1899 though the 1879 flood pulse, during a period of extremely low stage, was recorded at both Aire Head and the Cove.

There is therefore a complex pattern of interconnected drainage routes in the limestone behind and beside Malham Cove. However, the strongly contrasting flow-through times and the tracing of the Smelt Mill water to only one of the Aire Head springs show that this pattern is of only individual interconnections between discrete drainage channels; there is no evidence to suggest that the limestone approaches the conditions of a sub-uniform aquifer. The rapid pulse transmission indicates that a large part of the Tarn–Aire Head route is flooded.

The valleys

Most of the valleys around Malham are dry, though the amount of surface run-off is dependent temporarily on the amount of rainfall. After crossing the North Craven Fault, the Tarn Water, Smelt Mill stream and the streams running into Cow Gill Cote soon sink into the limestone. Moor Close Gill and Cowside Beck flow over impervious rocks and only Gordale Beck normally flows over thick limestone for any distance. Underground drainage is favoured in this area where the joint systems are well developed. Thus the problem is how and when the valleys were formed.

It has been suggested that some of the north–south trending valleys are relics of a pre-glacial topography. Clayton (1966) put forward, as an example, the group to the east of the Tarn leading from the Parson's Pulpit out on to the little dissected surface of the 405m (1,300ft) platform at Great Close. Clark (1967), on the other hand, favoured a subglacial origin for these valleys. He examined the valleys on Highfolds, Knowe Fell and Black Hill, and, taking into account various features such as their abrupt start and finish, and the humps in their long profiles, concluded that they were parts of an englacial and subglacial drainage system. Several chutes were observed running obliquely down the slopes of Highfolds and Knowe Fell, and deposits associated with the chutes and channels were noted to the north of Tarn Moss. In particular, he mentioned the mound of poorly sorted material which projects into Tarn Moss and appeared to be associated with a subglacial channel to the north, through which the Pennine Way now passes.

Melt-water from the Tarn area appears to have flowed down the two prominent valleys in the area—Gordale, and the Watlowes, leading to the Cove. A large pothole was once excavated close to Tarn Sinks, but the origin of its very poorly sorted, largely limestone debris remains uncertain. Since it was refilled, water has again run some distance down the valley after unusually heavy rain. A thin trickle of water even reached the Cove in March 1969. Earlier reports about the stream reaching the Cove were summarised by Howarth *et al* (1900).

The Watlowes dry valley terminates at the top of Malham Cove just over 100m vertically above the large spring at the base of the Cove. The vertical and overhanging cliff which forms the Cove curves in a great concave crescent to merge into the more broken scars and slopes on each side. Essentially, Malham Cove is a waterfall retreated from the scarp of the Mid-Craven Fault just to the south (see Fig 82), and since abandoned. Its development is however rather more complex in detail. The importance of direct glacial action is debatable, though at least part of the Cove's morphology must be due to waterfall action in periglacial environments within the Pleistocene, when the main drainage from the Tarn was over the surface and down the Watlowes. In addition the water resurging at the base of the Cove may have influenced its form by spring-sapping; however, the massive cavernous nature of the Great Scar Limestone is such that this can only have had a trimming action.

Gordale Beck is an exception at the present time in that it flows over the limestone for more than 1km after crossing the North Craven Fault. It is clearly saturated with lime since it deposits a considerable amount of tufa. The lime content has been found to be regularly higher than that of the Tarn system, which includes in its catchment water from peaty areas. Gordale's catchment is almost entirely from limestone country and it is possible that precipitation is more vital than solution (Moisley, 1955). Drift may also be blocking any sinks. However, it has been shown that at least some of the water drains underground from the Beck to the springs at the foot of the scar.

Gordale has many puzzling features. The main gorge itself, 75m deep, has often been explained as a collapsed cavern. There is little direct evidence for this, and, though cave formation and subsequent collapse must act as a partial mechanism in the incision of any youthful stream into limestone, the valley taken as a whole has the features of normal river erosion—down-cutting and recession of the waterfalls. How far glaciation and glacial melt-water have caused or modified the form is only one of a number of unanswered problems. Several melt-water features can be seen close to the Beck head around Great Close, including one prominent incision through the limestone between Ha Mire and Great

Close Mire (Clark, 1967). On crossing the fault, the Beck flows approximately south-east through a narrow steep-sided valley apparently controlled by one of the major joint directions. Moisley (1955) draws attention to 'lacustrine deposits' in this stretch, calcareous sediments at various levels through which the Beck has cut, leaving fragments of river terraces. He suggested a lake could have been contained by the natural wall that now marks the first waterfall. The stream flows through an aperture in this limestone wall, over a tufa screen and over more tufa forming the lower part of the falls. Phillips (1836), and other authors, refer to the time when water burst into the gorge during a violent storm in 1730. Their accounts imply that the gorge had been dry and that during the storm water broke through the already weakened wall. However, just to the west of this aperture is an abandoned tufa screen indicating that water flowed over the natural wall, but there is no attempt to put this in sequence with the formation of the main gorge.

The caves

The general lack of impervious capping on the hills of the Malham region has resulted in a scarcity of caves, and most of those which do exist are very small. At the western end of the block, Victoria and Attermire Caves are well known for the wealth of archaeological remains which they have yielded. The former consists of a single large chamber, almost entirely excavated by the archaeologists, while the latter has nearly 200m of rift passage leading from the back of its entrance chamber. Nearby, just south of Stockdale, are a number of shafts, the deepest being 50m, but none lead to any significant horizontal development.

On Pikedaw Hill, an old mine shaft gives access to the Pikedaw Calamine Caverns. These are a series of large, 10m diameter, phreatic tunnels which contain considerable amounts of glacial debris and also some zinc mineralisation. Nearly 1,000m of these tunnels are known, but all the ends are choked and they really only indicate the scale of cave development which remains unexplored in the Malham area.

The drainage of Malham Tarn and Cove has been described above. The Tarn sinks are impenetrable, excavations by generations of cavers in the Watlowes valley have all been unsuccessful, and divers have so far only penetrated 50m into the constricted flooded passage of the Malham Cove Rising. Nearby, Greygill Hole was once thought to be an ancient rising for the Tarn water, but it too is impenetrably choked after a few metres.

THE CRACOE–THORPE AREA

Just south of Grassington, where the River Wharfe crosses the Craven Fault zone, lie a spectacular series of reef knolls. Between the villages of Cracoe and Thorpe, the Skelterton, Swinden, Butter Haw, Elbolton and Kail Hills are each formed of reef limestones. The hills are up to 75m high and 600m in diameter, and the problems concerning their origin are the same as those concerning the reefs at Malham (see above).

All the knolls have been the scene of past lead-mining activities but only Elbolton Hill has any known natural caverns. In all 250m of passage is known in four small systems, the deepest being Elbolton Pot (41m). Little is known of the drainage of the knolls except that water sinking in Elbolton Hill resurges on the banks of the Wharfe over 1km away and 185m lower in altitude. Nearly all the known caves are formed along unmineralised NW–SE fractures. A feature of Elbolton Pot and Escoe House Hole is that parts of them reveal distinct signs of bedding in the limestone; clearly the reef knolls are not purely organic features but do contain patches of bedded limestone within the overall reef structure.

GREENHOW HILL

In the Greenhow area the Lower Carboniferous is represented by a series of limestones, at least 350m thick, exposed on the crest of an anticline trending roughly east–west. To the south the anticlinal ridge is cut off by the North Craven Fault zone whilst to the north and east the dip, of 10–30°, carries the limestone beneath the Grassington Grit. At the surface, the strike swings round the margin of the anticline from nearly east–west at Stump Cross to north–south at Mongo Gill Hole entrance. The change in the direction of dip has exerted an obvious influence on the cave system, seen most clearly in the way the main Mongo Gill–Stump Cross Passage closely follows the strike around the northern limb of the anticline (Fig 83). Many mineralised fissures have been worked for galena over a period of hundreds of years, and the miners have not only broken into the caves and partially excavated them, but have also considerably modified the local underground drainage.

Mongo Gill Hole is entered by the 15m North Shaft entrance and commences as a series of boulder chambers close to the fault zone, connected by a maze of passages which combine to form a single deserted, well-decorated gallery heading northwards. There are occasional short branches, including one connected with a streamway sumped at both ends, and the main cavern becomes

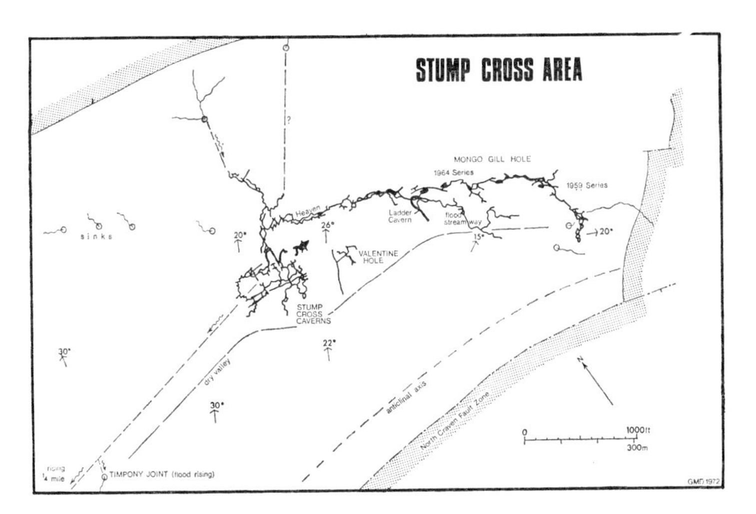

Fig 83 Plan of Stump Cross Caverns and Mongo Gill Hole

larger until the East Hade collapse and crawl leads to the 1964 series. The upper passages in this series comprise large rounded caverns and tunnels linked by relatively small galleries and modified in places by blockfalls. Below this deserted series, which contains substantial deposits of mud and sand fill, is a stream passage, prematurely deserted other than in wet weather as a consequence of the interception of its feeder streams by deeper lead-mining activity.

Upstream of the point of intersection between upper and lower levels—a high natural rift in Willie Waters Vein—the streamway is a lengthy and impressive passage obstructed by sumps which disappear in dry weather, and fed by several inlets. Downstream the water, when flowing, deserts its old course in favour of a complex of immature crawls ending in a flooded pot. The old waterway is a series of dry sandy caverns ending at an unstable clay choke formed by slumping from one of the clay-filled north–south joints which were frequently encountered in mines in the locality, and on the other side of which is the continuation of the passage in Stump Cross Caverns.

Mining also provided the entrance to Stump Cross Caverns, and these were almost immediately opened in part as a show cave. The system is very complex and at least five distinct levels of passage development, mainly strike orientated, can be discerned in this multiple-branched system. The deserted upper levels, many well adorned with formations, are extensively occupied by varying quantities of clayey fill which obscure large parts of the original galleries and block many side passages. The lowest level in the cave is a small vadose streamway, fed by sinks and seepage on the Grit/limestone boundary to the north, and impeded by numerous sumps. These are a probable consequence of its course cutting through the beds in the up-dip direction, but fortunately have a series of by-passes associated with them.

The flood stream from Mongo Gill flows beneath the deserted 1949 Series (Heaven), which equates with the former streamway in Mongo Gill Hole, and enters the Stump Cross watercourse at a diffuse junction area centred on deep phreatic rifts. From the downstream sump of Stump Cross the water flows westwards along the strike to rise up-dip from springs up to 1km away. The Timpony Joint outlet is restricted so that in severe floods ponding causes a series of springs to become active at progressively higher levels over almost the full distance between cave and rising.

Several impenetrable sinks behind Nussey Knot probably share these resurgences but, with the exception of some choked dry pots about 30m deep, Grenade Shaft and Valentine Hole are the only other accessible caves associated with the 5½km long Stump Cross–Mongo Gill system. In the former a 15m shaft enters a continuation of the ancient Show Cave level terminating at a massive collapse chamber with fine formations, whilst the latter is a 20m lead mine shaft re-opened to give access to a short length of boulder-choked passage. To the east, at Greenhow Hill itself, there are sinks now feeding streams in mine levels, and nearby is Pendleton Pipe, a pre-mineralisation solution cavern formerly worked for lead and now in danger of being quarried away. West of Stump Cross, a lost cavern encountered by mining was connected with Nape Well rising in Troller's Gill, and feeding springs in the same gill is Hell Hole, a 120m long and 40m deep pothole developed in north–south rifts. In Greenhow Hill there is still plenty of potential for new exploration which will help to solve many of the problems still remaining.

21

The Caves of Wharfedale

M. H. Long

At the head of Wharfedale, here known as Langstrothdale, the two major streams forming the infant Wharfe have their confluence at Beckermonds. The northernmost, Oughtershaw Beck, has very little of speleological interest as it flows largely on drift, with its source high on the slopes of the Yoredale Series. The other, Greenfield Beck, has incised a deeper, broader valley into the drift and down to the underlying rock, and in conditions of low flow the entire stream goes underground for two stretches. Although the possible extent of these systems is about 750m each, only a few metres have so far been explored, being largely wet bedding caves.

Along the south side of the Dale, from Beckermonds to near Deepdale, a number of minor streams sink on reaching the Great Scar Limestone. Of the several small caves and pots known, none is more than 150m long and 20m deep, and all are characterised by vadose canyon passages of small dimensions. Also in this part of the Dale the Wharfe again has an underground course, although apparently very immature as the total flow is only engulfed in times of severe drought.

The large rising on the south bank of the Wharfe at Deepdale marks a distinct change in the subterranean flow in the area as the submerged passage has been dived to a depth of 9m with no sign of a floor. Sinks some 100m higher in altitude have been tested to this resurgence, but it is also thought that part of the flow is (*right*) Deluge Pot in Out Sleets Beck Pot (*below*) Easy Street passage in Out Sleets Beck Pot

(*left*) A small rift passage in Scrafton Pot, Coverdale (*below*) The main phreatic tunnel in Sleets Gill Cave

derived from sinks in the Wharfe itself, indicating the presence of a shallow phreatic zone, possibly controlled by a small fault. Within 50m of the most easterly upper sink to this system lies the entrance to Langstroth Pot. Entered in the Hardraw Scar Limestone, this pot is some 1,400m long and 90m deep. It is strongly controlled by two parallel faults and consists of large lengths of vadose passage with numerous small pitches, the deepest being 14m. Part way down is a large inlet passage now carrying only a small misfit stream, and this fine swirling canyon cave most likely originated from sinks in Hagg Beck now deprived of water by down-cutting of the stream. Two short sumps below the final pitch are the only signs of phreatic development, and the continuing passage can be followed to emerge at a dug entrance within a few metres of the rising.

On the north side of the Dale, opposite Langstroth Pot, are a number of small pots up to 10m deep, together with Yockenthwaite Pot, which is something of an enigma. A large fault rift is choked some 15m below the surface, but a washed-out shale bed 60cm thick leads to two minor inlets and small pitches which provide a route back to the continuation of the entrance rift. A climb down in-situ breccia leads to the final choke at a depth of 45m. However, in one wall of the rift is an abandoned vadose passage which can be followed for 150m to an area of parallel faults, the way on being too tight. Almost straight below the pot is Yockenthwaite Cave consisting of 170m of low, wet bedding cave heading up-valley, and apparently unconnected with the pot.

Above Yockenthwaite is Pasture Gill, and sinks in this small beck have formed the pothole of that name, entered by way of a subsidiary shaft 50m north of the gill. At a depth of 20m is a thick shale bed where the main stream enters, and further pitches of 15 and 41m lead to a rift orientated along a fault. For no apparent reason the passage then leaves the fault in a small phreatic passage only to return to it some 80m further on. More fault rift passage gives way to a smaller vadose passage and the final pitch, to a sump at a depth of 104m. Throughout the system are a number of thick shale beds which exhibit strong control over the extent of all the vertical features. The rising for the system is on the north bank of the Wharfe further down-valley, and is also the outlet for water sinking in the river-bed at a number of points upstream of Yockenthwaite bridge.

At an altitude of 520m, the headwaters of Pasture Gill rise from a shattered scar at the base of the Main Limestone of the Yoredale Series. Entry cannot be gained, but it must drain a considerable area of the extensive bench, as also must another high-level rising at the head of the east branch of Deepdale.

Strans Gill Pot, in the next gill to the east, almost certainly has its entrance in the Hardraw Scar Limestone, as the Dirt Pot Grit outcrops only a few metres vertically above. Tight shafts and passages lead to a large rift pitch of 49m on a fault, where a number of inlets enter. Below, fault-controlled passages lead to a series of small pitches on another fault, and a sump at a depth of 105m, whilst several more inlets join the system. One of these extends almost to a point directly beneath Old Strans Gill Pot 2, a large surface opening choked with extensive collapse debris. The most impressive feature of the pot is an abandoned phreatic passage, The Passage of Time, containing magnificent formations, and also distinctive are the same thick shale beds seen in Pasture Gill and Yockenthwaite Pots, which again effectively control the limits of all the vertical features.

In all these three pots large individual calcite crystals have been found, many with traces of malachite present, and in several places bright green stalactites have been observed. Similar features also occur on the south side of the Dale in Bracken Cave; 460m of small bedding passage, the stream in which is probably derived from choked sinks in the lower part of Bouther Gill.

Straight opposite Buckden village lies the Birks Fell Cave system, some 3,200m long and 142m deep, and largely linear in extent. The entrance cascade is actually formed by the *Girvanella* Bed, for long held to be a marker bed between the Great Scar Limestone and the Yoredale Series, but now thought of as a rough guide only, occurring somewhere in the 'black' transition beds. Small vadose passages lead to large fault-controlled rifts with extensive collapse features, and long lengths of passage are very evenly graded with only small pitches and climbs to give the system its considerable depth. The detailed chronology of the cave is uncertain but it must be at least inter-glacial in origin, as at a sudden reversal in passage direction an abandoned upper level heads directly for an old resurgence cave only 20m away. This is Hermits Cave which lies some 30m above the present rising in the valley floor, thus giving some indication of the depth of excavation by just one glaciation in this part of the Dale.

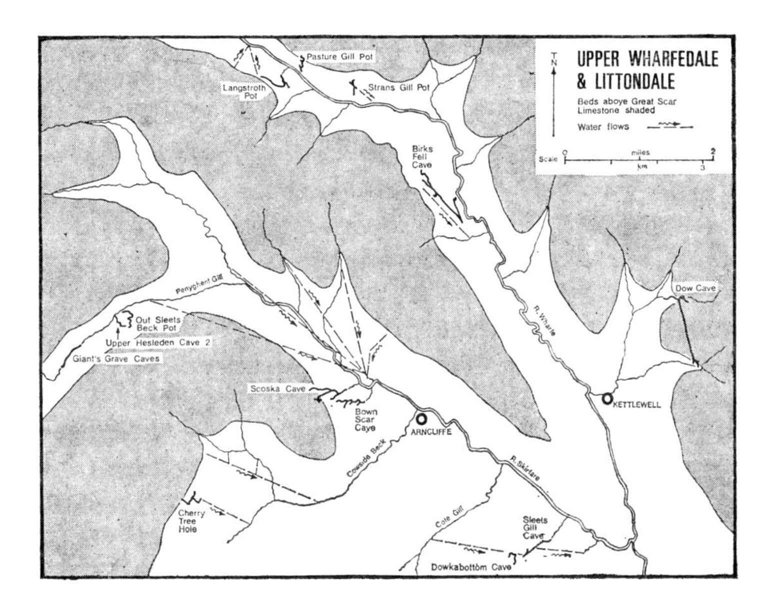

Fig 84 The major caves of Wharfedale and Littondale

Possibly unique in the Dale's hydrology is the fact that there exists a parallel system to the Birks Fell Cave, lying slightly further back into the hillside, and having its upper end in the Redmire Pot area and resurging at Birks Wood Cave. To date only short lengths of passage have been explored at either end, but it seems probable that this system acts as a 'cut-off', preventing water from entering the lower parts of Birks Fell Cave.

Immediately south of this drainage block is Firth Gill with a number of small caves and pots mostly associated with minor faults, and then the bench rises gradually to 420m in altitude. It is in this area that the only other known significant system lies on the western side of the Dale before the junction with Littondale is reached. The main sink is Foss Gill Pot which descends to a choke at 47m, whilst a number of other smaller streams to the south presumably feed the same resurgence at Foss Gill Cave, which can be followed in for 245m in bedding-controlled passages.

On the east side of the Dale very little of speleological interest is present until Caseker Gill is reached where lies the impressive Dow Cave system, 3,200m long, including the notorious Dowber Gill Passage. The majority of the main Dow Cave consists of a large stream passage alternating between stretches of canyon passage and wide bedding cave, all on a grand scale, and interspersed with several large chambers. One boulder choke can be passed to reach a series of waterfalls and the final choke, through which a large stream emerges. Although various sinks have been shown to enter the cave where it passes beneath Caseker Gill, the source for the main stream has yet to be established. Quite extensive mine workings can be entered from within the cave, and in one or two places thin veins of galena can still be observed.

Dowber Gill Passage consists of 1,300m of dead straight rift passage aligned along a major joint. For much of its length it is 20m tall, and the two-dimensional maze in a vertical plane, formed by extensive boulder chokes at varying levels, is the main cause of bewilderment to newcomers to the system. The limit of exploration upstream is a sump, shown to be fed in part by sinks high up in the Yoredale Series in Dowber Gill Wham, but as the end of the passage is almost under the gill it seems possible that the main source may be much further south. Providence Pot, situated in

Dowber Gill itself, extends for 330m and descends 30m to join Dowber Gill Passage 90m downstream of the sump.

The upper end of Littondale, like Wharfedale, has two major branch streams, which below their confluence at Foxup form the River Skirfare. The northern one, Cosh Beck, has a number of shallow caves at its head, none longer than 300m, whilst Foxup Beck has only two reasonably-sized systems, both about 350m in extent.

Further down valley, just above the confluence with Penyghent Gill (Hesleden Beck), a number of sinks in the river-bed engulf all the dry weather flow, leaving the valley completely dry for 4km. The multiple rising some 1·3km up-valley from Arncliffe is of considerable importance in the hydrology of Littondale as it is also the outlet for water sinking in Penyghent Gill (5km away in a straight line), and in Potts Beck and Crystal Beck, as well as sinks up on Old Cote Moor. It appears probable that large lengths of this extensively integrated system will consist of submerged bedding planes, but direct exploration may show this not to be the case, as was found in West Kingsdale.

Penyghent Gill is a steep-sided youthful valley fed by a number of streams draining the upper slopes of Penyghent. At the head of the gill the combined flow sinks on reaching the Great Scar Limestone at Giant's Grave Caves, an extensive through-system some 680m long, with mostly low passages, in places 12m wide. Below the rising the valley is cut down steeply and the bench along the south side of the gill contains numerous small caves and pots, together with three larger systems. The first two, Upper Hesleden Cave 2 and Out Sleets Beck Pot, both originate from sinks in the same stream; the first being entered at the rising and consisting of 520m of vadose streamway with four scalable waterfalls, over a vertical range of 64m, the other being entered at the uppermost sink with three short pitches and 975m of vadose passage. With a little effort both could be made into through-trips, and as such offer great scope for detailed observations of complete hydrological systems. The third system, Snorkel Cave, is again entered at the rising, and a route up three shafts allows a height of 40m to be reached, with 790m of awkward passages.

On the north side of the gill are a number of small caves up to 130m long, and although several streams must descend to the sub-valley system referred to above, no deep pots have yet been discovered. Of the numerous open entrances at the various risings a majority give the impression of having been beheaded by down-cutting of the gill and its main tributary, leaving the caves perched several metres up the sides. Also of some significance is the fact that with virtually no exception the caves are totally devoid of fill, thus lending weight to the theory that the majority of the caves and pots in the gill are entirely post-glacial in origin.

On the south side of Penyghent Gill at its junction with Littondale are a number of sinks known to feed the resurgence of Litton Fosse, as do a number of pots in Littondale itself. The deepest of these, at 43m, is Mustard Pot which descends in three pitches to a choke, and the area clearly holds potential for an extensive system. Unfortunately the rising, which can be entered for 30m to a choke, is utilised as a water supply, thus curtailing future explorations.

The exact watershed between the catchment draining to Litton Fosse, and that draining south to Bown Scar Cave, is uncertain, but is presumed to be approximately the boundary wall between Great and Little Scoska Moors. On Little Scoska Moor are a number of immature sinks, and the three known pots are all impassable at depths of 10m, although the Scoska–Bown Scar cave system lies some 100m lower in altitude. Scoska Cave, 1,400m long, and requiring mostly hands and knees crawling in bedding-controlled passages, has three main sections. The left-hand passage is obviously of considerable age, with extensive stalagmitic deposits at the far end which choke the way on, and a misfit stream which is the only water to utilise the cave entrance. The right passage soon splits, the right branch becoming a long crawl in a silt-floored canal, whilst the left branch meets the main stream, entering through a choke and rapidly sumping downstream.

It is this same stream which is presumed to be the main flow entering the Scoska Series of Bown Scar Cave some 230m away in a straight line. The entrance to the cave is an insignificant bedding, and the majority of its 1,150m of passage are covered by crawling, again in bedding-controlled passages. At the far end of a fine canal passage, a series of rift avens climb to a height of 32m and the ensuing cave consists of 245m of canals in a passage of small dimensions, to where the air space becomes unusable. A number of inlets join the main route below the avens, and these most likely have their origins on the moor to the south.

On the east side of the Dale there are no known extensive systems, but all the water sinking as far down-valley as Arncliffe ultimately resurges at the multiple rising 1km upstream of the village. Significant, however, are Potts Beck Pot with its large chamber, 30m by 22m and up to 25m high, Crystal Beck Pot with its six sumps in a highly faulted area, and the abandoned phreatic resurgence of Boreham Cave which beyond 50m of sumps ends at a choke of unsorted glacial till. Further down the valley, below Arncliffe, are a number of choked shafts up to 18m deep, and three unenterable risings, with as yet no known horizontal passage.

At Arncliffe, Littondale is joined by Cowside Beck, its largest tributary, in a steep-sided but evenly graded valley, at the top of which are a number of risings draining the eastern slopes of Fountains Fell. Of these only one has a known source, in Cherry Tree Hole, a complex system of great variety some 1,200m long and 41m deep. Three major inlets enter the system, which exhibits considerable control by a number of faults, giving rise to some extensive collapse features and spacious passages. In the same area is Darnbrook Pot, a largely vadose system some 335m long, which descends 41m in a fault zone to a choke of fine breccia situated almost beneath the large surface sinks known as the Cockpits. Further sinks in Darnbrook Beck combine with the Pot water to reappear some 3km away and over 100m lower, at a large rising on the south bank of Cowside Beck. Access to a part of this drainage route can be gained at Yew Cogar Cave, also situated on the south bank of Cowside Beck, but some 250m up-valley from the rising. Less than half the cave's 800m of passage are associated with the Darnbrook water, terminating in sumps in both directions, whilst the remainder forms an independent streamway which is believed to drain part of the Parson's Pulpit area.

Also draining a part of the extensive limestone uplands of the Parson's Pulpit area is the complex Cote Gill Pot/Dowkabottom Cave/Sleets Gill Cave system. At present unconnected, these three holes share a common rising at Moss Beck which lies a short way up-valley from the famous truncated spur of Kilnsey Crag. Cote Gill Pot, although less than 100m long, intersects a large choked passage below a small pitch, obviously a remnant of a far older system. Dye introduced here took only 10hr to cover the 3·2km to the rising, over 120m lower in altitude, indicating a route unlikely to contain long canals or sumps. Of more interest perhaps is the fine Dowkabottom Cave whose collapse entrance is adjacent to the tremendous closed depression of Dowkabottom. The majority of the 670m of passage are large, being associated with three faults, and in places contain extensive fill deposits including sandstone pebbles. As the entire catchment area has only two insignificant patches of Yoredale rocks, there is every likelihood that this cave is pre-glacial in origin.

With its superb phreatic main tunnels, Sleets Gill Cave is certainly the most impressive part of this drainage system. Entered by a sloping phreatic passage which descends 25m, this cave gives evidence of having been formed in a deep phreatic zone, a fact borne out more clearly by The Ramp, a unique sloping tunnel which climbs 60m in 90m of passage length. Unfortunately not all of this cave's 1,600m of passage are on such a grand scale, as extensive chokes block the way in places, leaving only small passages of recent origin as link routes. The hydrology of this system is largely unknown, as the only surface stream in the catchment area is Cote Gill, the rest of the flow being derived from percolation water; although just where the boundaries lie remains to be established. However, in times of heavy rainfall sufficient water enters the system to flood it all to the roof, except for the top portion of The Ramp which exhibits a fine array of pure white stalactites.

In conclusion it can be said that this part of the Dales contains caves the equal of any others in the north—long and deep, phreatic and vadose, and all worthy of more detailed examination. As the major proportion of the larger systems have only been discovered in the past few years, knowledge remains somewhat sketchy in places, although work is continuing in order to rectify this.

22

The Caves of the Black Keld Drainage System

D. Brook

The area of Great Whernside which acts as a gathering ground for the Black Keld resurgence is unique in Britain as a karst drainage unit. Most of the sinks within an area of $18km^2$ enter the Middle Limestone and penetrate two 6m thick bands of sandstones and shales to enter the massive Great Scar Limestone. Hence the caves display 'Yoredale cave' impervious floor characteristics in their lengthy initial stages, but become more complex after breaking through into the main limestone beds.

TOPOGRAPHY AND GEOLOGY

Behind Black Keld the land rises steeply in the classical scar morphology of the Great Scar Limestone to form one flank of the deep glaciated trough of Upper Wharfedale. The higher beds, of Yoredale limestones, form wide benches separated by steep scars, and the highest of these (the Middle Limestone bench) is pitted by shakeholes and partially covered by peat, sphagnum bog, clay and sandstone debris. The Langcliffe benches, west of the Great Whernside ridge, are the most prominent in the area and here the massive Middle Limestone is separated by black shale from the Top Limestone, only 5m thick, which may represent the Five Yard Limestone or be merely a local separation of part of the Middle Limestone sequence.

The top of the limestone series is marked by a prominent unconformity where a sudden lithological change occurs to the massive sandstones and shales of the Millstone Grit Series which rise to the summit of Great Whernside.

The sequence of strata described above is tilted gently to the south-east and extends without major dislocation into Upper Nidderdale. Southwards however a suite of east–west faults, subparallel to the Craven Fault system, have increasing downward throw into the Grassington Moor Mining Field. The Bycliffe Vein is on the most important fault in this suite and seems to define the southern limit of the Black Keld drainage block, since sinks on Grassington Moor emerge at Low Mill rising, near Grassington village.

Topographically the area is dominated by the bulky mass of Great Whernside, 705m (2,310ft) high. A deep valley drains its eastern slopes, and loses its headwaters at the Mossdale Scar sink. Below the alluvial flat of Bycliffe Moss the incised valley continues down to Hebden taking a course parallel with Upper Wharfedale.

GENERAL HYDROLOGY

The largest stream within the Great Whernside catchment is engulfed by the sink at Mossdale Scar which takes an average flow of 100l/sec. Myers (1950) showed that the rising at Black Keld (mean flow 150l/sec) was the only outflow in the vicinity which was large enough to account for the Mossdale water. This theory was proved by a dye test, and Myers also dye-tested the Gill House sink to the same rising. This vast area of high moorland draining to a common resurgence is further defined by a recent dye test which showed that Black Keld is the rising for the Rigg Pot water. Within the area delineated by these tested streams there are sixteen sizeable ones, whose most likely outlet is also Black Keld, but few of these enter known cave systems. Caves, in fact, are few and far between in this region, but two of them are large by any standard. Both are fine examples of litho-

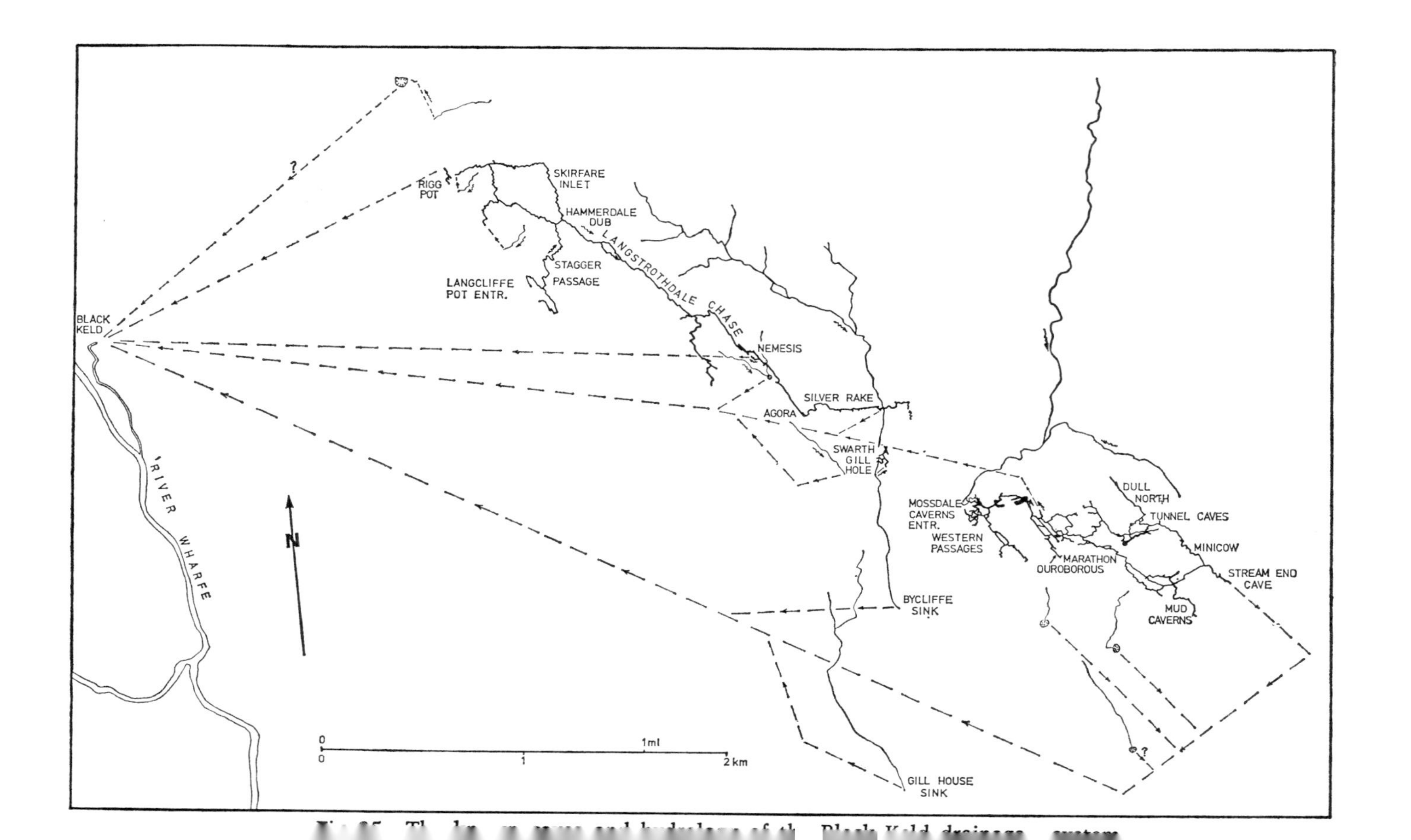

RIGG POT
SKIRFARE INLET
HAMMERDALE DUB
LANGSTROTHDALE CHASE
STAGGER PASSAGE
LANGCLIFFE POT ENTR.
BLACK KELD
NEMESIS
SILVER RAKE
AGORA
SWARTH GILL HOLE
RIVER WHARFE
N
MOSSDALE CAVERNS ENTR.
WESTERN PASSAGES
DULL
NORTH
TUNNEL CAVES
MINICOW
MARATHON
OUROBOROUS
STREAM END CAVE
MUD CAVERNS
BYCLIFFE SINK
GILL HOUSE SINK
0
1ml
0
1
2km
?

logical control on underground drainage, but, unlike the smaller examples found in the Yoredale limestones further north, these are channelling water away from its eventual resurgence. They terminate 180m above and several kilometres distant from Black Keld, and the hydrological problems will have to await the next phase of cave exploration before any attempt can be made to answer the questions 'how' and 'why'.

MOSSDALE CAVERNS

This major sink, 230m above Black Keld at an altitude of 425m (1,400ft), was first entered in 1941 (Leakey, 1947) and now leads to some 9·5km of explored and surveyed passages (Grandison, 1965). The system has a unique branching morphology with no less than six outlets. Beneath Mossdale Scar are several entrances, amongst a chaos of boulders, which all unite in Assembly Hall. This is a flat-roofed chamber floored by fallen blocks with a slope down to the south into a wide streamway leading to Blackpool Sands. The western limb of Assembly Hall is the gateway to Western Passages—a complex of abandoned and active passages uniting in the streamway of Broad Street whose eventual source is flooded fissures close to the main sink at Mossdale Scar. 2m above the Broad Street stream is the prominent elliptical entrance of Far Western Passages, which run south-east for 300m until they become floored by sandstone, subdivide, and close down.

Continuing down Broad Street the water deepens until it unites with the rest of Mossdale Beck at Blackpool Sands. Now the roof lowers over 1·5m deep water for 150m, through the Swim, into the large boulder fall under Boulder Hall which marks the end of the ponded section. Boulder Hall is the largest chamber in the cave, having a volume of 7,500m^3 beneath its flat roof; the stream emerges from the blockfall and cascades into a region of low, wet, confusing fissures. Down the stream the sandstone reappears and the roof rises into Broadway—the finest stream passage in the cave; but this is only short and the water turns off into Syphon Passage to flow north, against the dip, and disappear into flooded and choked fissures. Three dry passages, however, continue the down-dip, south-east, trend. Two are on the sandstone but the largest is developed 1m above it and all unite in Rough Chamber just beyond the abandoned outlet of Ouroborous Passage.

Rough Chamber marks the start of the long crawl-passages with smooth sandstone floors for which Mossdale is infamous. Beneath the floor the small stream overflowing from Syphon Passage enters the False Marathon Network, but the obvious exit from Rough Chamber is a high rift suddenly lowering to the crawl of Rough Passage. Higher in the rift, Shingle Passage connects with Oomagoolie Passage to form a 600m long loop back into Rough Passage near the point where the False Marathon Streamway can be entered by a short crossover. The smooth sandstone floor of Rough Passage is broken by a clean fault which causes a 60cm waterfall shortly before Kneewrecker Junction, where the cave divides.

To the east, Kneewrecker Crawl is a long low tunnel ending at Relief Junction where it intersects a much larger but deserted cave. Downstream the high levels become obstructed by boulders and sand at False Fourways, but a characteristic crawl on sandstone continues to Fourways Junction. Here the water runs into Far Syphon Passage to eventually sump while a fissure ascends into Fourways South Sand Caverns—a high-level series linking False Fourways and Tunnel Caves. Dull North Passage, north of Fourways, is a mud-choked inlet and Tunnel Caves runs south-east as a large arched main drain. Its water sinks into boulders, and a high-level continuation ends in a massive choke directly below the northern end of Minicow Passage.

The southern passage at Kneewrecker Junction is the start of Marathon Series, in which the water from the False Marathon streamway is soon encountered. The Marathon passages involve crawling and awkward walking for 600m; buckling of the sandstone floor is noticeable in the far reaches, which contain a lengthy, and strenuous, oxbow. Beyond a canal is the greasy chimney up into High Level Mud Caverns, which are an extensive series 10m above the sandstone and the present streamways. Beyond the chimney is the inviting Far Stream Cave leading on until Minicow Passage enters as a major inlet on the left. Minicow is a pleasant stream passage, but upstream collapse debris causes the caver to ascend into high-level flat-roofed caverns which choke above the Tunnel Caves collapse. Beyond the Minicow tributary, the streamway achieves impressive proportions before encountering the solid choke of Stream End Cave—the point in the system remotest from the entrance at Mossdale Scar.

LANGCLIFFE POT

As recently as 1968 Langcliffe Pot was a minor system some kilometres distant from any of the other caves in the region, but since that date some of the most notable explorations ever made in the Pennines have taken place (Rogers, 1969; Yeandle, 1971). These have solved many of the mysteries about the morphology and geological setting of the hydrological pathways in the area, but the far reaches of the cave pose yet more questions since the explorations have revealed a very complex hydrology.

The system now totals some 9·5km of surveyed passages of which 8km are in the Middle Limestone and 1·5km in the Hardraw Limestone. All the known tributary streams drop through the Middle Limestone as vertical shafts and communicate with the main system via crawls at the base of the limestone. Langcliffe entrance is a well-watered 25m shaft, the outlet being a short traverse dropping into the Craven Crawl. At No 1 Junction, Fools Inlet enters, its passage draining 300m from a boulder choke back under the Middle Limestone bench. Stagger Passage carries the water further under the gritstone ridge of Great Whernside for another 600m to Hammerdale Dub, passing the Strid Inlet en route. At Hammerdale Dub the passage finally achieves roomy proportions with the entry of Skirfare Inlet, which may be traced back until the streamway becomes impassable close to Thunder Pot, the main source of the water. A double level passage provides a crossover into Strid Inlet whose present upstream limits are a series of avens and a wet crawl.

Below Hammerdale Dub the Main Drain is 3m wide and 5m high until the large stream takes the smaller tunnel of Wetway which is an oxbow to the Kilnsey Boulder Crawl. These unite into the even larger streamways of Langstrothdale Chase continuing to Mile House where a complex of inlets includes the 600m long Gypsum Passage. 500m beyond Mile House, the water drops down sandstone steps into a flooded bedding plane. A dry by-pass rejoins the water in Boireau Falls Chamber where it has carved a canyon in sandstone and shale; a boulder choke in the floor guards access to the 20m pitch of Nemesis. This penetrates the Simonstone Limestone into a region of monumental boulder chokes—the gateway to Gasson's Series, which is totally developed within the Hardraw Limestone.

Beyond the Nemesis chokes, the streamway is 7m high and 3m wide but soon the water swings abruptly west under a wall into a short duck and descends through a mass of boulders to finally run into Poseidon Sump. The west wall of the sump is bedrock but the rest of Poseidon Series is within a chaos of boulders, which is associated with the Nemesis chokes. The total known dimensions of this chaos are 50m by 100m, by 40m high, and it resembles the Greenhow gulfs (Dunham & Stubblefield, 1945) and a similar, smaller feature noted in Dow Cave.

Before the duck the main rift continues south-east as the fossil passage of Sacred Way, floored by sand and containing a chamber with boulders covered in polygons of fungal mycelium. Sacred Way terminates after 450m at a traverse into the Agora, an impressive and well-decorated chamber which is the largest in the Langcliffe System. Now the cave suddenly changes direction, for after 3km of relentless south-easterly progress it turns due east down a stalagmite slope into Aphrodite Avenue, a handsome canyon with gour pools. Fallen boulders bridge the passage as it increases in size along Silver Rake, named after the vein of the North Mossdale suite which controls the direction of the cave

Page 429

Wharfedale, on the right, and Littondale, on the left, as seen from the air. The limestone benches of Old Cote Moor lie between the two Dales, and Great Whernside rises on the extreme right

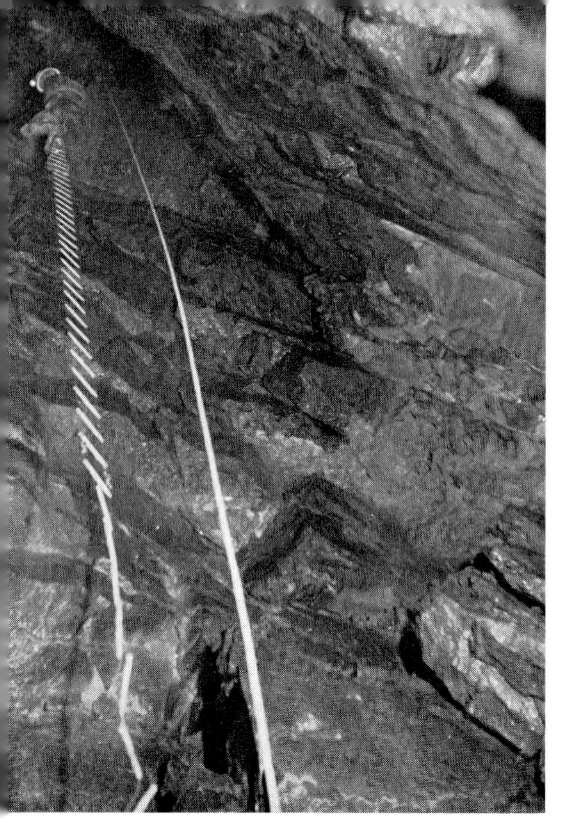

(*left*) The By-pass Pitch in Pasture Gill Pot. The roof on the right clearly shows the fault fractures which have located both the shaft and the passage below (*below*) Arch Chamber, a high joint-controlled rift in Robin Hood's Cave

beyond the Agora. The passage ends in a very solid choke of sand and boulders, but a lower route drops to an inlet passage which sumps downstream but can be traced upstream for 300m to a final choke of massive muddy boulders.

CAVE DEVELOPMENT

The Black Keld catchment also contains other smaller systems with many interesting features, but because of their great extent the Mossdale and Langcliffe caverns provide the most important illustrations of order and mechanisms of cave development.

Underground and surface deposits, and morphology, in Mossdale indicate that the stream has entered the limestone at several points during the cavern's history. These occur within the valley both above and below Mossdale Scar and at present only a thin remnant barrier of peat causes the beck to sink at the Scar and not continue down to Bycliffe Sink. Abandoned and choked passages, within the entrance complex and Western Passages, suggest that the Scar is an old sink, which has been re-invaded since the last Devensian glacier retreated up the valley and the 2m thick peat barrier was laid down in the bog of the deserted valley. Far Western and Ouroborous Passages are abandoned outlets for water sinking at the Scar and must have been initiated in the order stated because of their relationship to the present streamway. Syphon Passage, because of its restricted nature, seems to be a more recent development causing the fossilisation of the whole cave beyond it, in normal water conditions. The Shingle Passage–Oomagoolie loop poses questions since flow markings are conflicting, but a choked connection to North Relief Passage has provided water for the development of the latter at an earlier stage. The higher level of Relief Passage continues via Sand Caverns into Tunnel Caves, where it joins the old inlet of Dull North Passage which has been formed by Mossdale Beck once sinking much further upstream where the top of the Middle Limestone is now covered by streamwash. Tunnel Caves continue via Minicow to the Stream End Cave choke, but another (blocked) level is postulated to run parallel to Tunnel Caves into High Level Mud Caverns, which also have another abandoned feeder from the West.

The passages in Mossdale have a complex morphology due to the fact that they have been subject to considerable development during the tremendous floods which fill the cave to the roof. In spite of this, high-level remnants, and short lengths of active passage floored by limestone, indicate that much of the cave was initiated phreatically well above the sandstone which underlies the limestone.

Langcliffe Pot is morphologically much simpler than Mossdale Caverns within the Middle Limestone, and hence its mode of development may be interpreted and applied to the more complex history of cave formation in Mossdale. The details of stratigraphy in the Middle Limestone visible in Mossdale (shale, chert and fossil bands) are found to be remarkably constant throughout the whole of Langcliffe with its 8km of continuous exposures. Their effects on passage cross-sections are noticeable since they have determined both the ease of erosion and the collapse horizons. In Langcliffe the phreatic initiation at higher horizons than the base of the limestone is indisputable, since active inlets cut down to the sandstone only to rise above it further downstream. A fine example is Thunder Inlet, which at first is on the sandstone but rises stratigraphically on traversing Skirfare Inlet, whilst maintaining a fairly constant downstream gradient to Hammerdale Dub, where it encounters the sandstone once more.

Langstrothdale Chase repeats this phenomenon in a more impressive manner by climbing 4m up into the limestone whilst maintaining a constant gradient. A plan of the down-dip passages on the sandstone shows violent changes of direction under the gritstone/limestone boundary on the surface. This may be attributed to dip flexure under this boundary, possibly caused by warping as ice erosion removed the overlying gritstone. Stagger Passage is developed down-dip in a shallow syncline, but also reveals remarkable structures in the beds below the limestone. Here the smooth, unbroken sandstone familiar in Mossdale is replaced by thin rotten sandstone over shale which is contorted into a series of very sharp anticlinal ridges. Fractures in the sandstone also occur just before Boireau Falls Chamber. However, Langcliffe's unique breakthrough into the Simonstone Limestone is not achieved by way of such fractures, but via a canyon cut through massive undisturbed shale. A similar feature also exists between the Simonstone and Hardraw Limestones but here the exposures of argillaceous rocks are incomplete.

Within the Hardraw Limestone massive horizontal development has been discovered, not the great shafts which have been predicted (Myers, 1950), and the hydrology of Langcliffe, which was very simple in the Middle Limestone, becomes complex. Surface sinks above Gasson's Series do not enter Langcliffe but are diverted into separate caves by the umbrella of impervious strata above the Hardraw Limestone.

The brief observations recorded above cannot do justice to the unique problems posed by the challenging caves associated with the Black Keld system. Langcliffe Pot has indicated how other systems may break through into the Great Scar Limestone and enter extensive passages which exhibit different stages of development. If these characteristics are repeated in each of the widely scattered sinks of the region, the integrated system will present the greatest British challenge to both the sporting caver and the speleologist.

23

The Caves of Nidderdale

G. M. Davies

As a result of its steady dip east of Wharfedale, the Great Scar Limestone has been carried well below the surface in Upper Nidderdale and the caves are developed in higher beds. A consequence of the southerly overstep of the Grassington Grit group is that the Yoredale Series above the Underset Limestone is absent and only this bed, the Three Yard, Five Yard and Middle Limestones are exposed, all close to valley floor level. The Middle, in which virtually all the caves are formed, is exposed in the inliers of Limley, Thrope and Lofthouse. Faults, sometimes mineralised, cross these localities at frequent intervals and allow considerable local modification of dip to the extent that only the beds in the relatively unfractured Lofthouse inlier conform to the prevailing easterly downslope. Distinctive fossil and chert beds are a feature of the Middle Limestone and are well exposed in the caves.

The Limley inlier is an anticline with its axis aligned east–west, and on the northern limb, which is truncated by a large fault, the River Nidd sinks in its fissured bed to reappear underground in a number of collapsed inlets which coalesce into the main gallery of Manchester Hole. A single large river passage, some 3–4m square, trends southwards across the anticlinal ridge, in the process briefly exposing the sandstone top of the Simonstone cyclothem, passes the vast blockfall cavern of the Main Chamber, and follows the steepening dip to end in three sumps 17m below the entrance of the 450m long cave.

The divergence of streamways at the lower limit of Manchester Hole is continued in Goyden Pot, where the river reappears, and is further complicated by the presence of more inlets from the surface, a proliferation of which has produced the 3km long maze of passages explored in the cave. A phreatic origin is apparent, with subsequent vadose and blockfall modification, including the catastrophic collapses in the Main Chamber area which have not only diverted the whole of the Nidd into its present very large passage but also resulted in severe ponding and silting in the lower reaches of Manchester Hole. The river normally enters Goyden through various inlets near the Main Chamber, and continues south in the River Passage, leaving the old deserted streamway on top of the Bridge. The end of the known cave is the third sump, 35m below the entrance. East of the river is a virtually inactive, dip-influenced, network of crawls, chambers and high rifts, in many places wholly or partly choked by fill of various kinds. A small stream flows through this area, the Labyrinth, and into New Stream Passage, which ends at the deep Rift Sump.

The hidden courses of the river take it under the Grassington Grit, faulted down to valley floor level, to reappear in New Goyden Pot which lies in the Thrope inlier. Downstream of the low but deep inlet sump the river follows a relatively low passage to a long duck, and then quickly becomes a magnificent tunnel some 5m in diameter. Middle Sump can be by-passed via a large deserted gallery and the impressive blockfall chamber of the Planetarium to reach the downstream sump. Sinks above the 30m deep system have formed many avens and inlets, two of which are the entrances to the cave. Main Inlet is probably the point of re-entry of the Rift Sump stream from Goyden Pot, and this meets the main water just upstream of the Lower Sump—the present limit of exploration.

The combined waters then flow for more than 2km, beneath the surface outcrops of Grit, to emerge only 4m lower in altitude at Nidd Heads. There the river pours out of two entrances, behind which are separate networks of passages only a few hundred

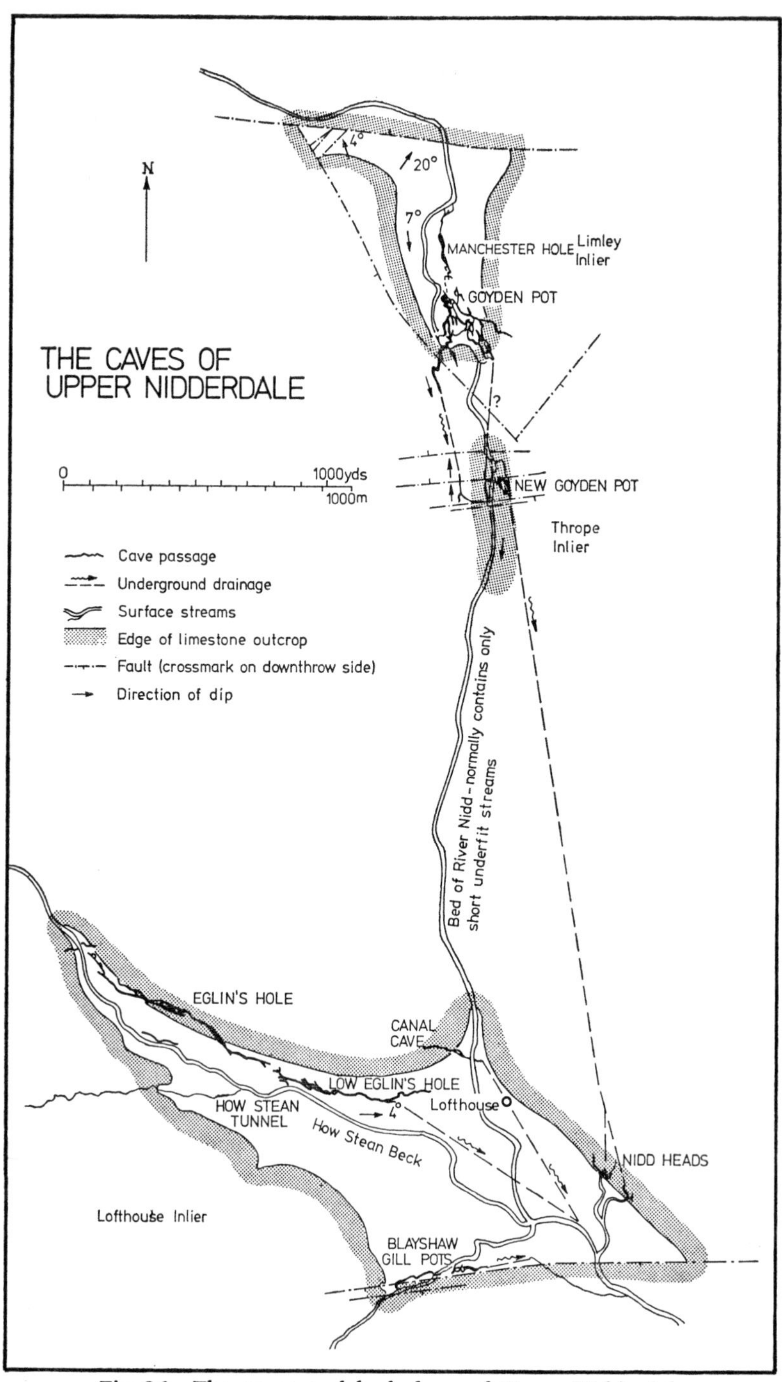

Fig 86 The caves and hydrology of Upper Nidderdale

metres in length before sumps are met. Most of these passages and chambers are at the top of the Middle Limestone, and local collapses have breached the base of the overlying shales.

The unique feature of the Nidd drainage system is that it is wholly developed beneath the floor of a major valley, yet has a sufficiently large fall from sink to rising to enable substantial parts of it to be explored. Furthermore the system is largely developed beneath an almost complete impervious cover, in limestone to which surface water has never had access except via the relatively small outcrops at sink and rising. Although the whole system at present floods to the roof, it has at one time been much drier, as indicated by large flowstone deposits which are now being eroded. The direction of the cave passages is strongly influenced by the varying dip, and modified in places, particularly in the larger passages, by joints. The numerous faults in themselves exert a limited control, and their importance seems to be in the abrupt changes of dip which they allow.

The second significant hydrological system of Upper Nidderdale is entirely within the Lofthouse inlier and commences at sinks in How Stean Beck which form the small stream occupying Eglin's Hole. Only a short cave, Cliff Wood Sink, can be entered at the sink area and there is no passable way through to the upper end of Eglin's Hole, where several restricted bedding planes and tubes join together and enlarge into a single wide bedding plane passage. One major inlet enters the deserted watercourse to give variety to its formerly well-decorated route with a total length of 1,300m.

A much greater water flow must have been carried at one time as the present trickle would have been incapable of creating the large and mainly vadose gallery which follows the 4° easterly dip to a boulder choke near the entrances to the cave, 20m below the sink level. The stream disappears under the ruckle and normally resurges 125m to the east where a daylight course of 10m ends at a drop into Low Eglin's Hole. In flood, however, part of the water penetrates down through boulder chokes into Tom

Taylor's Cave, a high vadose rift usually devoid of any appreciable stream, to emerge from this 200m long cave in How Stean Gorge, 17m lower.

Low Eglin's Hole is a 1,000m long cave commencing as a narrow vadose trench with a number of inlets all converging on the bedding plane which forms the roof of the main streamway. Normally the water disappears in a very small immature passage, and beyond this point is a muddy flood passage ending in a paraphreatic zone close to the resurgence level.

Forming part of the same hydrological system, Canal Cave, now fed by seepage, is a narrow cave whose upstream end is currently being steadily blocked by calcite flows. Downcutting in the Nidd river-bed has exposed the canyon passage, and consequent infilling with boulders and silt has produced the canal in the main cave and virtually closed the lower cave, into which sinks the underfit stream occupying the river-bed. A short course leads to a sump a little above the resurgence at Sandbeds Rising, where the combined streams flow out of alluvium in the river bank. A few short caves lie on the south-west side of How Stean Gorge, the major one being How Stean Tunnel, a short impressive through cave.

South-west of Lofthouse and also in the Lofthouse inlier are the two Blayshaw Gill Pots, forming the third cave system of the Nidderdale area. Part of Blayshaw Beck disappears into impenetrable sinks in the Five Yard Limestone close to a small fault. Underground the water emerges from a collapse pile, flows down a short well-developed passage and enters the Middle Limestone through a fault with a 15m throw. This, and a number of other small faults intersected by the cave, are slightly brecciated fractures, all throwing down to the south. The cave's single streamway is oriented down-dip to the east, but is solidly choked just downstream of the junction with the entrance inlet—a shaft down a fault plane.

Blayshaw Gill Pot No 1 is a little further to the east. It is a complex multi-level system of vadose passages, the lower ones containing the stream from the No 2 Pot. The final sump is at a level 30m below the top sink and only just above the resurgence—a pool in the alluvium of the Nidd floodplain.

Nidderdale contains a great variety of caves—perhaps surprising in view of the fact that they are almost all formed in a single band of limestone less than 40m thick. Lacking only the deep potholes when compared with the Ingleborough region, the caves clearly have a varied and complex geomorphic history.

24

The Chronology of the Caves

A. C. Waltham

The geomorphology of the karst and caves of north-west England poses a number of unsolved problems. The previous chapters in this volume contain descriptions of different regions by different authors, with, in many cases, slightly different interpretations of the patterns of cave development. There is, however, today reasonable agreement on the origins and environments of cave formation (chapter 4).

In contrast, there is no uniformity of opinion concerning the absolute chronology of cave development. This is because there is, as yet, virtually no evidence for the necessary cave dating. No one knows just how old the caves are. Eventually the answers will probably come from radioactive dating of either the caves' clastic sediments or the stalagmites. Suitable techniques have been developed recently but have not yet been applied to the karst of north-west England. When such work is started it will have its own problems; sampling will have to be done very carefully indeed in order to obtain material which is both uncontaminated and significantly related to the separate phases of cave enlargement.

Without this absolute dating of the caves or their contents, their chronology may only be determined by relating them to the events which have moulded the surface topography. On the surface, the majority of the landforms originate from the Pleistocene glaciations. In the upland areas, where most of the karst occur, the geomorphological effects of the successive glaciations were dominantly erosive, so that each ice advance tended to remove the results and evidence of the previous advances. The Pleistocene glaciations themselves are not therefore completely understood in this area, especially when compared to the detailed glacial chronology calculated in lowland Europe. Consequently, any postulated chronology must at present be considered rather conjectural, for it can only be based on a series of correlations between surface and underground features; many such correlations must remain dubious and even the final chronology is only relative.

Much of this however does not apply so definitely to the Morecambe Bay area. There, the results of Trassic karst processes can easily be distinguished (see chapter 11), though there is still considerable doubts about the effects of Tertiary erosion. Within the Pleistocene, the coastal geography of the area has meant that a long series of changes in sea level have left a variety of erosion surfaces and associated deposits. These are much better preserved, and give evidence of a sequence of events which is much more conclusive than any effects left by the successive erosion phases of the upland glaciers, each following the same pattern, in the Ingleborough–Malham region. Even more individual features, such as the layered screes (see chapter 3), are better preserved in this area. It should therefore be possible to construct the details of a long erosional history in the Morecambe Bay karst regions. Chapter 11 itemises much of the work so far completed, but also shows that most of the known cave development appears to originate from Devensian and Flandrian stages. The projection of this area's karst chronology even into the earlier stages of the Pleistocene is still proving difficult.

Malham is another region where there is some evidence for an absolute chronology, though even this is limited in extent. Pollen studies around Malham Tarn (Pigott & Pigott, 1959) have only shown that all the glacial drift is older than pollen zone I, that is, Middle Devensian or earlier. Victoria Cave contains a cave earth yielding a Middle Pleistocene fauna of hippopotamus, elephant and rhinoceros, overlain by a laminated clay deposited by glacial melt-water. Above this are further cave earths yielding reindeer and then bear, and human artifacts of the Upper Paleolithic period. Clearly Victoria Cave is very old, but its position and elevation mean that it cannot be related to the major cave systems further north, away from the fault zone. Tertiary karst erosion around Malham is indicated by some of the erosion surfaces, and also perhaps by the very large closed depressions around High Mark. It is not known just when, within the Tertiary, this erosion took place, but it does suggest a long geomorphic history of the present landscape. This does not necessarily infer a similarly long erosion history in the rest of the north-west England karst, for the limestone is higher in the Malham area than in the surrounding districts and should therefore have been first to be exposed to weathering.

The majority of the caves of north-west England lie in the limestones flanking the main Dales, from Nidderdale in the east to Barbondale in the west. It is here that the chronological problems are greatest, due to a lack of evidence on the surface which conclusively relates to known events in an absolute chronology. It is even more difficult to relate the rather better-known events of the Morecambe Bay and Malham areas back to the Dales; the contrasting environments provide too marked variations in the detail.

Within the Dales there are three major features of the geomorphology which must provide, and be incorporated in, a basis for a chronology. Firstly, the Dales themselves are classic U-shaped valleys, containing numerous glacial deposits; at least part of their morphology must be due to erosion by valley glaciers during the Pleistocene. Secondly, the patterns of cave development, everywhere they have been determined, show alternating sequences of erosion and deposition. Such marked changes must relate to strong climate variations, and, in turn, the strongest climatic changes that have affected the area have been through the glacial periods of the Pleistocene. Thirdly, some caves contain huge quantities of clastic sediments which can only be glaciofluvial in origin, and other caves are actively being eroded at the

Years	STAGE	ALPINE GLACIATIONS	GENERAL CLASSIFICATION	AMERICAN TERMINOLOGY	CULTURAL DIVISIONS	NW ENGLAND (LOCAL TERMINOLOGY)	POLLEN ZONES
0					Neolithic		VIIb
	Flandrian		Post-glacial		Mesolithic		IV-VIIa
5,000	Devensian: L	Wurm 2c	Last glaciation	Wisconsin	Upper Paleolithic	Corrie glaciation Valley glaciation	I-III
26,000	Devensian: M	Wurm 2b, 2a				North British glaciation	
50,000	Devensian: E	Wurm 1			Lower Paleolithic	Main Irish Sea glaciation	
75,000	Ipswichian		Last interglacial	Sangamon			
130,000	Wolstonian	Riss	Penultimate glaciation	Illinoian		Early Scottish glaciation	
150,000	Hoxnian		Penultimate (Great) interglacial	Yarmouth			
240,000	Anglian	Mindel	Antepenultimate glaciation	Kansan		Scandinavian glaciation	
310,000	Cromerian		Antepenultimate interglacial	Aftonian			
340,000	Beestonian	Gunz	Early glaciation	Nebraskan			

present time; the caves range in age, from before at least one glaciation, to the present day—post-glacial. In the first instance, therefore, a significant proportion of the cave and karst development must be related to the Pleistocene and post-glacial periods.

Pleistocene chronology provides its own series of problems. Table 13 summarises the relationships of various classifications of the later Pleistocene (Beestonian to Devensian) and Recent (Flandrian) periods. The stages are those which are now geologically accepted and have therefore been used elsewhere in this volume; their ages are the best approximations at present obtainable (West, 1968; Turekian, 1971). In the same table some of the more important other Pleistocene classifications are compared with the stages; the terminology developed in north-west England applies essentially to the Lake District, and it has not been extended successfully into the limestone Dales.

The Gunz and earlier glaciations have been recorded in the European Alps and across the lowland areas of Central Europe. Their effect in northern England, however, is not known for certain; the climate may have deteriorated sufficiently to give a full glacial stage, or alternatively only a periglacial environment. In the present context, the Cromerian and earlier stages are regarded as pre-glacial; this accepts that even periglacial phases in the Beestonian must have had some effect on the karst development, though the extent of this is at present impossible to estimate. During the Anglian, Wolstonian and Devensian stages, glaciers extended over much of north-west England. The intervening Hoxnian and Ipswichian stages were periods with warmer climates and are generally known as interglacials; evidence gleaned from lowland faunas indicates that the interglacial climates were so warm that there must have been complete glacial withdrawal from the highland areas such as the Central Pennines. On the other hand the Devensian has been divided into four glacial advances with three intervening, warmer, interstadials. The evidence has not yet been found for any sub-divisions of the earlier glacial periods. Surface deposits indicate the advance of at least two glaciations in the Dales, but the actual number could be considerably more. Similarly palaeontological evidence indicates many minor climatic variations within the Ipswichian and Hoxnian interglacials; whether or not these were marked enough to cause significant breaks in the karst processes is, as yet, unknown.

This very complex, and relatively unknown, sequence of climatic changes within the Pleistocene must now be correlated with the patterns of karst development. The solutional erosion of the caves and the initiation of new sinkholes largely took place during the warmer wetter phases. Incision of valleys on the limestone surface was mainly in periglacial environments, and the partial filling of the caves by clastic sediment was also a feature of the colder phases when melt-water streams were low in solutional aggressiveness yet overloaded with glacial detritus. Calcite deposits mark the breaks in the accumulation of the clastic fills. The rejuvenation of phreatic caves may be correlated with the lowering of resurgence levels; these in turn are probably connected with glacial excavation of the valleys. However, the surface evidence does not indicate which of the Pleistocene glaciations incised the valleys by how much; the best evidence concerning the valley deepening comes from the rejuvenation of the caves. The chronology of the cave development is therefore arrived at by only circular arguments, and can really only be regarded as an approach to the correct answers.

The most complete sequence of events in the geomorphic history is clearly going to come from the very complex cave systems such as Ease Gill Caverns, Pippikin Hole and Gaping Gill. In Ease Gill (Ashmead, 1967) a long sequence has been postulated, but in this system there are considerable problems in distinguishing series of passages developed through a number of climatic phases, from those developed by repeated stream diversions within the framework of the geological controls. Pippikin Hole has not yet been studied in detail, and Gaping Gill (see chapter 18) is a unique system with its own special problems, provided both by the great dominance of phreatic passages and also by the large number of unknown passage continuations.

Kingsdale should provide useful evidence of the karst chronology, with its well-preserved series of moraines and lake sediments, some of which can be related to stages within the development of the very extensive cave systems on the western side of the valley (see chapter 16). But an example of the difficulties of such relative chronology is provided by Keld Head—the flooded vadose resurgence passage level with the alluvial floor of the valley. Its flooding could be due to silting up of the valley floor behind the terminal moraine, in which case the passage may be entirely post-glacial. On the other hand, its vadose passage may have developed before the terminal moraine was deposited; the moraine held up a temporary lake and its sediments would have accumulated up-valley and flooded the resurgence; in this case, the passage pre-dates at least one glacial advance. Kingsdale also provides the only record of fossils which have been found deep in a cave and successfully dated. Unfortunately these reindeer bones can only be given an age of 'at least 8000 years'; this merely means they are not post-glacial and so contributes very little to the valley's chronology.

The relatively simple cave systems of Leck Fell contain three main stages of erosional development (see chapter 15). These three stages probably have the same total time span as the much larger number of stages indicated in the complex caves of Casterton Fell; they have merely not been sub-divided to the same extent. On Leck the end of the first stage is marked by the largest lowering of resurgence levels; this may then be correlated with a major glaciation and the deepest, glacial, incision of the valleys. It was followed by at least one more major glacial period. The glaciation responsible for the greatest valley excavation should therefore have been the Anglian or the Wolstonian. This assumes that all the phases of the Devensian glaciation provided only the relatively minor, late, modifications to the topography, and the Beestonian glaciations were of limited importance. At present it

Page 447

Straws and pillars in the Passage of Time in Strans Gill Pot

(*above*) The Passage of Time in Strans Gill Pot. The phreatic half-tube, wide bedding development and incised vadose trench are all magnificently developed (*below*) The Terminal Sump in Robin Hood's Cave

TABLE 14 *Tentative chronology of some cave passages in different parts of the Dales*

Area	Pre-glacial caves	Interglacial caves	Post-glacial caves
Casterton Fell	Montague East Passage Gypsum Cavern	Ease Gill Master Cave	Wilf Taylor's Passage
Leck Fell	Short Drop Cave Gavel Pot Main Tunnels	Lost Johns Main Drain	Lost Johns Old Roof Traverse
Kingsdale	Master Cave (except Canyon) Roof Tunnel	Swinsto Hole, Simpson's Pot	Rowten Pot Waterfalls
Gaping Gill	East Passage South East Passage	Far Country Series Disappointment Pot	Present (Flooded) Drainage Route
Fountains Fell	Fools Paradise in Gingling Hole	Out Fell Master Cave	Hammer Pot Entrance Series

is impossible to distinguish and compare the individual effects of the Anglian and Wolstonian glaciations; similarly the erosive features of the Hoxnian and Ipswichian interglacials cannot be separated.

There is some doubt about the magnitude of this early glacial excavation of the valleys compared with the effects of any subsequent or previous glaciation. However, throughout the Dales, the greatest change in the environment of cave development was one of the earliest changes. A broad threefold division of the erosional history does apply to all the known cave systems. Their passages can be classified as pre-glacial, interglacial or post-glacial, and are distinguished by the intervening deposition stages.

The pre-glacial passages date from before the Wolstonian or the Anglian—whichever stage saw the major glacial rejuvenation of the area. Some caves very probably date back to the beginning of the Pleistocene and into the Tertiary. The sub-division of this phase may relate to the various colder stages such as the Beestonian.

The interglacial phase must range from the beginning of either the Hoxnian or the Ipswichian up to the end of the Middle Devensian. It clearly has extensive scope for sub-divisions correlated with the interstadials of the Devensian and perhaps also the Wolstonian.

Post-glacial caves can only date from the retreat of the last glaciers, approximately at the beginning of the Flandrian.

On this basis, nearly all the caves of the Ingleborough–Malham region may be classified. Table 14 lists some representative examples from some separate areas, and the rather simple correlations in this table are widely acceptable, as is the chronology postulated above.

Further sub-division would be more and more controversial; it has therefore been avoided here and must await further research and a framework of absolute age determinations.

Acknowledgements

This volume has undeniably been the result of the hard work of a large number of people; besides the seventeen contributing authors and the photographers, many who have helped extensively will remain unnamed.

Throughout the long process of compilation Dr Trevor Ford has supplied continuous and invaluable criticism and assistance. It was Dr Marjorie Sweeting who started the project at the request of the Cave Research Group. She invited many of the authors to plan their contributions, and in the later stages has continued with useful assistance.

The book could never have been compiled if it had not been for the hundreds of unmentioned cavers who have explored the 'underground half' of the north-west England karst. They too have provided the descriptions and surveys of the caves. In most cases this work has been published in the caving club journals and most of these are now almost unobtainable. Consequently they are not included in the Bibliography with the exception of the small proportion of articles in the better-established journals which have a reasonably permanent library distribution. The assistance provided by all these unquoted writers must not be forgotten.

Also too numerous to mention individually are the relatives and friends who have been so helpful to each author, both in the field, down the caves, and during preparation of the manuscripts.

Much of the limestone area has some agricultural value, and access to the karst and caves cannot help but involve some form of intrusion even by the careful majority who have visited and worked in the area. The co-operation, assistance and friendship provided by the many farmers and landowners has been invaluable.

It should also be noted that there has been extensive co-operation between the various authors. Some of the chapters are really the joint works of a number of contributors. Local cavers with special knowledge of certain areas have also assisted, and in this respect D. Brook, M. H. Long and J. R. Sutcliffe have been particularly helpful.

The technical staff of the Geology Department of the University of Leicester, notably Miss S. Ward, Mrs J. Westerman and Mrs N. Farquharson, have provided valuable assistance in typing and the preparation of diagrams. Similarly Mrs M. Fry has typed a large proportion of the manuscripts. To all of these the editor extends his sincere thanks.

A. C. WALTHAM

Trent Polytechnic
6 June 1972

Bibliography

ANDERSON, F. W. 'Possible Late-Glacial Sea Levels at 190ft and 140ft OD in the British Isles', *Geol Mag*, 76 (1939), 317–21.

ANDERSON, L. C. & VERNON, J. 'The Quality and Production of Lime for Basic Oxygen Steel Making', *Journ Iron & Steel Inst* (April 1970), 32.

ASHMEAD, P. 'The Origin and Development of Ease Gill Caverns', *Trans Cave Res Gp*, 9 (1967), 104–112.

ASHMEAD, P. 'The Origin and Development of Caves in the Morecambe Bay Area', *Trans Cave Res Gp*, 11 (1969), 201–8.

ASHTON, K. 'Preliminary Report on a New Hydrological Technique', *Newsletter Cave Res Gp*, no 98 (1965), 2.

ASHTON, K. 'The Analysis of Flow Data from a Karst Drainage System', *Trans Cave Res Gp*, 7 (1966), 161–204.

ASHTON, K. 'Western Kingsdale: Cave Developments', *Univ Leeds Spel Assoc News-sheet*, no 9 (1966), 3.

ASHTON, K. 'Artificial Flood Waves in Caves', *New Zealand Spel Bull*, 4 (1967).

ATKINSON, F. 'Oxford Hole', *Cave Sci*, 2 no 14 (1950), 231–7.

ATKINSON, F. 'New Pots for Old—Bar Pot, Ingleborough, Yorkshire', *Cave Sci*, 2 no 11 (1952), B6–8.

ATKINSON, F. 'Some Notes on the Formation of Caverns in the Craven Area of NW Yorkshire', *Proc Brit Spel Assoc*, no 1 (1963), 67–78.

ATKINSON, T. C. 'The Earliest Stages of Underground Drainage in Limestones—a Speculative Discussion', *Proc Brit Spel Assoc*, no 6 (1968), 53.

AVELINE, W. T. *et al.* 'The Geology of the Neighbourhood of Kirkby Lonsdale and Kendal', *Mem Geol Surv* (1872), 44pp.

BARNES, B. *et al.* 'Skeleton of a Late Glacial Elk associated with

Barbed Points from Poulton le Fylde, Lancashire', *Nature,* 232 (1971), 488.

BENSON, D. & BLAND, K. 'The Dog Hole, Haverbrack', *Trans Cumberland and Westmorland Antiq & Archeo Soc,* 62 (1962), 61–76.

BLACK, W. W. 'The Carboniferous Geology of the Grassington Area, Yorkshire', *Proc Yorks Geol Soc,* 28 (1950), 29.

BLACK, W. W. 'Diagnostic Characters of the Lower Carboniferous Knoll-Reef in the North of England', *Trans Leeds Geol Assoc,* 6 (1954), 262–97.

BLACK, W. W. 'The Structure of the Burnsall-Cracoe District and its Bearing on the Origin of the Cracoe Knoll-Reefs', *Proc Yorks Geol Soc,* 31 (1958), 391.

BÖGLI, A. 'Kalklosung und Karrenbildung', *Zeit Geomorph,* Supplement 2 (1960). Internationale Beiträge zur Karstmorphologie, 4–21.

BÖGLI, A. 'Karrentische, ein Beiträg zur Karst morphologie', *Zeit Geomorph,* 3 (1961), 185–93.

BÖGLI, A. 'Corrosion by Mixing of Karst Waters', *Trans Cave Res Gp,* 13 (1971), 109–114.

BOTT, M. H. P. 'Geophysical Investigations of the Northern Pennine Basement Rocks', *Proc Yorks Geol Soc,* 36 (1967), 139–168.

BRETZ, J. H. 'Vadose and Phreatic Features of Limestone Caverns', *Journ Geol,* 50 (1942), 675–811.

BRETZ, J. H. 'Genetic Relations of Caves to Peneplains and Big Springs in the Ozarks', *Am Journ Sci,* 251 (1953), 1–24.

BREUIL, H. 'Observations on the Pre-Neolithic Industries of Scotland', *Proc Soc Ant Scot,* 56 (1921), 261–91.

BRINDLE, D. 'Car Pot Breakthrough', *Journ Yorks Ramblers Cl,* 7 (1949), 248–52.

BRINDLE, D. 'The Dow Cave–Providence Pot System, Wharfedale', *Journ Craven Pot Cl,* 2 no 1 (1955), 4–10.

BRODERICK, H. 'The Stream Bed of Fell Beck above Gaping Gill', *Journ Yorks Ramblers Cl,* 4 (1912), 44–53.

BRODERICK, H. 'Excavations at Foxholes, Clapdale', *Journ Yorks Ramblers Cl,* 5 (1924), 112–116.

BROOK, D. 'The Recent Events at Gaping Gill', *Univ Leeds Spel Assoc Rev,* no 3 (1968), 1–7.

BROOK, D. 'Sleets Gill Extension, Littondale', *Univ Leeds Spel Assoc Rev,* no 4 (1968), 5–9.

Brook, D. 'Caves and Caving in Kingsdale', *Journ Brit Spel Assoc,* no 48 (1971), 33–47.

Brook, D. & Crabtree, H. *Exploration Journal* (1969). Univ Leeds Spel Assoc.

Bullock, P. 'The Soils of the Malham Tarn Area', *Field Studies,* 3 (1971), 381–408.

Calvert, E. 'Gaping Ghyll Hole', *Journ Yorks Ramblers Cl,* 1 (1899–1900), 64–74 and 123–33.

Carter, W. L., Dwerryhouse, A. R. *et al.* 'The Underground Waters of North-West Yorkshire. Part II. The Underground Waters of Ingleborough', *Proc Yorks Geol Soc,* 15 (1904), 248–92.

Chappuis, P. A. 'Über die Fauna der Spaltengewasser und des Grundwassers Fauna', *Acta Sci Math et Natur Kolozsuar,* 6 (1943), 3–7.

Chubb, L. J. & Hudson, R. G. S. 'The Nature of the Junction between the Lower Carboniferous and the Millstone Grit of North-West Yorkshire', *Proc Yorks Geol Soc,* 20 (1925), 257.

Clark, R. 'A Contribution to Glacial Studies of the Malham Tarn Area', *Field Studies,* 2 (1967), 491–7.

Clayton, K. M. 'The Origin of the Landforms of the Malham Area', *Field Studies,* 2 (1966), 359–84.

Collingwood, R. G. *The Archaeology of Roman Britain* (1969).

Corbel, J. 'Les Karstes du Nord-Ouest de l'Europe', *Institut des Études Rhodaniènnes d'Université de Lyon. Mém et Doc,* no 12 (1957), 541pp.

Crabtree, H. 'The Use of Optical Brightening Agents in Water Tracing', *Journ Brit Spel Assoc,* no 46 (1971), 33–9.

Cubbon, B. D. 'Flora Records of Cave Research Group of Great Britain from 1939 to June 1969', *Trans Cave Res Gp,* 12 (1970), 57–74.

Cuttriss, S. W. 'The Caves and Potholes of Yorkshire', *Journ Yorks Ramblers Cl,* 1 (1899), 54–64.

Cuttriss, S. W. 'Gaping Gill Again', *Journ Yorks Ramblers Cl,* 3 no 8 (1908).

Cvijič, J. 'Hydrographie Souterraine et Evolution Morphologique du Karst', *Rec Inst Geog Alpine,* 6 (1918), 375.

Dakyns, J. R., Tiddeman, R. H., Gunn, W. & Strahan, A. 'The Geology of the Country around Ingleborough with Parts of Wensleydale and Wharfedale'. *Mem Geol Surv* (1890), 103pp.

DAVIS, J. W. & LEES, F. A. *West Yorkshire* (1878).

DAVIS, W. M. 'Origin of Limestone Caverns', *Geol Soc Am Bull*, 41 (1930), 475–628.

DAWKINS, W. B. 'Observations on the Rate at which Stalagmite is being Accumulated in the Ingleborough Cavern', *Rep Brit Assoc* (1873), 80.

DAWKINS, W. B. *Cave Hunting* (1874).

DENNY, H. 'On the Geological and Archaeological Contents of the Victoria and Dowkabottom Caves in Craven', *Yorks Geol Polyt Soc*, 4 (1860), 45–74.

DIXON, J. M. 'Starbotton End Mine (Springs Wood Level): A Preliminary Biological Report', *Mem North Cav Mine Res Soc* (1965), 57–74.

DIXON, J. M. 'Biological Survey of Springs Wood Level, Starbotton, Yorks', *North Cav Mine Res Soc, Individ Surv Ser* no 1 (1966), 9–14.

DOUGHTY, P. S. 'Joint Densities and their Relationship to Lithology in the Great Scar Limestone', *Proc Yorks Geol Soc*, 36 (1968), 479–512.

DREW, D. P. 'The Water Table Concept in Limestones', *Proc Brit Spel Assoc*, no 4 (1966), 57–67.

DREW, D. P. 'Limestone Solution within the East Mendip Area, Somerset', *Trans Cave Res Gp*, 12 (1970), 259–70.

DREW, D. P. & SMITH, D. I. 'Techniques for Tracing of Subterranean Drainage', *Brit Geomorph Res Gp Tech Bull*, no 2 (1969).

DUNHAM, K. C. 'Epigenetic Mineralization in Yorkshire', *Proc Yorks Geol Soc*, 32 (1959), 1–30.

DUNHAM, K. C., HEMINGWAY, J. E., VERSEY, H. C. & WILCOCKSON, W. H. 'A Guide to the Geology of the District round Ingleborough', *Proc Yorks Geol Soc*, 29 (1953), 77–115.

DUNHAM, K. C. & ROSE, W. C. C. 'Geology of the Iron Ore Field of South Cumberland & Furness', *Geol Surv Wartime Pamphlet*, no 16 (1941), 22pp.

DUNHAM, K. C. & ROSE, W. C. C. 'Permo-Triassic Geology of South Cumberland & Furness', *Proc Geol Assoc*, 60 (1949), 11–40.

DUNHAM, K. C. & STUBBLEFIELD, C. J. 'The Stratigraphy, Structure and Mineralization of the Greenhow Mining Area, Yorkshire', *Quart Journ Geol Soc*, 100 (1945), 209.

DWERRYHOUSE, A. R. 'Limestone Caverns and Potholes and their Mode of Origin', *Journ Yorks Ramblers Cl,* 2 (1907), 223–8.

EARP, J. R. *et al.* 'Geology of the Country around Clitheroe and Nelson', *Mem Geol Surv* (1961).

EASTWOOD, T. *British Regional Geology: Northern England.* Geol Surv (1953), 71pp.

EYRE, J. & ASHMEAD, P. F. 'Lancaster Hole and the Ease Gill Caverns', *Trans Cave Res Gp,* 9 (1967), 61–123.

FARRER, J. 'Further Explorations in Dowkerbottom Caves in Craven', *Proc Yorks Geol Polytech Soc,* 4 (1865), 414–22.

FOLEY, I. 'Lost Johns Cave', *Journ Yorks Ramblers Cl,* 6 (1930), 44–59.

FORD, D. C. 'Features of Cavern Development in Central Mendip', *Trans Cave Res Gp,* 10 (1968), 11–25.

FORD, D. C. 'Geologic Structure and Theories of Limestone Cavern Genesis', *Journ Brit Spel Assoc,* no 45 (1970), 35–46.

FORD, D. C. 'Alpine Karst in the Mt Castleguard-Columbia Icefield Area, Canadian Rocky Mountains', *Arctic & Alpine Res,* 3 (1971), 239–52.

FORD, T. D. 'The Goyden Pot Drainage System, Nidderdale, Yorkshire', *Trans Cave Res Gp,* 6 (1963), 79

FORD, T. D. *Guide to Ingleborough Cavern* (1971 *a*). Clapham.

FORD, T. D. 'Structures in Limestones Affecting the Initiation of Caves', *Trans Cave Res Gp,* 13 (1971 *b*), 65–71.

FOX, C. *A Find of the Early Iron Age from Llyn Cerrig Bach, Anglesey* (1946). Nat Museum Wales, Cardiff.

GARDNER, J. H. 'Origin and Development of Limestone Caverns', *Geol Soc Am Bull,* 46 (1935), 1255.

GARROD, D. 'The Upper Palaeolithic in the Light of Recent Discovery', *Proc Prehist Soc,* 4 (1938), 1–36.

GARWOOD, E. J. 'The Lower Carboniferous Succession in the Northwest of England', *Quart Journ Geol Soc,* 68 (1912), 449–586.

GARWOOD, E. J. & GOODYEAR, E. 'The Lower Carboniferous Succession in the Settle District', *Quart Journ Geol Soc,* 80 (1924), 184–273.

GEMMEL, A. 'Simpson Pot, Kingsdale, Yorkshire', *Cave Sci,* 1 no 2 (1947), 47–51.

GEMMEL, A. & MYERS, J. O. *Underground Adventure* (1952). Clapham.

GEORGE, T. N. 'Lower Carboniferous Palaeogeography of the British Isles', *Proc Yorks Geol Soc,* 31 (1958), 227–318.

GEORGE, T. N. 'Tectonics and Palaeogeography in Northern England', *Sci Progr*, 51 (1963), 32–59.

GLEDHILL, T. & DRIVER, D. B. '*Bathynella natans* Vejdovsky (Crustacea : Syncarida) and its Occurrence in Yorkshire', *The Naturalist*, no 890 (1964), 104–6.

GLEDHILL, T. & SERBAN, E. 'Concerning the Presence of *Bathynella natans stammeri* Jakobi (Crustacea : Syncarida) in England and Rumania', *Ann Mag Nat Hist*, Ser 13, 8 (1965), 523–32.

GLENNIE, E. A. 'Biological Supplement (Records of 1938–9)', *Newsletter Cave Res Gp* (1955).

GLENNIE, E. A. 'Biological Supplement (Records of 1940–46)', *Newsletter Cave Res Gp* : nos 58/59 (Sect 1) and nos 60/61 (Sect 2) (1956).

GLENNIE, E. A. 'The Distribution of Hypogean Amphipoda in Britain', *Trans Cave Res Gp*, 9 (1967), 132–6.

GLENNIE, E. A. 'The Discovery of *Niphargus aquilex aquilex* Schiodte in Radnorshire', *Trans Cave Res Gp*, 10 (1968), 139–40.

GLENNIE, E. A. & HAZELTON, M. 'Cave Fauna and Flora'. *In* C. H. D. Cullingford (Ed), *British Caving* (2nd edn) (1962), 347–410.

GLOVER, R. R. 'Optical Brighteners—A New Water Tracing Reagent', *Trans Cave Res Gp*, 14 (1972), 84–8.

GOODCHILD, J. G. 'The Glacial Phenomena of the Eden Valley and the Western Part of the Yorkshire Dale District', *Quart Journ Geol Soc*, 31 (1875), 65–99.

GRACE, G. & SMITH, F. H. 'Some Observations on the Glacial Geology of Furness', *Proc Yorks Geol Soc*, 19 (1922), 401–19.

GRAINGER, B. M. 'Survey and Report on the Geological Structure of Hensler's Passage, Gaping Ghyll', *Caves and Caving*, 1 (1938), 110–12.

GRANDISON, N. O. 'Mossdale Caverns', *Proc Brit Spel Assoc*, no 3 (1965), 43.

GREENWOOD, W. H. 'Flood Entrance: Gaping Ghyll', *Journ Yorks Ramblers Cl*, 3 (1910), 167–73.

GRESSWELL, R. K. 'The Glacial Geomorphology of the South Eastern Part of the Lake District', *Liv & Manch Geol Journ*, 1 (1952), 57–70.

GRESSWELL, R. K. 'The Post-Glacial Raised Beach in Furness & Lyth, North Morecambe Bay', *Trans Inst Brit Geog*, no 25 (1958), 79–103.

GRESSWELL, R. K. 'The Glaciology of the Coniston Basin', *Liv & Manch Geol Journ*, 3 (1962), 83–96.

HAZELTON, M. 'Biological Supplement (Records of 1947)', *Newsletter Cave Res Gp*, 72/77 (1958).

HAZELTON, M. 'Biological Supplement (Records of 1948–9)', *Newsletter Cave Res Gp*, 79/80 (1959).

HAZELTON, M. *Biological Records* (1960, pt 5 for 1950–53; 1960, pt 6 for 1954–6; 1961, pt 7 for 1957–9; 1963, pt 8 for 1960–2). *Cave Res Gp*.

HAZELTON, M. 'British Hypogean Fauna and Biological Records'.
Trans Cave Res Gp, 7 no 3 (1965)—for 1963
„ 9 no 3 (1967)—for 1964–6
„ 10 no 3 (1968)—for 1967
„ 12 no 1 (1970)—for 1968
„ 13 no 3 (1971)—for 1969

HEAP, D. *Potholing: Beneath the Northern Pennines* (1964).

HEM, J. D. 'Study and Interpretation of Chemical Characteristics of Natural Water', *U.S. Geol Surv Water Sup Paper*, 1473 (1959), 269pp.

HICKS, P. F. 'The Yoredale Rocks of Ingleborough, Yorkshire', *Proc Yorks Geol Soc*, 32 (1959), 31–44.

HILL, C. A. 'Clapham Cave', *Journ Yorks Ramblers Cl*, 4 (1913), 107–127.

HODGSON, A. *The Great Scar Limestone around Littondale and Upper Wharfedale* (1950). Imperial College, London (Unpublished Ph.D. thesis).

HODGSON, E. 'The Moulded Limestones of Furness', *Geol Mag*, 4 (1867), 40.

HOLLAND, E. G. *Underground in Furness* (1968). Clapham.

HOLLINGWORTH, S. E. 'High Level Erosion Platforms in Cumberland and Furness', *Proc Yorks Geol Soc*, 23 (1937), 159–77.

HOWARTH, J. H. *et al.* 'The Underground Waters of North-West Yorkshire', *Proc Yorks Geol Polyt Soc*, 14 (1900), 1–44.

HUDSON, R. G. S. 'The Carboniferous of the Craven Reef-Belt, the Namurian Unconformity at Scaleber, near Settle', *Proc Geol Assoc*, 41 (1930), 290–322.

HUDSON, R. G. S. 'The Scenery and Geology of Northwest Yorkshire', *Proc Geol Assoc*, 44 (1933), 228–55.

HUDSON, R. G. S. 'The Lower Carboniferous South of Carnforth', *Rep Brit Assoc Geol Trans* (1936 *a*).

HUDSON, R. G. S. 'Sudetic Earth Movements in the Craven Area', *Rep Brit Assoc Trans Section C (Geology)* (1936 *b*).

HUDSON, R. G. S. 'A pre-Namurian Fault Scarp at Malham', *Proc Leeds Phil Lit Soc (Sci Sect)*, 4 (1944), 226–32.

HUDSON, R. G. S. 'The Carboniferous of the Craven Reef-Belt at Malham (Abst.)', *Proc Geol Soc*, no 1147 (1949), 38–41.

HUGHES, T. McK. 'Exploration of Cave Ha, Giggleswick', *Journ Anthrop Inst*, 3 (1874), 383–7.

HUGHES, T. McK. 'On Some Perched Blocks and Associated Phenomena', *Quart Journ Geol Soc*, 42 (1886), 527–39.

HUGHES, T. McK. 'Ingleborough Part I', *Proc Yorks Geol Soc*, 14 (1901), 125–50.

HUGHES, T. McK. 'Ingleborough Part VI. The Carboniferous Rock', *Proc Yorks Geol Soc*, 16 (1909), 253.

JACKSON, J. *Personal Diary* (1870). Pig Yard Club Collection, Settle.

JACKSON, J. W. 'Dog Holes Cave, Warton Crag', *Trans Lancs Cheshire Antiq Soc*, 28 (1910), 59–82.

JACKSON, J. W. 'Note on a Javelin Point from Victoria Cave', *Antiq Journ*, 25 (1945).

JACKSON, J. W. 'Archaeology and Paleontology', *In* C. H. D. Cullingford (Ed), *British Caving* (2nd edn) (1962), 252–347.

JACKSON, J. W. & MATTINSON, W. K. 'A Cave on Giggleswick Scars near Settle, Yorks', *The Naturalist* (1932), 5–9.

JOHNSON, P. 'The Early History of Gaping Gill', *The Speleologist*, 2 no 9 (1966), 16–18.

JONES, R. J. 'Aspects of the Biological Weathering of Limestone Pavement', *Proc Geol Assoc*, 16 (1965), 421–33.

KENDALL, P. F. & WROOT, H. E. *The Geology of Yorkshire* (1924). Leeds, 995pp.

KING, A. 'Romano-British Metalwork from the Settle District of West Yorkshire', *Yorks Archaeol Journ*, 62 (1970), 410–17.

KING, C. A. M. 'Trend Surface Analysis of Central Pennine Erosion Surfaces', *Trans Inst Brit Geog*, no 47 (1969), 47–69.

KING, W. B. R. & WILCOCKSON, W. H. 'The Lower Palaeozoic Rocks of Austwick and Horton-in-Ribblesdale, Yorkshire', *Quart Journ Geol Soc*, 90 (1934), 7–31.

Lavell, C. (Compiler). *Archaeological Site Index to Radiocarbon Dates from Great Britain* (1970), 3J.1. Council for British Archaeology.

Leakey, R. D. 'The Caverns of Mossdale Scar', *Cave Sci,* no 1 (1947), 7–18.

Lees, A. 'The Waulsortian "Reefs" of Eire: A Carbonate Mudbank Complex of Lower Carboniferous Age', *Journ Geol,* 69 (1961), 101–9.

Long, M. H. 'Recent Caving Developments in Langstrothdale/Wharfedale', *Newsletter Cave Res Gp,* no 117 (1969), 6–16.

Long, M. H. 'Dentdale—Present and Future', *Journ Brit Spel Assoc,* 6 no 46 (1971), 9–17.

Malott, C. A. 'The Invasion Theory of Cavern Development (Abst)', *Proc Geol Soc Am* (1937), 323.

Manby, T. G. 'The Long Barrows of Northern England: Structural and Dating Evidence', *Scot Archaeol Forum* (1970), 1–27.

Manley, G. 'The Climate at Malham Tarn', *Ann Rep Council for Prom Field Studies (1955–6)* (1957), 43–56.

Manley, G. 'The Late Glacial Climate of North West England', *Liv & Manch Geol Journ,* 2 (1959), 188–215.

Marr, J. E. *The Geology of the Lake District* (1916). 220pp.

Mason, E. J. 'Ogof-yr-Esgryn, Dan yr Ogof Caves, Brecknock, Excavations', *Archaeologia Cambrensis,* 117 (1968), 18–71.

Matson, G. C. 'Water Resources of the Bluegrass Region, Kentucky', *U.S. Geol Surv Water Supply Paper,* no 233 (1909), 42.

McConnell, R. B. 'The Relic Surfaces of the Howgill Fells', *Proc Yorks Geol Soc,* 24 (1940), 152–64.

Meisler, H. & Becher, R. E. 'Hydrogeological Significance of Calcium–Magnesium Ratios in Ground Water from Carbonate Rocks in Lancaster Quadrangle, SE Pennsylvania', *U.S. Geol Surv Prof Paper,* 575 (1967), 232–5.

Miller, A. A. 'Pre-Glacial Erosion Surfaces round the Irish Sea Basin', *Proc Yorks Geol Soc,* 24 (1939), 31–59.

Milner, A. J. *The Caves of the Alum Pot Area* (1972). 14pp. Univ Leeds Spel Assoc.

Mitchell, A. 'When the Ice Cap covered Gaping Gill—Some Post-Glacial Problems', *Journ Craven Pot Cl,* 4 no 1 (1967), 36–40.

Mitchell, G. H. 'The Pleistocene History of the Irish Sea', *Rep Brit Assoc,* 17 (1960), 313–25.

MOISLEY, H. A. 'Some Karstic Features in the Malham Tarn District', *Ann Rep, Council for Prom Field Studies (1953–4)* (1955), 33–42.

MOORE, D. 'The Yoredale Series of Upper Wensleydale and Adjacent Parts of North-West Yorkshire', *Proc Yorks Geol Soc,* 31 (1958), 91.

MORRIS, J. P. & SMITH, H. E. 'Limestone Caves of Craven and their Ancient Inhabitants', *Trans Hist Soc Lancs and Cheshire,* 5 (1865), 199–231.

MOSELEY, C. M. 'The Metalliferous Mines of the Arnside–Carnforth Districts of Lancashire and Westmorland', *North Cav Mine Res Soc, Individ Sur Ser,* no 3 (1969), 32pp.

MOSELEY, C. M. 'The Fauna of Caves and Mines in the Morecambe Bay Area', *Trans Cave Res Gp,* 12 (1970), 43–56.

MOSELEY, F. 'The Namurian of the Lancaster Fells', *Quart Journ Geol Soc,* 109 (1954), 423–500.

MOSELEY, F. & AHMED, S. M. 'Carboniferous Joints in the North of England and their Relation to Earlier and Later Structures', *Proc Yorks Geol Soc,* 36 (1967), 61–90.

MUNN-RANKIN, W. 'The Peat Mosses of Lonsdale', *The Naturalist* (1910), 119, 153.

MYERS, J. O. 'The Formation of Yorkshire Caves and Potholes', *Trans Cave Res Gp,* 1 no 1 (1948), 26–9.

MYERS, J. O. 'The Mossdale Problem. The Problem of the Underground Water Flow', *Trans Cave Res Gp,* 1 no 4 (1950), 21–30.

MYERS, J. O. & WARDELL, J. 'The Gravity Anomalies of the Askrigg Block South of Wensleydale', *Proc Yorks Geol Soc,* 36 (1967), 169.

NEWSON, M. D. 'Erosion in the Limestone Stream System—Some Recent Results and Observations', *Proc Brit Spel Assoc,* no 7 (1969), 17.

OLDFIELD, F. 'Late Quaternary Changes in Climate, Vegetation and Sea Level in Lowland Lonsdale', *Trans Inst Brit Geog,* no 28 (1960), 99–117.

PARRY, J. T. 'The Erosion Surfaces of the South Western Lake District', *Trans Inst Brit Geog,* no 28 (1960), 39–54.

PHILLIPS, J. *Illustrations of the Geology of Yorkshire. Pt II. The Mountain Limestone District* (1836).

PIGGOTT, S. (*Proc Prehist Soc* Editor's Note). 'Secondary Neolithic Burials at Churchdale, near Monyash, Derbyshire', *Proc Prehist Soc,* 19 (1953 *a*), 228–30.

PIGGOTT, S. 'Three Metalwork Hoards in Southern Scotland', *Proc Soc Antiq Scot*, 87 (1953 *b*), 3–54.

PIGOTT, C. D. 'The Structure of Limestone Surfaces in Derbyshire', *Geog Journ*, 131 (1965), 41–4.

PIGOTT, M. E. & PIGOTT, C. D. 'Stratigraphy and Pollen Analysis of Malham Tarn and Tarn Moss', *Field Studies*, 1 (1959), 84.

PITTY, A. F. 'The Estimation of Discharge from a Karst Rising by Natural Salt Dilution', *Journ Hydrol*, 4 (1966 *a*), 63–9.

PITTY, A. F. 'An Approach to the Study of Karst Water', *Univ Hull Occ Papers in Geog*, no 5 (1966 *b*), 70pp.

PITTY, A. F. 'Calcium Carbonate Content of Karst Water in relation to Flow-through Time', *Nature*, 217 (1968 *a*), 939–40.

PITTY, A. F. 'Some Notes on the Use of Calcium Hardness Measurements in Studies of Cave Hydrology', *Trans Cave Res Gp*, 10 (1968 *b*), 115–20.

PITTY, A. F. 'Rates of Seepage in Poole's Cavern, Derbyshire', *Proc Brit Spel Assoc*, no 7 (1969), 7–15.

PITTY, A. F. 'Evidence related to the Development of Avens from Karst Water Studies in Peak Cavern, Derbyshire', *Trans Cave Res Gp*, 13 (1970), 53–5.

PITTY, A. F. 'Rate of Uptake of Calcium Carbonate in Underground Karst Water', *Geol Mag*, 108 (1971), 537–43.

PITTY, A. F. 'Contrast between Derbyshire and Yorkshire in the Average Values of Calcium Carbonate in their Cave and Karst Waters', *Trans Cave Res Gp*, 14 (1972), 151–2.

RAISTRICK, A. 'Some Glacial Features of the Settle District', *Proc Univ Durham Phil Soc*, 8 (1930), 239–51.

RAISTRICK, A. 'Roman Remains and Roads in West Yorkshire', *Yorks Archaeol Journ*, 33 (1934), 214–23.

RAISTRICK, A. 'Excavations at Sewells Cave, Settle, West Yorkshire', *Proc Univ Durham Phil Soc*, 9 (1936), 191–202.

RAISTRICK, A. 'Mineral Deposits in the Settle-Malham District, Yorkshire', *The Naturalist* (1938), 119–25.

RAISTRICK, A. 'The Calamine Mines, Malham, Yorkshire', *Proc Univ Durham Phil Soc*, 11 (1954), 125–30.

RAUCH, H. W. & WHITE, W. B. 'Lithological Control of the Development of Solutional Porosity in Carbonate Aquifers', *Water Resources*, 6 (1970), 1175–92.

RAYNER, D. H. 'The Lower Carboniferous Rocks in the North of England—A Review', *Proc Yorks Geol Soc*, 28 (1953), 231.

RICHARDSON, D. T. 'The Use of Chemical Analysis of Cave Waters as a Method of Water Tracing and Indicator of Types of Strata Traversed', *Trans Cave Res Gp*, 10 (1968), 61–72.

ROGERS, M. 'Langcliffe Pot', *Univ Leeds Spel Assoc Rev*, no 5 (1969), 7–15.

ROGLIČ, J. 'The delimitations and morphological types of the Dinaric karst', *Nase Jama* (Ljubljana) 7 (1965), 12–20.

ROGLIČ, J. 'Morphological Concepts—A Historical View'. *In* M. Herak & V. T. Stringfield (Eds), *Karst: Important Karst Regions of the Northern Hemisphere* (1971), 1–18.

ROSS, A. *Pagan Celtic Britain* (1967).

SCHOFIELD, P. C. S. 'Notes on Recent Fauna Studies in Yorkshire', *Proc Brit Spel Assoc*, no 2 (1964), 63–9.

SCHWARZACHER, W. 'The Stratification of the Great Scar Limestone in the Settle District', *L'pool Manchr Geol Journ*, 2 (1958), 124–42.

SHELLEY, A. E. 'Analysis of Two Coals from the Great Scar Limestone near Ingleton, Yorkshire', *Proc Yorks Geol Soc*, 36 (1967), 51.

SHEPPARD, E. M. '*Trichoniscoides saeroeensis* Lohmander, an Isopod New to the British Fauna', *Trans Cave Res Gp*, 10 (1968), 135–6.

SIMPSON, E. 'Notes on the Formation of the Yorkshire Caves and Potholes', *Proc Univ Bristol Spel Soc*, 4 (1935), 224–32.

SIMPSON, E. 'Gaping Ghyll Hole', *Caves and Caving*, 1 (1937–8), 27–31, 112–16, 140–1.

SIMPSON, E. 'Caves and Potholes in the Hull Pot Area, Ribblesdale, Yorkshire', *Cave Sci*, 1 no 8 (1949), 290–307.

SIMPSON, E. & ATKINSON, F. 'Lancaster Hole, Casterton, Westmorland', *Cave Sci*, 1 no 6 (1948), 202–17.

SMITH, B. 'Pre-Triassic Swallow Holes in Furness: A Glimpse of an Ancient Landscape', *Geol Mag*, 57 (1920), 16–18.

SMITH, C. R. *Collectanea Antiqua I* (1844).

SMITH, D. I., HIGH, C. & NICHOLSON, F. H. 'Limestone Solution and the Caves'. *In* E. K. Tratman (Ed), *The Caves of North-West Clare, Ireland* (1969), 96–123.

SMITH, D. I. & MEAD, D. G. 'The Solution of Limestone', *Proc Univ Bristol Spel Soc*, 9 (1962), 188–211.

SMITH, H. E. 'Limestone Caves of Craven and their Ancient Inhabitants', *Trans Hist Soc Lancs and Cheshire,* 5 (1865), 199–231.

SMITH, M. A. 'Iron Ores: Haematites of West Cumberland, Lancashire and the Lake District. Special Reports on Mineral Resources of Great Britain', Vol 8. *Mem Geol Surv* (1924), vi + 236pp.

SWEETING, G. S. & SWEETING, M. M. 'Some Aspects of the Carboniferous Limestone in Relation to its Landforms', *Mediterranée,* 7, (1969), 201–9.

SWEETING, M. M. *The Geomorphology of the Carboniferous Limestone of the Ingleborough District* (1948). University of Cambridge. (Unpublished Ph.D. thesis.)

SWEETING, M. M. 'Erosion Cycles and Limestone Caverns in the Ingleborough District', *Geog Journ,* 115 (1950), 63–78.

SWEETING, M. M. 'The Weathering of Limestones'. *In* G. H. Dury, *Essays in Geomorphology* (1965), 177–210.

SWEETING, M. M. 'Some Variations in the Types of Limestones and their Relation to Cave Formation', *Proc 4th Internat Cong Spel,* 3 (1968), 227–34.

SWINNERTON, A. C. 'The Origin of Limestone Caverns', *Bull Geol Soc Am,* 43 (1932), 663–93.

THORNBER, N. *Pennine Underground* (1965). Clapham.

THRAILKILL, J. 'Chemical and Hydrologic Factors in the Excavation of Limestone Caves', *Geol Soc Am Bull,* 79 (1968), 19–46.

TIDDEMAN, R. H. 'On the Evidence for the Ice-Sheet in North Lancashire and Adjacent Parts of Yorkshire and Westmorland', *Proc Geol Soc,* 28 (1872), 471–91.

TIDDEMAN, R. H. *In The Craven Herald,* 29 January 1892 (quoted at length in Kendall, P. F. & Wroot, H. E., *Geology of Yorkshire* (1924).

TONKS, L. H. 'The Millstone Grit and Yoredale Rocks of Nidderdale', *Proc Yorks Geol Soc,* 20 (1925), 226–256.

TROTTER, F. M. 'The Tertiary Uplift and Resultant Drainage of the Alston Block and Adjacent Area', *Proc Yorks Geol Soc,* 21 (1929), 161–180.

TUREKIAN, K. K. (Ed). *The Late Cenozoic Glacial Ages* (1971) Yale.

TURNER, J. S. 'Notes on the Late Palaeozoic (Late-Variscan) Tectonics of the North of England', *Proc Geol Assoc,* 46 (1935), 121–51.

TURNER, J. S. 'The Deeper Structure of Central and Northern England', *Proc Yorks Geol Soc,* 27 (1949), 280–97.

TYLECOTE, R. F. *Metallurgy in Archaeology* (1962).

VANDEL, A. *Biospeleology* (1965).

WAGER, L. R. 'Jointing in the Great Scar Limestone of Craven and its Relation to the Tectonics of the Area', *Quart Journ Geol Soc,* 87 (1931), 392–424.

WALKDEN, G. M. 'The Mineralogy and Origin of Interbedded Clay Wayboards in the Lower Carboniferous of the Derbyshire Dome', *Geol Journ,* 8 (1972), 143–60.

WALKER, D. 'A Site at Stump Cross, near Grassington, Yorkshire and the Age of the Pennine Microlithic Industry', *Proc Prehist Soc,* 22 (1956), 23–8.

WALTHAM, A. C. 'Cave Survey Interpretation', *Trans Cave Res Gp,* 12 (1970 *a*), 185–95.

WALTHAM, A. C. 'Cave Development in the Limestone of the Ingleborough District', *Geog Journ,* 136 (1970 *b*), 574–85.

WALTHAM, A. C. 'Shale Units in the Great Scar Limestone of the Southern Askrigg Block', *Proc Yorks Geol Soc,* 38 (1971 *a*), 285–92.

WALTHAM, A. C. 'Controlling Factors in the Development of Caves', *Trans Cave Res Gp,* 13 (1971 *b*), 73–80.

WARWICK, G. T. 'The Origin of Limestone Caves. *In* C. H. D. Cullingford, *British Caving* (1953), 55–82.

WARWICK, G. T. 'Caves and Glaciation: Pt 1—Central and Southern Pennines and Adjacent Areas', *Trans Cave Res Gp,* 4 (1956), 125–60.

WEST, J. *Antiquities of Furness* (1774).

WEST, R. G. 'Problems of the British Quaternary', *Proc Geol Assoc,* 74 (1963), 147–186.

WEST, R. G. *Pleistocene Geology and Biology* (1968), 377pp.

WHITE, W. B. & LONGYEAR, J. 'Some Limitations on Speleo-Genetic Speculation Imposed by the Hydraulics of Groundwater Flow in Limestone', *Nittany Grotto Newsletter* (NSS), 10 (1964), 155.

WILLIAMS, P. W. *Aspects of the Limestone Physiography of Parts of Counties Clare and Galway, W Ireland* (1964). University of Cambridge. (Unpublished Ph.D. thesis.)

WILLIAMS, P. W. 'Limestone Pavements with Special Reference to Western Ireland', *Trans Inst Brit Geogr,* 40 (1966), 155–72.

WILSON, A. A. 'The Carboniferous Rocks of Coverdale and Adjacent Valleys in the Yorkshire Pennines', *Proc Yorks Geol Soc,* 32 (1960), 285–316.

WOOD, A. 'Algal Dust and the Finer-grained Varieties of Carboniferous Limestone', *Geol Mag,* 78 (1941), 192–200.

WRIGHT, K. A. 'Borrins Moor Cave, Horton in Ribblesdale, Yorks. Micro-organisms: A Preliminary Investigation', *Mem North Cav & Mine Res Soc* (1965), 52–4.

YATES, H. 'Goyden Pot, Nidderdale', *Journ Yorks Ramblers Cl,* 6 (1934), 216–28.

YEANDLE, D. 'Langcliffe Pot—Gasson's Series', *Univ Leeds Spel Assoc Rev,* no 8 (1971), 15–19.

ZÖTL, J. *Steirische Beiträge zur Hydrologie* (Graz 1959), 125.

Location Index

A fully comprehensive index is not included because of the systematic chapter treatment of the various subjects. Not included in this index are the names of passages within individual caves, and the named caves in the biospeleological appendix on pages 164–8.

Page numbers in italics indicate plates